200 TRAILS TO GOLD

200 Trails to Gold

A Guide
to Promising Old Mines and Hidden Lodes
Throughout the West

Samuel B. Jackson

Doubleday & Company, Inc., Garden City, New York 1976

Printed in the United States of America
First Edition

Library of Congress Cataloging in Publication Data

Jackson, Samuel B
200 trails to gold.

Includes index.
1. Gold mines and mining–The West. 2. Silver mines and mining–The West. I. Title.
TN413.A5J3 622'.342
ISBN 0-385-09945-2
Library of Congress Catalog Card Number 75-14827

To the memory of
Jason
and the crew of the *Argo*—they followed
the first waybill to a lost mine

Foreword

Throughout the great days of American mining, the price of gold was an unvarying $20.67 an ounce, which was the amount paid for it by the United States Mint.

As this book goes to press, the price is fluctuating around $165 an ounce, both in this country and in the principal gold-trading centers of Europe. The advance began in 1973 and was associated with the considerable confusion in international finance of that year. Before settling back to its present level, gold had been traded at more than $200 an ounce.

The eightfold increase in the value of the precious metal suggests that many gold deposits previously unworkable may now prove profitable. The higher returns must of course be weighed against today's greatly increased cost of operating a mine.

The greater value of gold also has been an incentive for increased prospecting in the Western states, some of it of a hobby nature, some of it a serious quest for sudden wealth.

This report is a description of known mines, lost mines, and of promising mining districts and is not concerned with the weighty economics of the metals market. Nevertheless, a few major events in that field may be listed:

In the depression year of 1934 the United States went off the gold standard and a White House conference set the price of gold at $35 an ounce. Private ownership and trading were forbidden.

In 1942, with the United States involved in World War II, the War Production Board ordered all gold mines shut down as non-essential industries. At the close of the war the order was vacated.

In 1973 occurred the meteoric rise in the gold price already noted.

In 1975 Congress restored the right of American citizens to own, buy, and sell gold bullion. A free market, with gold finding its own level of value, appears assured for the foreseeable future.

Silver is commonly associated with gold, both in nature and in the mind of the prospector, and it will be associated with gold throughout the descriptions and narratives which follow. Its climb to a historically record price has been quite as spectacular as that of gold.

The lure of the native gold or silver deposit has thus been intensified. The newcomer in the field may well inquire: What preparation is needed for one to start into desert or mountains in search of a rich strike?

Almost all the great gold and silver discoveries in the American West were made by men with no formal technical training. They had something better, the same endowment that Napoleon hoped to find in his field marshals. It is said that when considering such an appointment the emperor would probe into the officer's military record, discuss his personal attributes, and then come to his crucial question: "Is the man lucky?"

For the prospector, too, there is no better qualification.

Contents

200 TRAILS TO GOLD

1
California's Outer Desert

If you should be driving toward Southern California and using the official AAA road map you could be puzzled by several very large blank spots that loom up on your route.

California is the most populous state in the union and most of its people live in its southern part. Why then does the map show such great areas of apparently empty country?

The mysterious blank spots represent the mysterious California deserts.

On a more detailed local map you still will find little evidence of human habitation in these areas, but a number of rugged, isolated mountain ranges will be identified. They are not linked in a chain like the Rockies or the Appalachians, but rise alone from wide, barren flatlands as if separated from each other by some strange geologic hostility. Hostile to man they certainly are, with their searing summer heat and their almost complete absence of water.

All these mountain ranges contain gold.

In the Oriflammes, an Indian youth, Julian Cabrias, walked away from a cattle drive for an hour or two and returned with a handful of gold nuggets. They were a gift for a cowboy who had befriended him, Harry Yarnell. The white man was advised never to waste time looking for the source, for the ancient tribal mine was well concealed.

In the Sheep Holes there someday will be found a leather-bound Bible and beside it the tall, bleached skeleton of an old desert character known as Hermit John. Somewhere along his faint trail is a rich Spanish mine which had been rediscovered by the hermit. The white-bearded wanderer once shipped five sacks of dazzling gold ore from the lonely railroad station at Amboy, but was never to pick up his remittance from the smelter. Like many another old-timer, he led his burros off into the desert and was seen no more.

In the Clippers, Thomas Scofield came upon a long abandoned little excavation in which a series of quartz veinlets sparkled with gold. Beside it was a mound of the rich ore which represented a small fortune in itself. Scofield gave up a profitable work contract with the Santa Fe Railroad, formed a partnership with a trusted friend, and went back to the Clipper range to claim his wealth. Seeking to retrace his steps, he time and again found himself hopelessly lost, a common experience in the Clippers and in many another desert mountain range.

The Old Woman Mountains take their name from the old woman herself, an Indian squaw who trudged into the remote little Santa Fe station at Essex, nearly dead of thirst but clutching a deerskin pouch filled with nuggets as large as marbles. She recovered, professed to speak no English, waved an explanatory arm toward the distant peaks, and walked away down the tracks.

In the Turtle range, a natural stone arch rises far back in the jumbled, waterless canyons, the landmark of a phenomenal concentration of placer gold in the sand below. The miner Kurt Amsden, who chanced upon these riches, carried out enough wealth to retire in comfort. From his home in Chicago he generously wrote directions to a friend in California, intended to guide him to the ample gold which remained. But the desert, with the active malevolence which some of its denizens believe in, confused the clues and turned the searchers back, destitute of gold and feeling lucky to get out with their lives.

In the New York Mountains, or possibly on vast, sprawling Kokoweef Peak nearby, there is a less orthodox bonanza. Several reliable miners, one in a sworn statement presented in court, have told of a great cavern with an underground lake which rhythmically ebbs and fills. When the black sands of its shores are scooped up and panned for gold they prove fabulously rich.

Then we have the Vallecitos, the Chuckawallas, the McCoys, and other ranges, even the forbidding Algodones sand dunes, which Indians—often the squaws—have secretly entered and from which they have secretly come forth carrying little bags of dust and nuggets to spend in a frontier trading post or village store. Yes, the gold is there.

This report will deal largely with gold or silver lodes which are known, not lost or hidden, such as those just described. From the business standpoint, the possibilities in a presently inactive mine, or in exploring a somewhat neglected mineral district, represent the strictly practical opportunities. Yet, in the company of mining men, one need not feel backward about admitting that he is looking for a lost mine. Great corporations have financed searches for such legendary bonanzas, as have kings of Spain and viceroys of Mexico. Even professional mining engineers have followed the gleam.

There is this to be said about following a waybill, or *derrotero* as the Mexicans call it, to a lost mine: it will lead into mineralized country. The ledge in question may prove elusive, but the searcher who keeps his eyes open may at least hope that the erosion of the years has exposed a new and even richer bonanza.

In examining the possibilities in California's outer, or eastern, deserts we will start at the Colorado River, the crossing at Yuma, Arizona, to be precise, and work westward. It will be helpful, however, to insert a brief historical note, for it was in the southern desert country that there occurred California's first gold rush, all but ignored by history. It antedated the stampede of the forty-niners by seven years.

As all the mining world knows, the discovery of gold is credited to James W. Marshall. He detected the yellow flakes as water and sand flowed through a water-powered sawmill which he was building for Captain John Sutter in the central part of the state. The year was 1848. The California Division of Mines, however, details how a certain Francisco Lopez, major-domo of a Southern California cattle ranch, made a discovery either late in 1841 or early in 1842.

The scene was San Feliciano Canyon, about thirty-five miles northeast of what was then the sleepy little pueblo of Los Angeles. The story is that Lopez casually pulled up some wild onions by the roots and noted bright particles in the clinging soil. Real gold has a faculty of identifying itself unmistakably, as opposed to sulphides and other fool's gold, and Lopez knew what he had found.

California was under Mexican rule at the time. Life was leisurely and easygoing. It was an era of subsistence farming, trade with the outside world being largely in cattle hides. Communication was quite limited. The excitement following Lopez's discovery was therefore confined to the scanty local population plus a limited number of experienced miners from the neighboring state of Sonora. Without fanfare, without notice in the American press, something in the amount of $100,000 was recovered from the San Feliciano placers. They afforded a small band of miners profitable work for more than two years, then the deposit was exhausted.

A shipment of 18.34 ounces of gold from these workings was sent around Cape Horn and is of record as being deposited in the United States Mint at Philadelphia on July 8, 1843. It was credited to Abel Stearns, a resident of Los Angeles and presumably an American citizen. The government in Washington also was informed of the California gold operations in an official report by Thomas O. Larkin, the United States consul at Monterey. These developments went unnoticed in the East and entirely failed to have the electrifying effect of Marshall's discovery a few years later.

San Feliciano Canyon was desert country. Lacking water, the miners were reduced to "dry washing" to recover gold from the auriferous sand and gravel. As this method still will be required in many locations, at least in the early prospecting stage, it may be well to describe it.

"As a method of mining it was simple, crude and inefficient, but it had the advantage of being inexpensive," writes Donald C. Cutter, a California mining engineer. "After the dirt was dug, it was sun dried on a large canvas and then pulverized into dust. The next operation was to throw the dirt by the panful into the air in order to allow the wind to blow away the lighter elements and let the gold dust fall back into the pan. Thus the old agricultural procedure of winnowing grain was the first method extensively used in California for mining; for not only was it used in the Los Angeles area, but also it was introduced by many of these same Sonorans into the mines of the Sierra Nevada after 1848."

The traditional utensil used in dry washing was the *batea,* a shallow wooden bowl. The same process could be carried out with a standard metal gold pan, but the batea was considered superior. It is possible that the rougher surface of the hand-hewn wood helped retain the gold.

Considerable ingenuity has been expended in trying to mechanize the dry washing process. Such machines, or plans for building them, are offered for sale throughout the Southwest. They all involve a sizable diaphragm which is agitated and tosses the gravel in the air at rapid intervals. Normally the apparatus includes a bellows to provide streams of air to winnow away the lighter particles. The system of gears, tie rods, pulleys, etc., to effect all this varies with the individual mechanism. The typical dry washer is hand-cranked, but for any sustained operation it is practicable to hook it up to a small gasoline engine. The power takeoff with which some jeeps are equipped also would serve.

In addition to the substantial recoveries in San Feliciano Canyon made by secular miners, tradition has the Spanish padres gleaning considerable gold from placer deposits convenient to the missions. An early report of the state mineralogist refers to this activity in a factual way and names San Francisquito, Placerita, and Casteca canyons as having been worked by the fathers and their Indian converts.

One legend of the missions, unsubstantiated, has had a lasting appeal. It concerns a large accumulation of gold at San Diego de Alcala, founded in 1769, from which the city of San Diego takes its name. The gold is said to have been buried at the time of some unspecified danger and, in the checkered history of church and state relations thereafter, never to have been dug up. A hopeful treasure hunter may still be encountered now and then in the wide, dry expanse of Mission Valley, which cuts through the San Diego urban area.

These minor placers were worked in days long before the population boom in and around Los Angeles. Today the old workings are engulfed in the vast urban sprawl that is Southern California and offer little even to the weekend hobby prospector. There remain, however, several other gold-bearing districts that are within easy driving distance–one might almost say commuting distance–of Los Angeles and its neighboring cities. Later we shall have a look at these close-in possibilities, but first it seems more important to examine the immensely greater prospecting fields offered by what we have called California's outer desert.

A start may be made in the southeastern corner of the state. This would be in Imperial County, which lies along the Mexican border. The county's eastern boundary is the Colorado River, which separates California from Arizona.

"Gold is widely scattered over the desert region of Imperial County," says a state report by R. J. Sampson and W. B. Tucker. "It probably has been mined to some extent in every mountain range."

The ranges of immediate interest are the Picachos, the Chocolates, and the Cargo Muchachos. All of them are fairly accessible from Yuma, Arizona, or from the towns of the rich agricultural empire known as Imperial Valley.

The first entry into these lands was by Spaniards, and by way of what is now the site of Yuma. Here various expeditions from Old

Mexico paused for the welcome water and forage that the Colorado River and its flood plain afforded. Across the river, on the California side, the Spaniards established a presidio and along with it the ill-fated mission of San Pedro y San Pablo de Bicuner.

Most of the early Spanish arrivals, soldier and cleric alike, were slaughtered in the surprise Indian uprising of 1781. The hoard of gold accumulated at Bicuner Mission was hastily buried, and none of the desperate friars who knew the hiding place survived the massacre. Treasure hunters have dug up a great deal of ground in their unavailing search for it.

The settlement at Yuma was not destroyed in the outbreak, and as the years went by it became increasingly a rest and supply point. Travelers of many kinds paused here before striking into the California desert, the last stage of their journey to the Coast. The United States, after its war with Mexico, took over the region, and with the discovery of gold in California the little river station saw more and more activity. Freight wagons rolled back and forth to the California settlements, and passengers jolted and sweated in the hard-riding coaches of the historic Butterfield Stage Line. Amid this later traffic another type of traveler was to be seen at times, the prospector plodding along with his burros, his gaze sometimes on a distant horizon, sometimes hopefully and appraisingly resting on such pieces of iron-stained quartz as might lie along his path.

One such wayfarer found the gold-spangled rock that led him to the Picacho lode and the richest mine in this far corner of the state. Another found an equally golden ledge but, sick and exhausted, had to abandon it, leaving us the legend and the lure of the Lost Mule Shoe Mine. Such episodes of discovery, of success, and of tragedy will multiply as we go through this parched and rugged country, but first we should take a general view of the country itself.

When one crosses into California from Yuma, he has two gold-bearing ranges directly before him. To the west are the Cargo Muchachos. To the northwest are the Picachos, which some map-makers do not designate by name but include in the Chocolate mountain range. Closer at hand are the narrow flatlands along the river, once heavily populated by miners working the Pot Hole placers. These alluvial deposits were the source of the store of gold, accumulated at Bicuner Mission and hidden—from the treasure hunter's viewpoint—with such deplorable ingenuity by the doomed priests. Many miners took small fortunes from the Pot Holes. The site is of no interest to today's prospector, for not only was the river bank worked to exhaustion but much of the old field has been inundated by an irrigation project.

The Picacho range is readily identified by the commanding landmark of Picacho Peak. This sheer mass of rock towers into the sky like an ancient ziggurat, and can be seen by travelers on Interstate 10, which is eighty miles to the north. Beyond the Picachos lie the Black Mountains, the Peter Kane Mountains, then the Chocolate Mountains proper, and still farther a succession of little ranges so isolated and so barren as to have little discoverable history. Yet from all of them have come gold and rumors of gold.

All this is to the northwest. Looking directly west, past the Cargo Muchachos, we find the virtually impassable Algodones Sand Dunes. Beyond the dunes one is out of mineral country, for the land flattens out into Imperial Valley.

If you want the Picacho region to bestow a quick and easy fortune on you, you have only to locate the Lost Mule Shoe Mine. The ore is so heavily charged with gold that it is like ready money and it is lying out on the desert in plain sight. But where in plain sight? Does the massy gold catch the sun's rays in Gavilan Wash, in Julian Para Wash, in Carrizo Wash with its rare desert spring, or in one of the various ravines that on maps are discouragingly labeled "Unnamed Wash"?

It is too bad that we can no longer talk with Ed Rochester, who spent his adult life between the eastern escarpment of the Picachos and the green strip along the Colorado River. Rochester was the acknowledged oracle of the far eastern desert and also maintained, in effect, a one-man, year-round rescue and first-aid station. He pulled struggling men and burros out of the muddy Colorado, revived men who were helpless from dehydration or heat exhaustion. He saved even more lives by living up to the acknowledged duty of all old desert rats—that of keeping tyros from starting too long a trip into the wasteland or starting any trip at all without a large—at least a half gallon—canteen of water. Rochester was often consulted by writers for *Desert* magazine, and much of his information and advice is preserved in the files of that highly useful publication. He contributed his opinions as to the Mule Shoe Mine problem and obligingly tried to untangle the tangled Picacho country in the minds of lost mine hunters.

To understand how the Mule Shoe Mine was found and lost it is necessary to trace the route, or rather the northern of two possible routes, which an early adventurer would take to get from Yuma to the California coast. Such a trek through this scorched land eventually reached the water wells and settlements of what is now the Coachella Valley, but for mile after weary mile the traveler was tramping from one to another of four known water sources. These were the Peter Kane water hole, Chuckawalla spring, Tabaseca tank, and the little settlement of Dos Palmas. It was Dos Palmas and its human companionship that the Mule Shoe traveler hoped to reach, for he was a very sick man.

We do not know his name. We know a great deal, however, about a man who befriended him in the veterans' hospital at Sawtelle, California, and heard the story of his discovery. This friend was William M. Smith. The discoverer weakened and died in a hospital bed, but the patient Smith recovered, and he spent years trying to find the little marker of the mule shoe and the fortune in gold which lay beside it. The story told Smith was as follows:

The ailing man had left Yuma riding a horse and leading a pack mule. He was adequately equipped with food, water, and blankets. He was on the trail out of Yuma about four and a half hours, according to his own estimate, when he felt so ill that he had to dismount and lie down. The place was at the crest of a mild rise, which he described as forming a saddle between two low hills. When somewhat recovered, he observed a nearby outcrop of quartz and its liberal showing of native gold. He knew he had made a real discovery. An assured fortune was clearly in sight. Too weak to build the standard location monument of stones or to contrive anything else in the way of a landmark, he took a spare mule shoe from his pack. This was to be his marker. Trying to make it more conspicuous, he folded his vest and placed it under the iron shoe. Then he picked up a few pieces of the loose ore, climbed on his horse, and continued on his way.

The weakened man made it to Dos Palmas. After a rest he continued to the coast, but at once had to enter the hospital. There he lingered on for two years with little improvement in health or stamina. On good terms now with William M. Smith, who appears to have been some family connection, and quite unable to stand the rigors of another desert trip, he drew a map and gave such directions to the discovery as he could. The search was definitely entrusted to Smith.

The latter found the story, and particularly the ore samples, so convincing that he bought a tract of land on the Colorado River as a headquarters. Perhaps he had a prescience of a long and difficult search. Perhaps, more confident, he wanted the ranch property for storing and milling his ore. The flat shore country along the Colorado always has been fairly good grazing land for cattle, and Smith's ranch, which he named the 4S, was probably a self-sustaining enterprise. He lived and farmed here for at least ten years, searching the desolate back country in such time as he could spare. Even after electing to move back to the Los Angeles area he made repeated trips to the Picacho country and during the mild winter weather continued to hunt for the Lost Mule Shoe. On these occasions he approached the search area not from the Colorado River side, as previously, but from the southwest, where the rugged complex of hills and ravines faces out on a wide plain traversed by the Southern Pacific railroad. He often was at Gold Rock Ranch, near the now abandoned station of Ogilby, and talked freely with Carl and Margaret Walker, the owners. A brief search in 1910, so far as is known, was his last. His desert friends concluded that, like the first sojourner, the man who deposited the mule shoe, Smith had passed that shadowy line of physical vigor when sustained tramping and climbing no longer were possible.

River and desert people have discussed Smith's quest on many occasions, and they know pretty well where the sickly discoverer must have traveled. They say that the rich outcrop must have been reached by way of one of the following dry desert washes: White, Bear, Carrizo, Gavilan, or Julian Para. It is the "by way of" that confuses the matter. All these washes have branches, and Julian Para in particular spreads out into several long ravines which in their turn have branches.

A reasonable way to start a search today would be to go to Gavilan Wash. There is an established dirt road that leads north from the remains of old Ogilby and eventually reaches Blythe. Taking off from this route at a point near Gold Rock Ranch is a fainter road on which an automobile can proceed as far as Indian Pass. The pass forms an entry to the rugged, broken hill country sloping toward the Colorado River. A jeep trail, none too easy and maintained by nobody, winds all the way down Gavilan Wash and eventually to the river. Along the way other dry ravines branch off. Nearly all of them have some mild rises which might be described as "saddles." On which one did the sick man dismount and rest?

This area lies northwest of Picacho Peak. Southwest of the peak and much closer to Yuma is another lost ledge, this one of quartz of an unusual yellowish tinge flecked with visible gold. The discoverer, a lone German prospector, either perished of thirst or was killed for the valuable ore he was packing out toward civilization on his burro.

The outcrop is described as being in a small basin, high in the hills, and with the advantage of containing one of this country's rare and widely spaced water holes. A clue to identifying the basin is the fact that from one point the prospector could see the green strip of foliage running along the Colorado River. This observation was made in 1910 or 1911. Today the view would be of a larger sheet of water created by reclamation projects–either Ferguson Lake or Martinez Lake. The search area for the German's lost mine recently has become more accessible. There is now a Ferguson Lake Road branching off the paved Yuma-Imperial Dam thoroughfare, and it takes one right alongside the area where the high basin and its rich ledge should lie. The Ferguson Lake Road crosses Senator Wash and it is there that one may start looking. A line running north-northwest from this crossing would cut through the middle of what existing clues indicate is the most promising territory. It would be desirable to set a course by compass, and the recommendation would be to explore at least three miles on each side of such a line. This, of course, is a big order.

It was Ed Rochester who first heard the old German's story. He saw the rich ore, in fact was paid with some of it for helping the man get his two burros across the Colorado River. The prospector was planning to pass through the forbidding western Arizona desert to the old mining center of Quartzsite, there to settle accounts with his grubstaker and to outfit for working his great strike. He never was seen along the Colorado again, nor did he come back to reclaim some articles he had left with Rochester. A sinister note was detected by the river man in a bit of news that came back to him through the miners' grapevine. An Indian on the Arizona side had been observed pounding up some rich gold ore in a mortar. Like the few pieces paid to Rochester, the Indian's ore had a rare and distinctive yellow tinge.

There are alternative quests to those identified with the mule shoe and the yellow quartz. If one follows the Southern Pacific tracks a mile or so out of Yuma he is close to the Cargo Muchacho Mountains. This not very extensive range rises southwest of the Picachos. A great deal of gold has been taken out of these mountains and a great deal remains. Mineralization is demonstrably widespread. State reports tell of a number of idle mines with ore running $15 to $25 a ton at the old $20 gold price. The most businesslike approach to the Cargo Muchachos probably would be an evaluation of these old workings in the light of today's conditions.

There was only one camp in this district, Tumco, which now consists of only a few ruins and the town cemetery. The Golden Cross group of claims at Tumco produced $3 million in gold. Another rich property, the American Girl Mine, shut down when values thinned out at a depth of 740 feet.

There is a road through this region roughly paralleling the railroad tracks and the turnoff for Tumco is two miles beyond Ogilby. In the Tumco area dirt roads lead into various parts of the Cargo Muchachos and scenes of much early mining. The eastern slopes of this chain appear the most promising for virgin discoveries, since they have not been nearly as thoroughly prospected as the western side. There is a straggling desert trail starting at the agricultural inspection station on Interstate 8, a trail which runs ten miles to the base of Stud Mountain. This would place one in a starting position on the eastern slopes.

Nine miles northeast of Glamis, another little station on the Southern Pacific, is the old Mesquite placer field. A great deal of work already has been done here, but modern excavating machinery could conceivably permit a profitable revival. Engineering reports show that at a depth of three hundred feet there exists a layer of gold bearing gravel about six feet thick. The value per cubic yard has not been established.

The Mesquite diggings are in the southern reaches of the Chocolate Mountains, and just beyond them is the Gold Basin district, which has a history of both gold and silver mining. One of its last residents was Desert John, who kept a hopeful vigil over his Bay Horse Mine. It was a shallow shaft on a vein said to run forty to sixty ounces of silver to the ton. John was never to profit by it. There was a rumor going the rounds that the aged man had accumulated a hoard of placer gold. In 1930 he was found murdered, his cabin thoroughly ransacked.

It was ten years before this tragedy that a mining engineer, L. Harpending of Long Beach, California, called on Desert John and to his own surprise produced another lost mine puzzle. Harpending was there in his professional capacity at the behest of a friend, a small investor who was considering buying the Bay Horse and another of the gray-beard's properties. The location was in the southern reaches of the Chocolate Mountains.

The investigation left Harpending and his friend some free time and they decided on a little exploring. As the engineer later recounted it, they made their camp under some paloverde trees in a wide wash, presumably near Desert John's. He said a number of small ravines branched out from here "like the fingers of your hand." This is the nearest one comes to a clue as to the location of their camp.

The two men took separate routes, each going up one of the ravines, then crossing over and coming down another. They discovered two tanks, or natural rock catchments, that promised a water supply for all or most of the year. Harpending discovered something more. It was a strong ledge of quartz, about three feet wide, and extending clear across the ravine for a distance of some forty feet. This would make it an exceptionally large outcrop, indicative of a large ore body if the quartz carried any values. No gold or silver was visible, but to Harpending's practiced eye the formation looked promising. Never without his pick and sample sack, he carried away a few pieces.

Several days after he returned home, the engineer had a sample assayed and found that it ran thirty-two ounces of silver to the ton, plus a little gold. This was confirmed by a second assay. Such a result certainly raised interesting prospects. At the silver price then prevailing, Harpending knew that he had no overnight bonanza, but he foresaw a profitable, long-term operation. He had a living to make, and accepted an offer to manage an ore reduction plant in Plumas County. He remained at this well-paid job for three years. When the mill closed down he was at liberty to consider working the big silver lode and confidently returned to California's southeast corner and its desolate mountains.

Harpending believed he had a clear memory of his tramp up and down the dry ravines, a sharp picture in his mind of his campsite, and particularly of the massive ore outcrop. Yet, in an unhurried, painstaking exploration, he failed to locate that three-foot-wide vein which had stood out, plainly visible, for a distance of forty feet. Twice again during the next ten years, between professional assignments, he returned to the search. Always without success. With the soaring silver prices of the 1970s, the Harpending lode has taken on the complexion of a true bonanza, though also a truly lost one. The location is in Imperial County, and today the paved Ben Hulse Highway runs through the general area. A spot marked "Gold Basin" on road maps would be a reasonable takeoff point.

The engineer's experience is one that has been repeated in many parts of the Southwest. The desert seems to have an uncanny way of confusing the lost mine hunter. There are few features that can correctly be called landmarks. Each little peak, or gully, or landslide looks much like another. The canyons and other depressions that cut through a desert range are multitudinous. If one climbs to the top of a ridge he has achieved a view of perhaps a dozen similar ridges. More than one searcher has given up because of a firm conviction that this silent land is exerting some mystic power of deliberate concealment.

In Harpending's case, the engineer admitted that he was not certain that he ever found his way back to the original discovery ravine. Later, as if for consolation, he theorized that he might have reached there, but that a flash flood had washed enough sand, gravel, and broken rock down the gulch to alter the appearance of things and possibly even to have covered the silver ledge completely. Such an effect is certainly possible after one of the characteristic desert cloudbursts. Even if this explanation is accepted, the silver ledge is still there, and what one flash flood can cover up, another can reveal again.

In recent years it has become easier to drive into this area. The paved Ben Hulse Highway, designated State Route 78, provides a paved road very close to the old diggings associated with Desert John and his fellow miners. Away from the road, the prospector is in country just as arid and desolate as it ever was. The twentieth century has even added an extra hazard. If one goes too far north of the old mines he exposes himself to bombing in the Chocolate Mountain Aerial Gunnery Range.

All the trails out of Yuma thus far described eventually reach the ample water supply and other amenities of the Coachella Valley. Midway on the northern route there is dependable water at Chuckawalla Spring. Its clear little pool is protected by a masonry wall constructed by state fish and game officers. From this base the gold hunter can explore the Chuckawalla and Little Chuckawalla mountains. The spring and mountains take their name from one of the largest of the desert lizards. The chuckawalla is quite harmless, but the commotion as it suddenly darts out of one's path can be a little startling.

A little way down the canyon from Chuckawalla Spring one finds the remains of a curious little cave colony. Placer gold is found in a number of the ravines of the surrounding mountains. No rich concentrations are known to occur, but during the depression of the 1930s a few miners eked out a little income here. An earth bank of fairly soft yet stable composition made it easy to dig caves, and these became the miners' homes. The flow of Chuckawalla Spring is too slight for sluicing. Recovery of placer gold in any sustained operation in this region would necessarily be by the dry washing method.

The Little Chuckawallas extend eastward from the spring for about fourteen miles. Along their southern side runs the trace of the old Butterfield Stage Road, which we first encountered near Yuma. In the dry desert such trails are well preserved. This one can easily be traversed by four-wheel-drive vehicles and possibly by standard passenger cars. An old-time landmark on this route is Wiley Well. It could be made the base for exploration of the nearby Mule Mountains, definitely a gold-bearing range, as are the more distant Palo Verdes. The mining in the latter range was largely in the neighborhood of Thumb Peak. A fair dirt road leads south from Wiley Well to the peak and its abandoned mines.

As to the main Chuckawalla range, it has seen extensive lode mining, and state reports accent the mineralization of its northwest section. They note, in addition to gold, occurrences of silver, copper, and lead.

Chuckawalla Spring can be reached most easily by a drive of about fifteen miles southward from Interstate 10, after carefully getting directions at Desert Center. It is more adventurous, and could be vastly more profitable, to approach from the south, starting along the route to the Mule Shoe gold. During such an approach you not only can search for that bonanza but also for the Lost Bull Ring Mine.

The latter may have been California's very first gold operation, for it apparently antedates even the San Feliciano placers. Discovered by a Frenchman, and this was said to be in 1836, its ore was so rich that it could be taken overland to the Colorado River for a complicated and expensive haul to Swansea, Wales. After the one shipment was made, the shallow shaft was covered with logs. Whether, like the later western emigrant trains, this early miner was using oxen for hauling is something we do not know. At any rate he had in his possession a brass "bull ring." This he firmly attached to the wood cover of the shaft, a small marker but a distinctive one.

The Frenchman never returned. He must have talked, however, because a couple of decades later, with California mining well under way, his story appears to have been widely known around the gold camps. The only rediscovery of record was an accidental and ineffective one.

This interesting sequel is related by Philip A. Bailey in his *Golden Mirages* and by another chronicler of lost mines, R. W. McAllister, although neither can supply a name for the frustrated paterfamilias concerned. In 1900 a family was traveling by wagon to Southern California. On the old Ehrenberg-San Bernardino road, which still can be traced in many places, they made a late afternoon camp. One of the children, a girl of about ten, did some exploring in the sandy stretches between the scattered desert brush. She came back with the news that she had seen a covered "well" and that it had a shiny ring on the top. The fact meant nothing to the weary parents and the next morning the party pushed on.

Settled on the Coast, the family made friends and, at that period, found that some of the conversation dealt with mines and mining. In due course they heard of the Lost Bull Ring Mine. It was an excited father who now took his daughter back along their old trail. He found what he hoped was their old campsite, carelessly observed at the time but now vitally important. It was now up to the child. She doubtless tried hard. Her father tried harder still. But the Lost Bull Ring Mine remains lost.

We now have passed out of Imperial and into Riverside County, which is bisected by Interstate 10, a major transcontinental artery. This is the most direct route leading eastward from Los Angeles and it carries heavy traffic. Yet, along a course of two hundred miles through the desert, the only signs of life are a few, a very few, refueling stops and a small residential development at Desert Center. On each side of the highway there remain hundreds of square miles of arid, unpopulated wasteland, with steep, picturesque mountain ranges punctuating the landscape.

At the eastern edge of the county, a short jog from the Colorado River, is the town of Blythe. It is often mentioned in a weather report as registering the nation's highest temperature for that particular summer's day. Blythe is the natural supply and departure point for exploring the desert region along Interstate 10. Although not far from the amazingly rich La Paz placers on the Colorado River, worked during Civil War days, Blythe never was primarily a mining town. It is associated rather with a successful irrigated farming development in Palo Verde Valley.

It was in Blythe that the late Randall Henderson published a small weekly newspaper as a young man. His enthusiasm for the gaunt, vacant country stretching off to the northwest led him into many exploring trips and ultimately to the founding of *Desert* magazine. This publication has been the repository of an immense amount of desert lore and is faithfully read by rock hounds, amateur archaeologists, searchers for abandoned mines and lost mines, and by various other desert enthusiasts.

The closest mineralized mountain range to Blythe is paradoxically one of the most difficult to explore. The McCoy Mountains are clearly visible from Interstate 10 but can be approached only by a sketchy dirt road which tapers off to a mere trail. This route passes through a wide basin on the eastern side of the McCoys but veers farther and farther away from the mountains. There is also a jeep trail, not shown on maps, along the western flank. It will take one fairly close to McCoy Spring, nine miles north of the highway, which appears to be the only dependable source of water in the entire range. Along with McCoy Mountains we may consider the Palen range, which is the next one to the west and even more isolated than the McCoys.

A 1917 report of the state mineralogist tells of veins in both the McCoy and Palen ranges running twenty, fifty, even one hundred feet in width and carrying gold, silver, and copper. These would be extremely large ore bodies, and if they are of commercial grade could be extremely valuable. The usual excellent and circumstantial reports of California's official mining agency are not very illuminating in this instance, for it appears that the field geologist did not actually examine the deposits.

County records, however, give an indication of widespread mineralization. Discoverers of the so-called Crescent group at the north end of the Palen Mountains staked the amazing number of 50 contiguous claims, each of them of course measuring 1,500 feet in length and 600 in width. Since it is required that there be demonstrable values on each claim, this would seem evidence of very extensive outcrops of promising ore. Most of these claims were later allowed to lapse and after a space of half a century it is possible that all of them have been abandoned. The location is seventeen air miles northwest of Blythe.

The Palen Mountains also contain an immense deposit of gypsum, but it appears to be of no immediate commercial value. The trouble lies in fragments of marble which make the material difficult and costly to process. No areas of pure gypsum so far have been found.

In the McCoy Mountains a promising early development was the Mountain King Mine on their eastern side. Old-timers along the Colorado River felt that the McCoys also were the location of a rich surface deposit of either loose nuggets or crumbling, easily broken vein matter from which an Indian family obtained gold. As late as 1921 a

Mojave Indian named Chinkinnow was bringing such gold into Blythe to pay for supplies.

The mine was not general tribal knowledge. The secret reposed with a middle-aged Mojave squaw who, for reasons unknown, was something of an outcast. When a fugitive Papago named Papuan married the woman and thus gave her some status, she gratefully led him to some mysterious fastness where gold, apparently plenty of it, was easily obtained. Their takings were generously shared in the Mojave colony and much of the gold ultimately found its way into the trading post at Ehrenberg owned by Bill McCoy.

After Papuan's death his widow was cultivated, or perhaps harassed, by a local miner named Hartmann. Like many another white man, he found the Indian mind impregnable. The secret of the gold deposit went only to an adopted son, the above-mentioned Chinkinnow. This young red man made the mine pay extra dividends. According to Dorothy Robertson, who accompanied her husband on prospecting trips out of Blythe and gathered such local history as she could, the Indian frequently took gifts in exchange for a promise to show the diggings, then, when out on the trail, would manage to get confused and forgetful. He continued to bring in enough gold for his own needs.

Another mineralized range near Blythe is the Big Maria. The most promising metal appears to be copper, and ore running 41 per cent occurred in the St. Joseph Mine. There is a major highway running north from Blythe along the Colorado River, U.S. 95, and at intervals there are short dirt roads leading to the old gold and copper workings on the eastern side of the Big Marias. On the west side of these mountains the chief activity seems to have been quarrying for highway or other construction purposes. The quarry work has resulted in several good roads being run into the canyons. These could be a real convenience to the prospector. They are reached by taking the paved Midland Road running northwest out of Blythe.

Midland Road will take one also to the Little Maria Mountains. The hamlet of Midland is near their southern tip and the old Victor Mine marks the area of known mineralization. Both mining reports and local information indicate that the higher elevations of neither the Big nor Little Maria mountains have been very extensively explored. The northeastern slopes of the Little Marias in particular, as well as the twelve-mile uplift known as the Granite range, the Little Marias' near neighbor, would seem to offer much virgin prospecting ground.

Keeping in the far outer desert and fairly close to the Colorado River, we move north to two other mountain ranges of interest.

The Whipples occupy the inside of a giant bend in the Colorado south of the rapidly developing resort areas on Lake Havasu. Boaters and water-skiers on this man-made lake are doubtless aware of the broken peaks that loom up fifteen miles away on the southern skyline. It is unlikely that any of these vacationers will hike into the interior of this range. It is unlikely, in fact, that white men have ever made any intensive examination of the innermost Whipples. An almost complete absence of historical and mineral information prevails regarding the eighty-odd square miles that make up the lofty core of the range. Around the more accessible circumference, the lower levels, there are a number of mines, some engineering works connected with water supply, and even an Indian reservation.

The greatest concentration of mines is on a rib extending west from the main uplift and terminating in Pyramid Butte, a landmark on U.S. 95 thirteen miles north of Vidal Junction. Here, at the American Eagle, the Gold Hill, or other old workings, some survivor of early mining days may have something to tell about the interior of the Whipples. Do the gold formations suddenly end as the mountains rise higher? Or is this roadless, almost trail-less region something that was neglected in early prospecting days?

Two points of reference should be kept in mind by anyone driving in this vacant land; Vidal and Vidal Junction. The former is a small but long-established settlement on the Santa Fe Railroad; the latter, six miles to the northwest, is near the Colorado River Aqueduct, which carries water to the thirsty cities of Southern California. Vidal Junction looks out on one of the most extensive blank spots on the California map. Not completely blank geologically. For the space is partly occupied by a sprawling, desolate complex of cliff and canyon known as the Turtle Mountains. And sheltered far in the recesses of this range is the fabulous gold deposit known, hunted, and sometimes cursed as the Lost Arch Mine.

Besides the lost bonanza, there are known gold-bearing veins in the Turtles and a few known areas of indifferent placer ground. The quartz ledges so far detected carry only modest values in gold and have been regarded as too narrow for profitable mining. This judgment was made in the days of horse-drawn transport and the $20 gold price. A new, open-minded approach might result in more optimistic findings.

The odd name of the range may have derived from native fauna, for there is a turtle which has adapted itself to this almost waterless land. It is known as the desert tortoise. Another explanation is based on the strange concretions found in great numbers in some localities which, when they don't resemble cannonballs, often have a domelike shape resembling a turtle. The more accessible canyons of the Turtles receive a variety of intermittent visitors. The range's craggy rock formations attract photographers, Indian petroglyphs attract archaeologists, and a minor Jungfrau known as Mopah Peak attracts mountain climbers.

Lying a short distance within the southern rises of the Turtles is the Lost Arch Inn, which has been unwittingly listed as a tourist accommodation in at least one travel guide. In reality, the inn is a one-room cabin built of nondescript lumber which long was occupied by a desert character named Charlie Brown, and later by a Jesse Craik. Something of a wag, Brown spent some hard-won gold dust to have a large, neat sign painted and the legend "Lost Arch Inn" was thereafter spread across the front of the shack.

The Lost Arch term refers to the much discussed natural rock formation which is the pointer to a rich concentration of gold nuggets and a possibly richer ledge of gold-bearing ore from which the nuggets came. John D. Mitchell, the pioneer collector of lost-mine stories and a successful mining man himself, places the arch and the nugget hoard near the north end of the Turtle range, but does not give any basis for this choice. He believes the gold was first discovered by a small party of Mexicans on their way to the sensational placers at La Paz, Arizona. The Mexicans were said to have built an adobe shelter near the spot. Then, it is asserted, they returned to Mexico, not with the riches of La Paz but of their lucky discovery en route. In 1900 a prospector named Peter Kohler, who at that time was not aware of any bonanza in the Turtles, observed the crumbling walls of such a structure.

The account which carries weight concerns the year 1883, a prospector from the eastern states named Kurt Amsden, and Amsden's unidentified partner. Of their previous experiences in the desert and the circumstances of their journey into the Turtles nothing is known. It is known, however, that the two men found a fortune under the arch. In a letter to a friend, Dick Colton of Barstow, California, Amsden described the nuggets as lying on or close to the surface. They were to be had just for the trouble of picking them up, he wrote. Amsden and his partner, rich with gold and perhaps a bit giddy with their good luck, suddenly found that they were not at all rich in water and food. They were in a remote spot, far from human habitation, utterly without water or any edible game or plants. Sobered by a realization of their danger, they buried half their takings. With lighter loads of gold than they had first planned to carry, and with such food and water as remained, they set forth. The nearest refuge from thirst and heat which was known to them was forty miles to the north, a station on the Santa Fe Railroad known as Goffs. Amsden's partner died during the desperate trek but the easterner staggered into the little whistle-stop settlement, still weighted down with a heavy bag—or would it now be two bags?—of coarse gold.

His hardships and bare survival had given Amsden all he wanted of the desert. He took his new wealth back home and unselfishly tried to direct his friend Colton to the treasure. In company with Herb Witmire and Mort Immel, Colton tried to follow the map and directions sent him. The men looked primarily for the stone arch, which was and is the vital landmark. Amsden, the best authority, accents this and the arch also figures, if less positively, in the story of the Mexicans' discovery. The Colton trio could find no arch and they could find no gold.

The Turtle Mountains cover a great deal of territory. In his earlier exploring years, John Hilton, a well-known western writer and artist, had come across a natural arch in that range. Like the prospector Kohler, he was not aware of its significance. Years later, he and Walter Ford, another authority on desert travel, tried to follow up Hilton's earlier discovery and locate the nugget field. If experience and detailed knowledge of the California desert counted, these men should have found it. But luck, the essential ally, was not with them.

Did a long-time Turtle Mountain denizen known as Pack Rat Joe chance upon the gold that Amsden had buried? "Half a tubful," the easterner had written. Joe lived in and roamed the Turtles for many years. For years he also was a friend of James Harrigan of Los Angeles, a successful illustrator who often explored the desert photographing Indian petroglyphs. In many long trips the two made together Joe never mentioned any important mining activity and Harrigan did not get curious.

There came a day when the old desert man was out in civilization to buy supplies. There was some tragic pile-up of cars on the highway near San Bernardino and the Pack Rat was among those killed. Back in the Turtles, friends of old Joe found piles of useless auto parts, broken machinery, and other items of junk that the solitary man seemed driven to pick up and carry home. This habit had given him his nickname. The friends also found that Joe had $40,000 in the bank. It was not derived from junk. At some time he had made a gold strike. It may have been a new one. He may have found the cache buried by Amsden and his unlucky partner. He may have found the Mexican-Amsden nugget field itself. Even if Pack Rat Joe did discover the deposit and mine it completely out, the Lost Arch holds promise of wealth. Accounts agree that above the concentration of nuggets there existed a red formation, possibly the edge of a blanket vein, that appeared to consist of hematite. This is a familiar ore of iron which also can carry gold. Those who have studied the Lost Arch puzzle surmise that the gold found in the sands must have eroded from the higher formation. In view of the bountiful amount below, they argue, the red ore above would constitute a real bonanza, probably more valuable than the placer itself.

A great arc of almost level desert curves around the Turtle Mountains on the north and west. So flat is the country to the west that a graded dirt road runs straight as an arrow for a distance of forty-eight miles. The area through which it passes is called Ward Valley. There is no town along its entire length, but halfway along the lonely course one can look west and see some relics of human activity. These represent old mines in the Old Woman Mountains.

The range is about twenty-five miles long and has a north-south axis. Mining has been extensive enough to create many roads into the various canyons. Fourteen of the routes are prominent enough to be shown, along with mileages, on maps of the Automobile Club of Southern California. Additional jeep trails exist.

Mineralization in the Old Woman Mountains was commented on favorably by the state mineralogist as early as 1896. At that time he reported "veins of gold and silver ore and also silver-lead ore." Among the mines operated with some success under the old $20 gold price were the Golden Fleece, Enterprise, Florence, and Oro Plata. Activity was amost entirely in the northern half of the range. The southern half was formerly much more difficult of access, but now can be approached by a semi-improved road branching off U.S. 66 and following the Santa Fe Railroad. Chubbuck and Milligan stations would be takeoff points for this southern sector.

The original "old woman" has been referred to briefly earlier in this report. There is little amplification of the story possible. The year of her almost fatal journey into the mountains was 1890 and the spot where she finally obtained water, food, and assistance was Essex, then an extremely isolated place serving some purpose for the Santa Fe line. Today one can drive to Essex on U.S. 66 and from there can drive well into the mountains, for the trackless sand flats over which the squaw struggled with her deerskin pouch full of large nuggets is now traversed by a graded road to Sunflower Springs, where a group of old mine workings will be found.

This route must come very close to the secret Indian hoard. No hints were received by the railroad men who helped the weary woman. One account appropriately has her giving several nuggets to them before she stolidly walked off down the tracks, but giving no information. Nor did a watchful station agent observe any other Indians going to or from the mountains by way of Essex.

The most northerly gold territory to be considered in this chapter is that in and around the well-developed Ivanpah district. This brings us near the Nevada state line. The area is reached by Interstate 15, the busy highway that runs northeast from Los Angeles to the gambling and entertainment center of Las Vegas, and thence to Salt Lake City and major arteries to the Middle West and East.

At Mountain Springs Pass, fifteen miles from the Nevada border, Interstate 15 is in the midst of many old and once prosperous mines. Some of these were abandoned during extreme slumps in the market price of silver. With silver now worth at least five times as much as its old-time average, they would seem well worth re-examination.

South of this concentration near the pass there are workings of varied value and varied history over a considerable area, the greatest activity apparently having been in the New York Mountains. In the middle of that range are found the ghost towns of Vanderbilt and Barnwell. Down on the flat and now served by a paved road stands what is left of an early supply and shipping town, Ivanpah. It will be convenient to refer to the whole area under consideration as the Ivanpah district, although official mining reports differentiate a number of districts with individual names.

Gold and silver occur together and in separate deposits throughout the Ivanpah district. In the case of gold, we shall look into a persistent story of a lavish concentration of gold around the shore of an underground lake.

There have been about eighty mines of substantial production in the Ivanpah district. The earliest digging was for turquoise. Old excavations a few miles north of Halloran Springs on Interstate 15 contain stone hammers and scoops made of shells which archaeologists connect with the pueblo tribes of New Mexico and Arizona. These Indians are believed to have made long excursions to mine the prized turquoise, which is used in the ornaments of southwestern Indians even today. It is evident that the forty-niners were not the first Americans to travel a long way in search of precious minerals.

The Indians apparently were not interested in rich deposits of silver, which were to be seen in the Clark Mountains. Consequently these proved a bonanza to the first white men to enter the area. The discoverers were soldiers from Fort Mojave who employed their spare time prowling around the surrounding country. The veins they discovered were very rich. The Allie Mine had some ore running 4,400 ounces of silver to the ton. Much of the north slope of the Clark Mountains, rising up from Kearny Pass, was and is an area of extensive mineralization, and the Beatrice, Lizzie Bullock, and Jackson mines rivaled the Allie in high values.

The ore had to be rich to be worked at all. "The early mines were isolated, far out at the extremities of a long communication system," writes William E. Ver Planck, a California state geologist. "Supplies came from San Francisco by way of San Bernardino and a long haul by freight wagon over Government Road. Burros were used for the last stage of the journey where wagons could not go. The outgoing ore had an even longer journey, because most of it was shipped to Wales for smelting. Freight rates were hundreds of dollars per ton. . . . Workings were necessarily small and shallow. Only the richest deposits could be considered."

As would be expected under these conditions, mines were abandoned as soon as the richest, easily worked parts of a vein were exhausted. Today's miner might well find ore in place in these properties that could be worked at a profit. Transportation difficulties are now minimal. It has been noted that Interstate 15 cuts right through the area, and the main line of the Union Pacific is a short distance away.

In the New York Mountains to the south, values were in gold. The amalgamation process of recovery used during the active years was successful only with free gold, that is, gold which could be freed from the ore by crushing and grinding. Engineering reports affirm that large tonnages of ore in the sulphide zone, where gold is held in chemical composition, remain undisturbed.

The Castle Mountains, east of the New York range and containing the ghost town of Hart, have a history of what a state report calls "small but rich gold veins." Clay has been mined from steeply dipping fractures in the Castles' volcanic rocks, and Ver Planck states that these same fractures contain the narrow, high-grade veins. The range extends across the state line into Nevada. To get into a known gold area look for the old Oro Belle and Valley View mines.

The Providence, Piute, and Old Dad mountains should be added to the list of prospecting possibilities in the general Ivanpah region. The record shows all of them to contain a scattering of old mines.

The Providence range has seen a great deal of activity and at least twenty roads lead up various ravines and canyons to old, largely abandoned mines. This otherwise neglected range boasts one tourist attraction, the Mitchell Caverns. Now part of the California state park system, they constitute a vast underground labyrinth which has been only partly explored.

Entering the Ivanpah district today, one will be limiting his opportunities if he restricts his interest to gold and silver. There is an amazing variety of mineralization. Copper, lead, and zinc were mined in some quantity in response to the demand and high prices during the two world wars. Geologists report significant occurrences of tungsten, vanadium, antimony, magnesite, and that rare component of North American ores, tin. A discovery in this area of "rare earths," the source of several little-known space-age elements, was quickly taken over by one of the great mining corporations. Engineers estimate that the deposit could supply the world's needs in this line for at least a hundred years.

Two minor ranges just south of Mountain Springs Pass remain to be mentioned, the Ivanpah and Mescal. There has been substantial gold and silver production in both. A landmark quite near Interstate 15 is Kokoweef Peak, rising to 6,038 feet above sea level. This brings us to the gold-encircled lake, for the weight of evidence, in fact the flat statement of one explorer, is that on Kokoweef is the inconspicuous little opening that leads into a strange cavern and its even stranger golden sands.

The account of the lake hardly qualifies as a factual report, yet any modern-day prospector who gets interested in the Ivanpah district is certain to hear of this subterranean mystery. He also will hear that several otherwise reliable miners have reported independently on the matter and in substantial agreement as to fact. The lake even has figured in a lawsuit, although the avowed object of the litigation was to obtain title to a more conventional mining property. The ascertainable information runs as follows:

The lake lies in a large cavern which extends beyond the lake for an unknown distance. The water rises and falls at intervals of several hours. When it is low, wide banks of black sand are exposed. Those who have scooped up some of the black sand say that it pans out as very rich in gold.

The first descent to the lake, or at least the first that the Ivanpah district heard about, was made by two Indian boys, brothers. These two established the fact of gold on the lake shore, exhibiting nuggets they had picked up to their own people. Eventually the tale sifted through to a few scattered mining camps. On a later trip by the two boys there was a misadventure of some kind and one of the brothers drowned. R. W. McAllister, who tried to trace down the facts for his *Lost Mines and Treasure of the West,* says that Indian superstition, or perhaps a natural revulsion against the eerie depths and their now tragic associations, thereafter kept the tribesmen away.

A prospector named John Delano was the first white man of record to enter a cave on Kokoweef but he found no gold and no lake. He may have gone through a different entrance. He may have been in a different cave. The existence of the far-reaching Mitchell Caverns to the south suggests something of a honeycomb nature for the country.

Better luck awaited two explorers who entered the cave in 1927. One was a Coloradan named E. P. Dorr, who said he had been told of the deposit by an Indian who worked on his father's ranch. He had enlisted a mining engineer as a partner. They came out with essentially the same story as the Indian boys. The rise and fall of the water, the presence of nuggets and fine gold in black sand were confirmed. Their high hopes ran into a legal obstacle. During the active mining period in the Ivanpah district a great deal of land was covered by mining claims, some promising areas being almost blanketed. The discoverers now found that the surface above the cavern already had been taken up. A small mining corporation was the owner of record. The validity of its claims was attacked in court and in this connection Dorr executed an affidavit describing the lake and its value.

The action against the corporate owners failed and soon the cave's entrance, which still was known to only a handful of people, was closed by a dynamite blast. The responsibility was attributed locally to the disappointed plaintiffs. The corporation, interested only in mining zinc on its holdings, made no attempt to reopen and explore the big cave.

There was another prospector who told of chancing on the cave, the lake, and the gold. He was Riley Hatfield, formerly of Raleigh, North Carolina, who originally had come to the desert for his health. In the year 1904 he appeared at Crescent, a gold-turquoise camp midway between the Ivanpah district and booming Searchlight, Nevada. The lone Southerner was befriended by a group of five mining men at Crescent, and these included John D. Mitchell, then a young prospector who already was gathering notes for the account of lost mines and buried treasure that he was to publish in 1933.

Mitchell and Hatfield soon happened to be in Searchlight at the same time. Hatfield had some placer gold to sell and Mitchell directed him to an assayer. A number of gold nuggets were exhibited and the sojourner, who was described as having a long white beard, told of finding them in a cave. His description of a lake and black sand tallied with the accounts which later were to come to light. The elderly Hatfield, who complained of a mild heart condition, was eager to take Mitchell as a partner.

The latter was agreeable but wished to remain at Crescent long enough to sell the mining claims he held there. Hatfield therefore took off with his burro toward the Ivanpah district for what he planned as a three weeks' trip. He said confidently that he would be bringing back more gold. Neither Mitchell nor Crescent saw the old man again.

At this point it is well to pause and get our bearings. The gold country so far described consists of the desert area lying against California's eastern border and extending from the Mexican boundary up to the latitude of Death Valley. The three major highways mentioned as access routes, interstates 15, 40, and 10, all converge on the Los Angeles area as one follows them westward.

Leaving the Ivanpah district on such a westward trip and remaining on Interstate 15, one comes to a limited but rather intensive area of mineralization surrounding the highway stop of Halloran Springs. Numerous old mines attest early discoveries of gold and silver on both sides of the road. Those on the north are interspersed with turquoise diggings. An easy entry into this region is afforded by a paved road leading to a microwave relay station on Turquoise Mountain. This is not the only remote spot where a relay station provides today's prospector with a good road.

Between Halloran Springs and the edge of the great metropolitan complex that is Southern California, Interstate 15 runs through almost vacant desert country for some forty-five miles. The best-known point in this stretch is Baker. On this route one observes the Soda Mountains quite close to the road on the north side. To the south is the long, narrow ridge of the Maverick Brothers Mountains. Mining activity in these ranges appears to have been minimal. A few old mines in the Sodas probably can be reached by rough dirt roads but careful local inquiry is recommended.

Dropping down to the next major arterial, Interstate 40, we are close to the Clipper Mountains. Hidden somewhere in this little range is a series of narrow, parallel quartz stringers which, if described accurately, are so impregnated with gold that they outshine the richest "seam mines" of the Mother Lode.

The fate of the original finder is unknown. That he met a sudden and possibly violent end seems probable. This is indicated by the fact that not only was a heap of high-grade ore left beside the mine shaft, but also a quantity of native gold was left stored in a Dutch oven. The latter utensil, still in use in many households, is a heavy, lidded, cast-iron kettle. Because of this distinctive item found at the scene the diggings are often referred to as the Lost Dutch Oven Mine.

The skeleton of the strangely missing miner may well be lying somewhere nearby. The rediscoverer, dazzled by the gold and critically short of water at the time, did not make a search.

The approach to the Clipper Mountains would be by way of the twin settlements of Danby and Danby Station, one on the Santa Fe Railroad, the other on U. S. Highway 66. The principal technical report on the Clipper range does not refer to the Dutch Oven Mine, but it does describe a promising deposit stated to be five miles northwest of Danby. The vein is said to be a massive one, with the values in free gold. It was being developed by a vertical shaft, but in 1920 a heavy flow of water was encountered and operations were suspended. When a state field man reported on the property in 1931 the mine appeared abandoned and no further mention is found in official records.

The facts about the Dutch Oven came from an experienced hard-rock miner, Thomas Scofield of Los Angeles. He was in Danby under contract with the Santa Fe Railroad, which had engaged him to drill a tunnel into a nearby spur of the Clippers in the hope of developing water. Drilling for water was an indispensable procedure as the early railroad lines were pushed through the southwestern desert.

On his Sundays off, Scofield the miner inevitably spent the day prospecting, and the dry washes and canyons of the Clippers were the only accessible ground for a one day's tramp.

One Sunday he was pleased to find a flowing spring in one of the parched ravines. Someone evidently had made good use of this water supply, for a faint but unmistakable trail led away from it and back into the mountains. The trail seemed an invitation to something of interest in an otherwise empty country. Scofield decided to follow it. Later he recalled only one distinctive feature along its twisting, ascending course, a pair of high-standing rock formations. They were about three feet apart at the base, he said, and the trail passed between them. He was to see them only once more—on his excited, unobserving trip homeward. He was singularly unobservant also regarding any identifiable natural features in the setting of the mine. He was too interested in the mine itself.

What he saw was a shaft going straight down on a body of "bluish quartz." Parallel iron-stained streaks occurred in the formation, and these all displayed gold in coarse, highly visible masses. After taking in this startling display his experienced miner's eye noted that the capacity of the shaft about equalled the volume of the excavated rock. In other words, none of the rich ore had as yet been hauled away. Had the owner gone out to civilization to get a string of pack burros? Would he soon be back to cash in on his great discovery?

Examination of the nearby camp answered these speculations in the negative. The place had long been deserted. A tent pole remained sunk in the ground and a few shreds of canvas stirred in the breeze. There was a bed of very dry branches of greasewood, and tattered remains of what may have been a couple of blankets. Short lengths of drill steel, a heavy hammer known as a singlejack, a shovel, an ax, and several other tools lay about. If further evidence were needed that something tragic and mysterious had closed down the mine and camp, it came when Scofield found the Dutch oven and lifted the lid. The vessel contained gold.

A story that later spread through the miners' grapevine had the oven quite full of dust and nuggets, had it still awaiting a lucky finder. This hopeful view was based on belief that the utensil had been entirely full of gold and that Scofield had transferred only a couple of handfuls to his pockets. The fact seems to be that it contained considerably less than $1,000 worth and that Scofield took it all. In any case our miner was more interested in where the real fortune lay, in the mine itself. He made a selection of what he considered average, representative samples of the already mined ore. Then, the dangers of desert thirst in mind, he started back on the long and unfamiliar trail. As indicated, he was none too observant on this hurried return and he made no written notes. He arrived safely at the spring and hurried on to Danby. Then, ignoring his tunneling job, he took the next train to Los Angeles.

An account of a lost mine necessarily ends in anticlimax. If there were a triumphant rediscovery the mine by now would have yielded up its treasure and its story would appear only in state or federal technical papers. The Dutch Oven Mine remained lost, a major disappointment and an incredible puzzle to Thomas Scofield. From the dust and nuggets taken from the oven and such gold as he pounded out of the ore he carried away, he realized a little more than $1,000. A sample of the vein matter which was assayed by John Herman in Los Angeles showed fantastic value, he said, though the precise figure was not revealed.

Aware of the dangers of solitary exploration deep in desert country, Scofield took in a trusted friend as partner. When the men tried to retrace the route into the Clipper range they could not even find the landmark spring, where the trail of the luckless early miner was first discerned. The desert played its old tricks of deception and confusion. Like other searchers we have met along the Colorado and still others that we shall meet later, Scofield and his partner tried again and again, only to be baffled.

At the end of the nineties, Scofield apparently had left the desert for good, but the alluring story of the Lost Dutch Oven was still told around many a campfire. Old-timers believed that if even the ready mined ore could be packed out to the railroad it was so rich that it might yield a small fortune in itself.

Does modern transport make a search more promising? A trail bike just possibly could make exploration faster in the various canyons leading into the Clipper range. This would facilitate search for the elusive spring. But is it still flowing after seventy years? A helicopter might help. It is excellent for mapping a route to be followed later on foot. Observation from above can usually spot a conspicuous mine dump, but as will be seen in connection with explorations in Arizona, positive identification of a particular mine from the air is rarely possible.

There is one modern development which might provide a new approach to the Dutch Oven riddle. It will be recalled that Scofield went into the Clippers from Danby on the Santa Fe Railroad. This means that he entered the southern side of the range. Now the *width* of that range, from the first rising ground on the south to where it sinks into Clipper Wash and flat desert on the north side is barely seven miles. Could the old trail, which was long enough to cause Scofield real concern with regard to distance, time, and thirst, have led clear across the crest? Is the mine actually on the north slope? Within recent years a stretch of Interstate 40 has been realigned and now runs along the northern base of the mountains. This side formerly was very difficult of access. If one leaves his car where Interstate 40 crosses Clipper Wash, he need walk only four miles up the wash to be at the western tip of the mountain chain. If he prefers to start at the east end, the departure point would be the old Goldhammer Mine. Both these approaches are speculative but look interesting. Once in the mountains and on foot, however, the searcher faces the same confusing and somewhat dangerous conditions as did Scofield himself.

Thirty-three miles west of Danby is Amboy, now served by U.S. 66. Although a small place, it was the focal point of much early mining activity because it was on the Santa Fe Railroad. It was a point from which miners could ship ore to the smelters or bullion to the mint. Through the accident of a broken bag, some of the richest gold ore in all the West was exposed to a group of men who were lounging, late one afternoon, on the Amboy station platform. The ore belonged to a white-bearded recluse known as Hermit John and it came from an old Spanish mine that he had found.

This was in the days that the Virginia Dale mines were being extensively worked, and though some forty miles away across the torrid expanse of Bristol Dry Lake, the Dale mines were employing Amboy as a shipping point.

On the afternoon in question the Hermit emerged from the heat waves and intermittent mirages of the dry lake and brought his string of five burros directly to the railroad station. There he shifted a number of heavy ore sacks off the animals and onto the loading platform. According to Bill Pine, the Santa Fe agent, this would be the third time that Hermit John had made such a shipment. The ore went to the American Smelting and Refining Company smelter at Selby, near San Francisco.

Noting that one of the sacks had a large tear in it, Agent Pine called John's attention to the ore spilling out on the platform. The old man fumbled in the gear packed on one of his burros, extracted another bag, and dumped out the ore to resack it. The little group of bystanders stared. The rock was a light gray quartz, with liberal iron stain and, as one of the observers related, "almost plastered together with gold."

The spectators, besides Station Agent Pine, were John Locke, the local storekeeper, Jim Walsh, the Santa Fe's section foreman, Pete Ring, a mining man, and John D. Mitchell, whom we have met before.

As he had on previous occasions, the Hermit withdrew to some mesquite trees a short distance from the tracks, removed the remaining packs from his burros, and prepared to spend the night. During the evening, John Mitchell paid him a friendly call. The tall, patriarchal figure had disposed himself so his campfire flickered on a large, leather-bound Bible, which, it developed, he always carried and in which he read and reread only the Book of Psalms. Mitchell, for his part, carried on his hip a flask of Taos Lightning. The evening became convivial and as usual in the California of those days the men's talk turned to gold.

Hermit John could not well disclaim "laying up treasures on earth." The scene at the station had been too revealing. His great strike, he told Mitchell, was definitely a Spanish or, possibly, a Mexican mine and very old. Several mining tools of antique make still lay around. There were three graves, but nothing in the way of carving on wood or stone to identify them. An arrastra, which is a primitive ore grinding plant, its stone floor fairly intact, showed how the miners had recovered their gold. The Hermit, of course, did not offer any clue to the location and even stopped short of saying how the deposit lay with regard to its surroundings. He did mention that there was a partly caved shaft that might once have been a well, but also might be an old mining excavation.

This was all the information imparted to Mitchell, even under the influence of the fiery Taos whiskey. We do not know whether the antiquated tools were left lying around and are there to this day, nor do we know what the forsaken graves look like. Even if the old arrastra is located, it might not be in the immediate vicinity of the mine. For various reasons, this crude but effective milling device could have been constructed some distance away.

This conversation with Mitchell over the evening campfire may have been the old man's last contact with humankind. Next morning, having posted a letter to the smelter, loaded his burros with supplies from John Locke's store, and filled his capacious water cans from the Santa Fe well, the gangling figure and his animals receded in the heat and glare of Bristol Dry Lake. He never was seen in Amboy again. We know this much in a negative way: an Armenian freighter who was making regular trips between Amboy and Virginia Dale and who formerly had sighted old John from time to time, never saw him from this occasion onward.

Where do we look for the ancient mine?

The first mountains the Hermit would encounter beyond the dry lake are the Calumets. They form a picturesque and little-explored range which has yielded virtually no mineral information, favorable or unfavorable. Toward the south the Calumets taper off into level desert. In this expanse there is a succession of small, scattered uplifts which seem to have escaped any names. Their eastern slopes look down on Cadiz Dry Lake and Cadiz Valley, which can be approached via State Route 62. If the old man had gone still farther he would have been in the Sheep Hole Mountains, which, as we shall see later, had their share of rich gold strikes. On the scanty evidence, the Sheep Holes seem favored.

One hint, possibly too vague to be of assistance, is recalled. Hermit John told Mitchell that north of his discovery, the distance not indicated, there was a very large outcrop of iron ore. He had worked the deposit long enough to determine that it contained gold, but not in such quantity that the ore could be mined, packed out on burros, and shipped to San Francisco under the conditions that faced him at the time. A search for the massive iron and gold outcrop might be a reasonable undertaking today. It might lead one on to the old Spanish mine and its wealth. Under modern conditions, it might prove a bonanza in itself.

The country north of Amboy also is of considerable interest. From near Amboy there is now an improved county road that runs north for eighty miles through quite barren desert country. It terminates at Baker, on Interstate 15. Baker is a small place but is well known to travelers between Los Angeles and Las Vegas, since it is a natural refueling and refreshment stop. Even the sparse desert vegetation gets sparser on this Amboy-Baker route. The road passes black cinder cones, vast lava beds, some sprawling sand dunes, and a dismal alkaline expanse known as the Devil's Playground. These natural features are all quite valueless as far as precious metals are concerned, except possibly some undiscovered spot in the Devil's Playground.

In 1933 a prospector was found dying of thirst on the northern edge of this wasteland, and in his pockets were a number of gold nuggets, the largest almost an inch in diameter. He died without regaining consciousness, but was recognized as a man who had bought supplies in Baker and had remarked that he was going prospecting in the Playground. He obviously found gold, but how important the discovery was, or where it was, never has come to light. The incident was tantalizing enough to send a local man, Albert G. Walker, and a companion on a search. They found no gold, but it might be argued that they nevertheless had good luck–they came out of the Devil's Playground alive.

Thirty miles north of Baker there is an old mine of historical interest, the Amargosa. Whether it is of practical mining interest is uncertain. One view is that its ore body is worked out. It is true that during the present century only intermittent leasers have shown any interest in the Amargosa and they have reported nothing of much value. Neither have they reported much serious additional exploration. Another view, also without contemporary supporting evidence, is that there are massive untapped reserves and that the next hundred years will be the Amargosa's finest. With perhaps faulty logic, the belated discovery of the Big Bonanza at Virginia City is cited.

The mine stands high on a very rugged peak, within a mile of State Route 127. This is a modern highway which runs north from Baker, serves the eastern side of Death Valley National Monument, and then connects with the Nevada road system. The name of the mine property comes from the Amargosa River, said to be the longest underground stream in the world. The elusive water surfaces in Death Valley for a short distance. In the flats below the lofty mine the flow is all below the desert surface. The only water in the area is the bitter trickle at Salt Spring, which even animals will not drink. Disappointment with Salt Spring is feelingly expressed in John C. Frémont's account of his expedition of 1844.

Mexicans may have mined the rich surface ore of the Amargosa in the early 1800s. The first work of historical record was in 1849, when Sheldon Stoddard, a Mormon frontiersman and guide, was in possession. He and his associates constructed an arrastra, which was quite adequate for making recovery from the gold-studded "picture rock." This is a miner's term for ore with highly visible gold in it. Such ore was bountiful in the mine's early days. The Stoddard operation was quickly broken up by Indians. Stoddard escaped alive and knowledge of the abandoned bonanza soon spread throughout the Southwest.

The story from then on was one of repeated attempts to mine the rich ore and of repeated Indian attacks. There are four recorded onslaughts by the red men and there probably were others. In each case some miners were killed and the remainder driven away. In the more peaceful era of the 1890s the Amargosa was worked by Adrian Egbert of Daggett. The place seems to have been deserted for a number of years and Egbert heard of it from an aged Mexican. The earlier operations, intermittent and fleeting as they were, had exhausted the bonanza ore at the surface. Egbert and associates set up a mill for working vein matter of more modest values. A fortresslike stone building was erected, not for defense but for storage. It remains there, an imposing landmark.

There apparently has been no occasion for a comprehensive study of the property by state field men, and the mine and its environs remain a puzzle. Pervasive mineralization in the rugged, barren mountainside, plus the phenomenal richness of the original outcrop, persuade some mining men that either the Amargosa proper or other sections of this gold-bearing zone have additional riches to offer.

There is authenticated placer ground in this general region. Silver Dry Lake is a conspicuous sight just north of Baker. Across the lake in a northwesterly direction, says the California Division of Mines, is a deposit of gold-bearing gravel traceable for five miles. An attempt to work it was given up, not because of lack of value but because the exceedingly fine gold could not be recovered by the methods used.

A much more concentrated deposit, a rich placer bar in the old tradition, was found in 1872 by Johnny McCloskey of Bishop, California. The discovery came at an unfortunate juncture. McCloskey had his wife and small daughter with him on a long homeward trek across the desert which had been marked by ill luck, hardship, and some danger. The location of the strike, which McCloskey never was able to find again, was thirty to forty miles above the Amargosa Mine on the dry Amargosa River course. At the discovery point, however, clear water came to the surface for a short stretch. It was a logical place to camp and a logical place to try some brief gold panning. He found the river sands very rich. Although the father felt he had a fortune in his grasp, the little family had to push on without delay.

Shortage of food? Threat of one of the desert's truly dangerous cloudbursts? A sick child? The reason for the haste remains unknown. The consequence, however, was that the family arrived home safely but a rich mine was lost.

We have been following an irregular course westward from the Colorado River and are drawing nearer the populous coastal region of Southern California. There will still be promising mineralized districts to explore, and they will have the advantage of being closer to supply and outfitting points. Before leaving the outer desert, note should be taken of several old placer fields which, on the basis of fairly recent information, still yield colors for the hobby gold miner and might encourage more serious operations.

These areas can best be described in relation to Barstow, California, an old-time railroad center and now the point where busy Interstate 15 and Interstate 40 converge in their course to Los Angeles.

The Coolgardie Placers, obviously taking their name from the rich Australian mining district, lie about fourteen miles north of Barstow amid a web of unpaved desert roads. Some local maps still identify the forsaken camp by name. The most recent state report on these diggings was that of 1930, at which time the place was virtually deserted after an estimated production of $100,000. This gold was obtained by the dry washing method. Some visitors in 1967 were interviewed by the press and reported getting colors in every pan even in this worked-over ground. The auriferous area is identified by a number of abandoned shacks and ample evidence of digging. Occasionally an old miner may still be encountered at Coolgardie, prowling about in a search for "hot spots."

Also north of Barstow and about four miles due east of the Coolgardie field are the Williams Well Placers. The country here is flat. Gold is not confined to channels, but according to engineers' reports is distributed in a mantle of earth covering several square miles. During the depression days of the 1930s, Jasper Dodson of Barstow operated six little dry washing plants powered by gasoline engines and took out several thousand dollars.

In the low mountains south and east of the two placer deposits are a number of old lode mines, but none appears to have been important enough to be noted in state or federal reports. Part of this region, known as Rainbow Basin, has been taken over as a public recreation area.

A third placer field in the Barstow district is found on the southwest slope of the Newberry Mountains. This range lies east-southeast of

Barstow, and the diggings may be reached by the graded Camp Rock Road branching off Interstate 40. The names Camp Rock Placers and Clark Diggings both have been applied to this site. Its attractive feature is that bedrock, where gold normally is concentrated in any placer deposit, is found here within two feet of the surface in some places and seldom deeper than twenty-five feet.

2
West of the Orocopias

The Orocopia Mountains are an unimpressive and little-known range, but they are a convenient point of reference for separating the large outer deserts from the mineral areas closer to the Pacific Coast.

East of the Orocopias, Southern California is mostly arid, sparsely settled country. To the west, that is between these mountains and the ocean, there also are large areas of desert but they are interspersed with tracts of irrigated land. Water supply and population increase as one moves nearer the Coast. Eventually the traveler is in the vast Los Angeles metropolitan complex, where the bulk of California's population has chosen to live.

Even in this congested and overdeveloped section there are a few gold-mining opportunities. When we have examined various intervening places of promise these close-in spots will be described.

The Orocopias will be feelingly remembered by any surviving drivers from the early cross-country trucking days. The principal route to Los Angeles, a narrow two lanes, passed through this range and pioneer truckers have a grim recollection of the grades and curves encountered. In case of mechanical failure the driver found himself in an empty, waterless region with any assistance, with even food and water, many miles away. The mountains themselves are as desolate as ever, but transport now moves along a wide straight freeway, Interstate 10, skirting the northern base.

Some of the most promising mineralization encountered as we move west is that of the Ord Mountains. The range rises just north of modern developments in Lucerne Valley, which is served by State Route 18. A good road connecting Lucerne with the thriving little city of Barstow runs close to the Ords, and there are several unimproved branch roads that will take you to old mines and very close to some barely touched land which would appear to be excellent prospecting territory.

Ultimately the major product of the little mountain chain may turn out to be copper, but official reports give gold almost equal billing. Early reconnaissances tell of veins of such impressive size that one wonders whether the engineers of the last century were using the term "vein" in the sense understood today. We are told that "fissure veins ranging in thickness from 5 to 25 feet occur along the west side of the mountains" and that they could be traced "for at least 16,500 feet [about three miles] and perhaps 19,500 feet or more." These were primarily copper deposits, with the gold content stated to be between two and three tenths of an ounce per ton. The information is from an official state report.

Another such document relates that gold-bearing sulphides can be traced on the surface for two and a half miles. There were several workings on the formation referred to, known collectively as the Brilliant Mine. A comparatively recent operation on this or a similar outcrop, and for which no name is given, terminated in 1934. The visible vein matter, stated to be on the southeast slope of the range, was described as eight feet wide and visible on the surface for 3,000 feet. Values were in free gold and the ore that was milled averaged $8.90 a ton at the old $20 value of gold. Another massive and barely exploited deposit is listed as the Ord Mountain Mine. This yielded copper and gold. It apparently required twenty-eight contiguous claims to take in this extensive mineralized zone. Yet, according to F. Harold Weber, Jr., a state geologist, operations that extended on an intermittent basis over a span of eighty years produced only 2,000 tons of ore.

Why a great mine was not developed in the Ord Mountains is a puzzle. It is uncertain whether the answer lies in the mountains themselves, rugged and remote as they are, or in exaggerated descriptions. The excellence of California's state mining reports already has been noted. Some of them, however, especially in the early days, do not make it clear whether information on a given mine was based on a personal, professional visit and examination or whether the writer obtained his facts secondhand. The massive unnamed lode cited above under the 1934 date is especially tantalizing. Using the engineer's data and the gold price prevailing as this paragraph is being written would give us ore averaging $71 a ton across a vein eight feet wide and traceable for half a mile on the surface of the ground. This would be a bonanza in anybody's book.

The Ord Mountains are almost encircled by other little desert ranges. The only break is due south, where one looks out on flat Lucerne Valley. On the record these peripheral ranges show lesser mineralization than the Ords, but there has been gold mining in all of them and they all could be of interest under today's conditions. Reading clockwise from the southwest, these surrounding ranges bear the names Granite, Sidewinder, Stoddard, Newberry, Rodman, and Fry. A network of old roads of dubious safety exists and a number of the more important mines are shown on local maps. A mining landmark directly east of the Ords and a little south of the old town of Ludlow is the Bagdad-Chase Mine, one of the richest and longest-lived producers in Southern California. Also in this direction are several black volcanic craters. One of these, or perhaps a similar cone farther away, was the scene of one of the most unusual exploits in all western mining.

It concerns two miners who certainly had very precise information on a rich surface deposit of gold nuggets. Prospectors had heard of the hoard in the early 1850s and, as we shall see, a few of them had made ill-starred attempts to reach it. The successful trip appears to have been made in the mid-1860s. The partners spent almost a year in logistical preparations for a fearsome trek into one of these volcanic wastelands. The careful planning enabled them to reach their goal and they came back with all the wealth they could carry.

Many another prospector has wanted to make his own try for this bonanza, and has tried in vain to talk with a surviving partner or his descendants. Unfortunately the names of the protagonists, even if known at their departure point, which was Twentynine Palms, have been lost with the years. This makes the account unsatisfactory, even open to suspicion. But the story has come down to us from various sources and always consistent as to facts. If you go prospecting in the California desert you are very likely to hear it, just as you are likely to hear of the underground lake on or near Kokoweef. It therefore seems permissible to marshal the available facts and set them down.

We know that one of the partners was an Indian. It is possible that the gold was tribal knowledge. The red man's hardships during the trip were so severe that he died soon after getting back to civilization, possibly from the effects of a strange sulphurous miasma that rose around the nugget deposit. The other miner was a white man and in his hard-won opulence he purchased a fine citrus grove in Riverside County. He resolutely refused to enter the crater country again.

The two men set out from Twentynine Palms in the reasonably cool autumn weather. Their takeoff point gives little clue to their destination, for that settlement, then known as Oasis, was the supply point for widespread prospecting and mining. They had been busy for months hiking far into the desert. Water in sealed containers and canned or packaged food had been buried at intervals. These caches were for their return journey. Outward bound, they were to carry water and provisions on two pack mules.

It is evident that the men had exact knowledge of where they were going. This knowledge is one of the mysteries of the case. Had the information been handed down to the Indian by his forebears, or had a white prospector once stumbled upon the gold?

As we know, the enterprise was a success. There appears to have been a steep climb encountered as the two neared their goal, a climb up the side of an old volcano. The pack mules were turned loose on the flat to fend for themselves and were not seen again. As planned, the partners made the final ascent on foot and necessarily walked the entire distance of the return journey. They were weighted down with gold. On the outward trip they had found several skeletons. It was assumed that the unfortunate men had been headed toward the same bonanza as themselves. Quite evidently the story of the nuggets had reached others.

The value of the gold carried out is usually given as $65,000, a substantial sum in those early days. It is pointless to try to figure out how much weight the two men could carry. They certainly were weary and they may have been ill. The lone survivor made much of the strange pollution in the air at or very near the gold deposit. It rose as a yellow dust on the volcanic slope, stirring from the earth with the slightest breeze. Settling on the perspiring skin, it proved intensely irritating. As an unknown and malignant chemical it must have inspired some fright and thus become a double hazard. Foul air in an old mine shaft or tunnel can be dangerous and miners know it. This pollution assailed them in open air. It appears to be the sole instance of the desert's adding a gratuitous surface poison to its already considerable perils.

As to what the place looks like, one account says the gold was found on "a small black mountain." This supports the general impression that it lay on or in a cinder cone. A larger, prominent landmark of this kind is Pisgah Crater, clearly visible from U.S. 66 and noted on many maps. It is owned by the Santa Fe Railroad, which at times has processed and shipped the volcanic ash. Such natural features are fairly common on the California desert. A survey by Bruce Vinson found about one hundred such craters, most of them looming up as symmetrical black cones. A discovery of precious metals is not to be expected in lava flows or volcanic remnants as such, but there are instances of other formations, older in time and mineral in character, being raised or shifted by vulcanism and consequently occurring in this unlikely association. The strange gold deposit found by the two partners may therefore be classified as an unusual but not impossible natural phenomenon.

Near Twentynine Palms there is an opportunity to examine an abandoned mining district which is easy to reach and, being fairly compact, is also easy to explore. To a prospective new enterpriser in the field of lode mining, a look at the old workings cannot fail to be interesting and instructive. There also may be a practical attraction. Few if any gold districts are completely mined out. Today, evaluated in the light of modern transport and technology and greatly increased prices for gold and silver, the ore left in place may offer a real prospect of profit.

The area is known as the Dale Mining District, taking its name from its first and perhaps its best mine, the Virginia Dale. It is reached by State Route 62, the turnoff being twelve miles east of Twentynine Palms and 144 miles from Los Angeles. This junction is the site of Old Dale. Later a New Dale grew up closer to the mines. Today Old Dale and New Dale are not even ghost towns, merely "historical sites." Their buildings have been torn apart for lumber, mine machinery has been carted away by scrap dealers, an abandoned safe blown open by hopeful cracksmen. Only the mines and their characteristic dumps of tailings remain.

It is a short drive from Route 62 to the Virginia Dale and another heavy producer, the Supply Mine. Charles M. Schwab, the steel magnate, was a heavy investor in the Supply and its sister mine, the O.K. With due care about deteriorating roads, one can drive to the Golden Crown, another major operation, and view perhaps a dozen more old workings. To reach the Brooklyn may require a climb on foot. A state mining report of 1931 estimated that the inactive Brooklyn still contained commercial grade ore worth $450,000 at the old gold price. Operations were at their height in the 1880s and 1890s, at which time the Virginia Dale property was considered to be worth $500,000. The

Dale mines kept going with varying success until World War I. Leaner or more complex ores were making most of them marginal or losing operations. Increasing costs during wartime, while the price of gold remained fixed, were a factor in their closing, and well-paid jobs in war industries drained away miners and even some owner-operators. Except for a little intermittent leasing and possibly some very recent activity, Old Dale, New Dale, and their surrounding mines have remained dead since World War I. The 1920 census gave the population of New Dale as "1" and this resident, Sam Joiner by name, has now gone to his reward.

Before getting interested in any individual mine, the investigator should determine whether it now lies in the Joshua Tree National Monument. Creation of this reserve lopped off a number of mines in the southern part of the mineralized zone. Most of the Pinto Mountains and the better of Dale's old mines fortunately are located to the north and outside the park boundaries.

At least two of the best mines went largely undeveloped because of litigation, which was the curse of many a western mining camp. A rich claim often seemed a vulnerable property to certain shrewd lawyers. The legal squabbles along the great Comstock Lode, if there were any point in retelling them, would fill a volume. In speaking of such troubles at Dale, the veteran mining man Ronald Dean Miller writes in his *Mines of the High Desert:*

"It is a sad commentary on the ways of miners that when things are not going well, while the digging is tough, they work well together to seek success. But when success appears close, or has arrived, the working harmony falls apart. Lawsuits start, injunctions issue and soon all work bogs down in a legal morass."

It perhaps should be recorded that in 1923 a fast and spectacular recovery of gold was made at New Dale—$25,000 in a single day. It was witnessed, but not shared, by David and Anna Poste, who were in New Dale in connection with a lease on one of the inactive mines.

A chauffeur-driven car arrived at the ghost camp, which in that year had most of its buildings empty but still standing. A well-dressed woman alighted, looked around as if getting her bearings, then went up a short trail to an old house standing apart from the main street. On a map of New Dale in Miller's book, this structure is designated "Red Light District." The visitor returned to the car, carrying with great difficulty a heavy iron box. Managing to get her burden into the rear seat, she climbed in after it and was driven away.

The Poste couple had witnessed the performance and described it to Sam Joiner, New Dale's lingering permanent resident. The old man chuckled and said he knew who the woman was and knew of her past connection with the isolated house. As for the heavy little chest, in 1910 gold bullion worth $25,000 had been stolen from one of the mines and, up to this episode at least, never had been recovered.

Mines of the Dale district lie on both sides of the line dividing San Bernardino and Riverside counties. San Bernardino was the largest county in all the United States until the admission of Alaska. Only a hundredth part of it is arable land. Local boosters might quarrel with the view that the other 99 per cent is desert, but the prospector, rationing himself with sips of water from his canteen, will have no illusions. The redeeming feature of the San Bernardino desert is the prevalence of gold. We have the word of the California Division of Mines that gold is disseminated in all or virtually all of the many little mountain chains scattered throughout the county.

When we return to the highway from the old scenes of Virginia Dale mining, the wide, flat expanse of Dale Dry Lake comes into view. There are ruins of an old salt works on the eastern edge and beyond that one sights the Sheep Hole Mountains.

The range is some fourteen miles in length. Several gold mines have operated on the fringe of the Sheep Holes, and it was the opinion of Virginia Dale miners that the interior was heavily mineralized. Phil Sullivan, an experienced and fairly successful desert miner, told friends that if he were again a young man and setting out to find a bonanza he would head direct for the Sheep Holes. The mountains are difficult country. There is only one known spring, and if there exist any of the life-saving *tinajas,* or natural rock water troughs, they are not of record. Those who have studied Hermit John's fragmentary story believe his old Spanish mine was in the Sheep Hole range. A state report tells briefly of two parallel gold-bearing veins "in the center" of these mountains, where ore had been worked in an arrastra. The information is extremely vague.

The one man known to have found a rich deposit in this inhospitable terrain endured great hardships on his return journey and died without benefitting from his strike. He was L. O. Long, believed to have lived in Fresno and other towns in California's central valley, and once to have been engaged in mining at El Dorado on the Mother Lode.

Long was found exhausted near Dale Dry Lake. He had a small

burden of gold nuggets. His water and food had been used up and the dehydration of the thirsting desert wanderer was far advanced. The man's condition worsened rapidly, but he did manage to tell of finding a surface concentration of placer gold. It was in a brushy canyon, he said, and this canyon contained a spring with a little green foliage around it. There was one definite marker at the diggings, his shotgun. On the long tramp from his discovery back to civilization, Long could carry just so much weight and he preferred gold to firearms. He therefore shed his gun, wedging it into a rocky formation at the discovery site. Within days the emaciated man expired.

The Sheep Hole riches missed rediscovery only by a quirk of fate. In 1925 another miner chanced into the little canyon, and he found the shotgun. Later, he estimated that the spot was about ten miles into the range from the eastern edge of Dale Dry Lake. The gun was of a fine make and still in good condition. Unaware at the time of its significance, the miner appropriated it and walked on, musing on his good luck. He was walking away from a fortune.

The Sheep Hole range can be penetrated a short distance by motor car, but only at the south end. A dirt road leaves State Route 62 two miles east of Clark Pass, then splits into several branches. All the branches come to dead ends.

Running clear across the desert country, in this area about fourteen miles south of the Sheep Holes, is the Colorado River Aqueduct. It extends from the river to its water distribution system for the coastal cities. It is of interest to the gold hunter because the road that accompanies it, with numerous connections with main highways, can prove useful. It also is of interest because of a gold-bearing ancient river channel that was uncovered during construction work.

The pipeline project involved tunneling through several small mountains. The rock usually was solid and self-supporting, but one bore in the Little San Bernardino Mountains struck a body of crumbling, loosely cemented gravel. Its character showed it to be the bed of an ancient stream, now left high and dry and embedded in a younger formation. Such channels, uplifted by some convulsion of the geologic past, are a feature of the Sierran Gold Belt and have been extensively mined. Workmen tunneled through this soft spot for about sixteen feet and were in solid rock again. To forestall any cave-in that might partially block the tunnel they constructed stout wooden retaining walls on each side.

A machinist on the project, who understandably gave a fictitious name, John Mason, in telling his story, knew something about the northern channel deposits. He found a place where he could squeeze behind the plank barrier. This he did one night and scooped up some gravel. Washed out, it left a trail of coarse gold in the pan. In the free and easy ways of a desert construction camp, he was able to go back again and again, into the tunnel and behind the shoring. By the time his crew moved on to the next assignment, he had a valuable little jar of gold.

There could be no return to the tunnel and its gold. For the protection of its water, the city of Los Angeles put locked steel doors on all the tunnels and enclosed these entrances with strong wire fences. In telling his story to Donald R. Hoff years later, the pseudonymous John Mason speculated on whether the same channel might be traced on the surface, and a mining location made outside aqueduct property. He decided he was too old for the search himself.

The start of such an undertaking would be at the old concrete foundations for the one-time Berdoo construction camp, which are still in place. Access would appear to be simple. The camp was in Berdoo Canyon on the southwest side of the Little San Bernardino Mountains. Local maps show a Berdoo Canyon Road which branches northeast off Dillon Road about five miles north of Indio.

In these investigations we have been on or near California's State Route 62, a very useful road that crosses the desert between the two coast-to-coast arterials, Interstates 10 and 40. It merges with Interstate 10 not far from the widely publicized resort of Palm Springs. Before leaving Route 62 we should take note of a gold-bearing region which, in spite of its proximity to the mass population of Southern California, contains much primitive country and some demonstrably promising prospecting ground.

Since Twentynine Palms has figured prominently in the previous pages, consider that we are traveling west from that point. After twenty miles we are at the community of Yucca Valley and a paved road leading northwest to Pioneertown. This apparently antiquated village, with its gold-rush architecture, was built as a movie set. It is now maintained as a resort, with emphasis on old west atmosphere.

A little beyond Pioneertown one can feel quite realistic about being in the old West. Pavement ends and dirt roads straggle off into various dry, empty ravines and canyons. Both roads and ravines slope up rather sharply, for they are on the eastern ascent into the San Bernardino Mountains, which rise to an elevation of nearly 10,000 feet.

Placer gold has been found in all of these drainage courses. By name the ravines are Burns, Pipes, Antelope, Water, Chapparal, Little Morongo, and Big Morongo. There are additional, smaller gullies on which no names have been conferred.

The best placer ground and the strongest indications of valuable gold veins in place have been recorded in Burns and Pipes canyons. These contain visible remnants of placer mining, dating largely from the 1930s. This was the time of the nation's great economic depression, and a number of able-bodied jobless men were having a try at gold mining for lack of any other way to earn a livelihood. The returns, in the San Bernardinos and elsewhere, were usually meager but it was a living of sorts. When an improved economy and, later, a wartime industrial revival opened up good jobs the placers were largely deserted. They are not made up of rich gravel, but neither, under these circumstances, did they get worked out.

One man who stayed on beyond the depression as a full-time miner was Charles V. McClure. He was fifty years old when the economic slump wiped out his job and his savings and caused him to take to the hills. Twenty-four years later he was still living in Burns Canyon, not rich but enjoying life and recovering enough gold for a modest income. In 1958 he had visitors from *Desert* magazine, and demonstrated to them that he could get color from any one of his numerous prospect holes. In McClure's opinion, the ravines of the entire east slope of the San Bernardinos were auriferous in some degree.

The chance for real wealth, he believed, lay in the veins from which the placer gold had weathered out. He had found float ore running $200 a ton. McClure had been unable to trace it to its source, although there had to be such a source, and if discovered it almost certainly would be of value.

The existence of at least one high-grade, eminently workable deposit is demonstrated by the luck of the flamboyant E. J. "Lucky" Baldwin. His Doble Mine, some distance northwest of McClure's diggings, turned out to be a rich one. Some surface structures at the presumably worked-out shaft are standing today. They are near the scattered rubble where the town of Doble once stood.

The trail to Baldwin's mine takes us into the San Bernardino National Forest via Burns Canyon. This graded dirt road is the best route for an exploratory trip through the district. At the old camp of Doble one is near the popular resort area of Big Bear Lake and also near the century-old gold-mining district of Holcomb Valley.

In 1860 rich gravel was discovered in Holcomb Valley. The area was close to Southern California population centers and consequently was worked so intensively that any further returns from placer mining would seem unlikely. The prospects for finding profitable ore in place may be slightly better and a principal objective would be the Lost Van Dusen Mine.

Jed Van Dusen and a partner whose name has not been preserved struck out from the diminishing placer field and prospected the narrow gorge now known as Van Dusen Canyon. They struck it rich, a quartz outcrop of substantial size showing visible gold. Working as secretly as possible, they accumulated a considerable stake, so considerable in fact that the partner took his share and left for home.

Van Dusen stayed on. Although the mine itself was hidden, there was no secret about Van Dusen's place of residence, and on weekends he usually was mingling with friends. When he had not been seen for some time, these friends investigated. He was found in his cabin, murdered.

The killer never was identified and the mine, known to be some distance from the cabin, never was found. Van Dusen Canyon is flanked on the west by Bertha Peak and by Gold Mountain on the east. Somewhere in that region the rich ledge awaits a later discoverer.

Before we leave this territory note may be taken of a possibly neglected field of exploration not a great distance from the placers worked by Charles McClure. It is an area dominated by Big Horn Mountain, and comprising between fifty and sixty square miles of arid, unpopulated country. It is served only by sketchy dirt roads, few in number. One such route, leading west from Ruby Mountain and coming to a dead end, penetrates the southern section. From the north it may be entered by several similar dead-end roads leading off the pavement in the vicinity of Timico Acres.

There is little information, favorable or unfavorable, about this country, for any early exploration seems to have escaped the record. Along its western edge are the Akron, Reef, Lester, and Santa Fe mines, while to the east, somewhat outside the area, are the Cholla, Los Padres, and Green Hornet. These peripheral operations may well encourage prospecting in the interior, using Big Horn Mountain as a central landmark. In a day of four-wheel-drive vehicles, of dune buggies and trail bikes, such exploration becomes much more practicable than in the past although to carry ample water is as necessary a precaution as ever.

North of this area are the Fry Mountains and the Rodman Mountains, both of which must be described as unknown quantities, since they figure briefly or not at all in mineralogical reports. Both of the ranges lie close to the Ord Mountains, which, as we have seen, have important deposits of gold and copper. We are now near the junction of Interstate 40 and Interstate 15 at Barstow, and also are near the old silver camp of Calico.

Calico is now reconstructed as a ghost town for tourists. Their striking colors and patterns explain the name of the Calico Mountains, which rise as a backdrop to the town. For those unfamiliar with underground workings, a tour of the tunnel and drifts of an old silver mine is available as one of the recreational features. (Another authentic view of underground mine workings may be had in the Nevada State Museum at Carson City.) The Calicos themselves probably have been fully explored but there is a promising area of unpopulated country spreading out behind them in a northeasterly direction. It embraces some six hundred to eight hundred square miles.

This is virtually all public domain and open to mining location. A large area is taken up by Coyote Dry Lake, and overlooking this great *playa* are several once productive mines. They include the Olympus at the north end of the lake and the Elvira on the south.

Seven miles east of Coyote Dry Lake are the Alvord Mountains, rising to 3,450 in height, and we are reminded of the curious black formation, heavily charged with gold, that has become known as the Lost Alvord Mine. Rather than a "mine," it is a high-grade outcrop waiting to be mined. As far as is known, only a single fantastic specimen has been chipped off the ledge.

It was in the boom discovery year of 1860 that Charles Alvord, an Easterner sixty years of age, joined a desert prospecting party. Among his new associates he was described by the then widely used term "dude." The group was headed for Death Valley in hope of locating the Lost Gunsight Mine, a bonanza left behind by a hard-pressed emigrant party in the initial gold rush. The Gunsight was a rich and authentic silver discovery. Its story will be told in a succeeding chapter, for it remains an alluring object of search even today.

Alvord's party took off from Los Angeles and the route was much the same as Interstate 15 now follows. Our interest is in the men's layover of several days in what is now known as Spanish Canyon. This shallow depression, a ravine rather than a canyon, lies five miles north of Interstate 15, the turnoff being fourteen and a half miles east of

Yermo. The camp was chosen because of a good water seepage, and this source has been pretty well identified as the present Mule Springs.

There was some spare time as men and beasts rested. In company with one Asahel Bennett, Alvord did a little exploring. As the Easterner later remembered it, the two men passed over a low saddle some distance up the ravine, then decided to separate. This was unexplored territory at the time, with the possibility of a virgin mineral discovery, and it was reasoned that their chances were greater if they returned to camp by different routes.

It was Alvord who had the luck. He came across a curious outcropping of black rock and even his inexperienced eye told him it was something unusual. He put a piece of the rock in his pocket and doubtless noticed its unusual weight.

At camp the Easterner either timidly failed to show his black specimen to his companions or if he did they were not impressed by it. The party moved on. How thoroughly they searched for the Lost Gunsight is unknown. Certainly they did not find it. Some such expeditions broke up long before getting even close to their destinations as they experienced the unfamiliar heat and thirst of desert travel and brooded on the weary miles that lay ahead of them.

The group returned to Los Angeles, with Alvord and his all-but-forgotten specimen. Only then did some knowledgeable friend show interest in the strange rock and suggest an assay. The return showed that Alvord had found a deposit of gold ore and that the ore was very rich. The mineral was described as being chiefly manganese but carrying gold in very large amount. A new expedition was immediately organized.

As has been seen in relation to other lost bonanzas, the desert has an almost mystic capacity for concealing things. For the eastern "dude" it may have been exceptionally secretive. Alvord could not–or would not, as some of his partners charged–lead the way to the rich gold ledge. It is probable that he was genuinely lost. But there was suspicion that Alvord, now that he had his bearings, was saving the mine for himself. There was gunplay of some kind and by one account Alvord was killed.

Much of the information on the case came from Joe Clews, who seems to have been the one real friend that Alvord had among the miners of the original Gunsight search. Clews had actually seen the rich black ore, and after Alvord dropped from the scene he made some ineffectual searches of his own.

To give directions to the search area involves a bit of analysis. The intense interest in the episode resulted in some optimistic misnaming, and detailed maps of San Bernardino County show an Alvord Peak, an Alvord Mine, and an Alvord Well. None has any direct connection with the man himself or with the black manganese outcrop, although they help identify the area generally accepted as the lode's location.

Not unanimously accepted, however. Harold O. Weight, whose writings on desert topics are backed up by a great deal of personal exploration, believes the discovery was made in the Panamint Mountains. This range forms the western boundary of Death Valley National Monument and will figure prominently in a later part of this report. The Panamints would be the logical objective of the party seeking the Lost Gunsight Mine.

Weight has definitely placed Alvord in the Panamints, and has given circumstantial accounts of the Easterner's wanderings, alone or with fellow searchers, of his near escapes from starvation, and of his trouble with an abscessed injury to his thigh. He quotes a Dr. S. C. George as relating that Alvord was shot to death somewhere in the Panamints by a man named Jackson, who was enraged because Alvord could not or would not take him to the rich ledge.

It is hard to reconcile the two search areas, for they are more than fifty miles apart. The turnoff from Interstate 15 to the water source now known as Mule Springs, already alluded to, remains the most promising approach. Why? Because Joe Clews, probably Alvord's closest friend, repeatedly said so.

Southeast of the Alvord region we find the Maverick Brothers Mountains, which already have been mentioned in passing. More than passing attention must be given to the "T" they form with the north-trending Cady range, because from that T came tantalizing clues of a geological puzzle, a puzzle of some scientific interest and of more than slight financial interest. It concerned a discovery of volcanic obsidian containing gold, and the value of the glassy, gold-charged substance was said to exceed $350,000 a ton. A ton of obsidian, at rough calculation, would be a cube measuring two and a half to three feet in each dimension.

For a long time the secret of this wealth, a secret that did not include its precise location, reposed with Ray S. Caldwell. A practical mining man, he developed one of the last producing silver properties in the Calico range. After making several searches for the strange obsidian deposit, he decided to open the field to others. He chose *Desert* magazine for his disclosures and told his story in the January 1968 issue.

The discovery was made by a member of a family by the name of Williams who had come from Missouri and were settled at Yermo. A son, aged fifteen, was too young to work in the nearby mines, but during a summer school vacation he found employment as companion and helper to a Los Angeles artist who painted desert scenes.

The two started their trip into the desert country at Afton, a station on the Union Pacific railroad which now can be reached by a seven-mile stretch of road leading off Interstate 15 at a point thirty-four miles east of Barstow. Williams said he and the artist went into the mountains "by way of Afton Canyon." This could take them into the Maverick Brothers Mountains and toward the T, but it also leads toward the Cave Mountains, a separate range that cannot be altogether eliminated as a site. They were well provisioned and spent a pleasant week. While the artist sketched or painted, the boy had time on his hands. He roamed about, got interested in picking up "pretty rocks," and carried his little store home with him. What made some of them pretty was a combination of black and yellow color.

When he was older, young Williams followed his father's example and worked in the Calico silver mines. As a mucker, hauling ore out every day, he learned something of minerals and picked up a knowledge of prospecting. Some lingering impression now led him to get out the souvenirs of his trip with the artist. He knew by this time that any yellow metal occurring in quantity probably was not gold but an iron sulphide, and he was not particularly hopeful. Nevertheless, he took one of the shining black and yellow specimens to the mine's assayer. The liberal amount of metal it contained was not iron sulphide. It was gold, and the value figured out at $352,000 a ton. A sample sent to San Bernardino returned a similar assay.

We are not told why there was not an immediate rush into the nearby hills. Young Williams evidently held his tongue, and went on earning his living. As to the assayer, the story is that he was often far gone in his cups.

When the tale resumes, Williams is now "Old Missouri," with nearly a lifetime of indifferent success behind him as a prospector and miner. He never had found his way back to the bonanza of his boyhood years. In the late 1950s he was to make one more attempt, in company of a new-found friend and believer. This was Ray Caldwell, the Calico silver miner, who had met the old man and casually teamed up with him while both were prospecting. Hearing Old Missouri's story over a campfire, Caldwell was understandably excited. He listened carefully to the reminiscences of that long-past outing and believed that the general area, at least, could be determined. In partnership, the two men went to Afton station.

Williams unhesitatingly led the way to a lofty, bowl-shaped valley and appeared certain that this was the place where the artist had kept busy with sketch pad and canvas. The brief exploring time available to the two men on this occasion—Williams was in feeble health and Caldwell had a stalled truck to cope with—did not turn up anything of promise, even plain obsidian. When the trip was over, Old Missouri had had enough. There was a verbal agreement on shares, and Caldwell was coached for other, later searches to be made on his own. By 1968 the younger man also admitted failure.

The technological argument as to whether gold could occur in volcanic obsidian is something outside our scope. One experimenter states that he has created such a gold-obsidian specimen in his laboratory under conditions that could be duplicated in nature. The mountain valley, at least the one which the aging Missourian insisted was the right valley, is oval-shaped, surrounded by rugged peaks, dotted with small, weathered hills, and is about four miles across.

Old Missouri's obsidian is not noted in government mining reports and the reports have little to say about either the Cady or the Maverick Brothers mountains. There also is a dearth of information about the extensive Bristol range, the next one east of the Cadys. These mountains, especially the section north of Interstate 40, are among the blankest of the blank spots to the map-maker. This northern part of the Bristols is eighteen miles in extent, but the wide-ranging scouts of the Automobile Club of Southern California could find only two roads, both old and in dubious condition, leading into the section. There is known to be a natural stone arch near the northern end, and also an old Indian mine which was discovered by a white prospector in 1907.

South of Interstate 40 the Bristols are a little more accessible, and one mine, the Orange Blossom, has been of some importance.

Here we are close to the Old Dad Mountains, only a few miles to the northeast. Although a small range, mineralization is liberal enough for considerable mining to have developed in the past. The mountains lie close to the highway and most parts are within a single day's hike. Just beyond are the Granite Mountains, which contain quite a network of old roads and trails and an unusual number of springs for a desert range. A microwave relay station has provided an improved road leading well into the higher elevations.

At Barstow the interstate highway system swings south toward Los Angeles and we continue west on State Route 58. Ahead is Edwards Air Force Base, often in the news in connection with the testing of new military aircraft. The northeast corner of the base was very much in the news in 1926, for just outside the government reserve is the town of Kramer Junction, the nearest populated place to the Kramer Hills. These hills were the scene of a roaring gold rush in the "roaring twenties." Newspaper accounts were perhaps too vivid, so we quote from a report of the California Division of Mines:

"Thousands of people from Los Angeles, San Bernardino and adjoining territory visited the scene and hundreds of mining claims were filed. Many shallow shafts were sunk, but none of this work resulted in a mine." (In professional usage an excavation is not a "mine" until it has either shipped or milled ore. Until then, even if the work is extensive, it is a "prospect.")

The report goes on to say that in the low Kramer range there is a schist belt running east and west which is about two miles long and one hundred to two hundred feet wide. "High grade streaks and pockets are found where the schist is in contact with narrow dykes of volcanic rock," the report continues. "Gold is evenly disseminated and is all free, even at depths."

There would appear to be some similarity to the so-called seam mines of the Mother Lode, to be described later. Fascinating high-grade specimen ore often is found in seam mines, but the veinlets are mostly too narrow to pay for sustained mining. The Kramer rush was 100 per cent failure as far as any immediate profit was concerned. Last word from the area was that a study was under way to see whether it would pay to mine the little veins and enclosing barren rock by blasting and use of a power shovel.

Due north of Edwards Air Force Base is the scene of some immensely profitable mining. Twenty-six miles along U.S. 395 brings us to Randsburg and its neighboring camps of Johannesburg, Red Mountain, Midway, and Atolia.

Randsburg owes its existence to a Brooklyn newspaperman named F. M. Mooers. Sent west to write up the California mining activity, he became infected himself. He teamed up with a footloose prospector named C. A. Burcham and the two spent weary, discouraging months wandering in the arid country south and east of the operating mines of the Mother Lode. They were among the many miners, latecomers to the California mines, who formed something of a backwash into the as yet unexploited desert. The Goler placer excitement, to be described later, occurred in 1893 and may have attracted Mooers and Burcham to the general area of their great strike, which was made in 1895. While the two men were camped on a hillside above the site of the present Randsburg, Mooers went out to gather some of the scanty burro brush for a campfire. This homely chore resulted in his great strike. By this time the Easterner could recognize promising mineralization and he had no doubts about the outcrop of gold ore which jutted up amid the burro brush. The deposit became the great Yellow Aster Mine. After only a few months of development work the partners were offered $400,000 for the property. They declined to sell and went on to make more than $10 million over a space of twenty years.

There of course was a stampede to the newly discovered gold field. The camp that resulted took the name of Randsburg and a nearby settlement was called Johannesburg. Both names derive from the great South African mining region, the most productive in gold mining history, although the local deposits had little in common with the vast reef formation known as "The Rand" and virtually a symbol of inexhaustible wealth.

Some of the newcomers attracted by the Yellow Aster strike did very well for themselves. Dan Kelsey, who had been working as a freighter hauling borax, made a discovery which he named the Blue Daisy. He quickly sold it for $170,000. A county constable, Cyrus Drouillard, resigned his post to join the Randsburg rush and located the St. Elmo. It was so rich that he refused an offer of $300,000.

One mine surpassed even the Yellow Aster in production. This was the Kelly. It was not a gold but a silver property and its fabulously rich ore held out for a long time. One of its miners, in his later years, was put in charge of the little "mining museum" maintained at Randsburg. He could regale visitors with a firsthand account of how a single round of shots in the face of the Kelly tunnel, about six by seven feet in dimension, brought down a pile of rock worth $40,000. Few concentrations of silver have equaled the best ore shoots of the Kelly Mine.

Some of the old-timers at Randsburg will point up to the immense waste dump of the Yellow Aster and to Government Peak that rises just beyond and tell you that there is plenty of mineral to be found on the other side. In truth, the Rand Mountains twist away in a southerly direction for a good fifteen miles, with only two primitive old roads to indicate that anyone ever was interested in them. Did the swarming prospectors drawn by the Yellow Aster strike thoroughly explore this range? An area of some sixty square miles north of the present Colgate Road would seem to be worth some attention. The chief landmark in this area is Sidney Peak.

The El Paso Mountains lie twenty miles west of Randsburg. A good paved highway runs along their southern base and jeep trails lead into them. At their western end is a scenic area now constituted as Red Rock Canyon State Park. Not scenic but definitely interesting is a man-made attraction which geologists declare will endure longer than the pyramids of Egypt. It was created by a single man, the labor of a lifetime. It is the tunnel driven through Copper Mountain by William Henry "Burro" Schmidt. While it may, just may, conceal a "Crystal Room" rich in gold, its stated purpose was more commonplace: to facilitate the shipment of ore from a mine of uncertain value.

Schmidt came from a New England family with a history of tuberculosis. Brothers and sisters had died in their early years, and at the age of twenty-three William was told that his only hope was to get into the warm, dry air of the western deserts. It was sound advice, for the once susceptible young patient lived to be eighty-three—though a cynic might suggest that life as a "human mole," as Robert Ripley once dubbed the old tunneler, was hardly worth prolonging.

It was in 1906 that Schmidt located gold and copper claims high on Copper Mountain. Haulage of ore was almost impossible because of steep-walled canyons between the mine and any road. The miner hit upon the idea of tunneling clear through the mountain. The exit would be in clear view of the level road along the base of the El Paso range and ore could be sent down a chute, he believed, to a loading bin below.

Once started, the tunnel itself became Schmidt's obsession. The mine it was intended to serve was neglected and it never has been determined whether it contained ore that was worth shipping. Schmidt worked on ranches in the San Joaquin Valley each summer to earn enough money for mining supplies and to sustain his frugal way of life for the rest of the year. The tunnel was driven by hand labor, a single jack pounding ceaselessly on drill steel in conventional mining fashion. He economized in many ways, including the use of short fuses to his dynamite charges. On one occasion he cut them too short and appeared at a neighboring camp bloody but not seriously hurt. The tunnel went straight through into solid rock for 1,600 feet, then turned at a right angle. Eventually, after years of labor, it came out into daylight. Visitors say the view from this opening, commanding a wide expanse of the Mojave Desert, is an inspiring one.

One account has Schmidt working on his tunnel for thirty-two years, another says thirty-eight. In any case, he finished it in his sixty-ninth year and had more than a decade of life ahead of him to reflect on his achievement. There was a flurry of newspaper and magazine publicity, then the old man and his curious project were forgotten. In his last years he had the companionship of another miner, Mike Lee, and the partners made some small ore shipments.

A turnoff at the ghost town of Garlock on the Red Rock-Randsburg road will take one into the El Pasos. The route twists among numerous old mine workings and signs will eventually direct one to a steep climb and a dead end at the Schmidt tunnel. At last reports the tunnel and the miner's old cabin were open to visitors.

A belief grew up among El Paso miners that Schmidt's work on the tunnel was a blind, that he was guarding a great gold concentration somewhere in the bore. For unknown reasons, the supposed deposit was termed the Crystal Room. The name has some affinity with the so-called Bridal Chamber in the Cresson Mine at Cripple Creek, Colorado. The latter was a subterranean hollow whose walls were covered with crystallized gold. Schmidt's long tunnel through Copper Mountain did indeed cut through some veins of gold and copper. Two such leads were explored by drifts of fifty feet or so, and they may have supplied the ore that he and Mike Lee shipped out. No evidence of anything corresponding to the fancied Crystal Room has come to light.

To some western miners the El Paso Mountains were associated chiefly with the lost–and found–Goler placers. Goler was a German who in 1867 was making a lone trek across the then unexplored range. He came upon a surface deposit of gold nuggets at a spring, identified as Mesquite Spring, in the long, shallow ravine which maps now show under the name of Goler Canyon. He gathered some of the gold and hurried on, fearful, he explained later, that there were Indians about.

In Los Angeles he had trouble getting anyone to join him in working the deposit, despite the impressive display of gold nuggets. Perhaps his English was too hard to understand. He certainly was not a good promoter. At length he interested a rancher named G. P. Cuddeback, and the two men made several trips into the desert country. They got nowhere near Goler Canyon but they did discover some medium-grade placer ground which was worth working for several years.

With increased traffic in this part of the Mojave—we have witnessed the excitement at Randsburg—the original Goler gold, so near an important water supply and exposed on the surface, was inevitably discovered. The immediate area was very rich. Miners found good values also in several gulches tributary to Goler Canyon. The gulches can still be identified today, and bear the names Colorado, Red, Mormon, Slate, Sand, Reed, and Benson. Judge Edward Maginnis, who once served as Wells Fargo agent at Garlock, said that $350,000 in gold was shipped through his station, and that he believed the total yield in Goler Canyon and its branches exceeded a million.

There is ground for belief that Goler himself shared in the takings from the extensive placer deposits around his original find. In the 1870s a freight wagon driver known as Slate Range Jack gave a ride to a man who said he was Goler and who told in a matter-of-fact way about his discovery years before and about his own share in the present workings.

West of the El Paso Mountains there is an immense area of sparsely settled land, with a few widely scattered old mines. The only proved mineral zone lies still farther west in the higher, pine-clad elevations of the Piute Mountains. Here some good placer ground and several productive lode mines gave rise to the town of Claraville, which for a brief period in the late 1860s had a population of 3,000. Not a structure is standing today, but scraps of machinery and other metal relics are found lying about. Since the district is now part of the Sequoia National Forest, roads to some of the old camps are well maintained.

Eight miles west of the Piute mines, but much farther by road, is the old camp of Havilah. Here the swashbuckling Asberry Harpending, prominent in early San Francisco real estate and finance, had a rich mine. Harpending was a Southerner, and during the Civil War some sharp operators in San Francisco saw a chance to dispossess him because of his Confederate sympathies. The litigants managed to get a court order for taking over the mine and even to get a detachment of federal troops sent to enforce the order. Warned barely in time, Harpending issued guns to all his workmen, hastily prepared some defenses, and defied both troops and lawyers. The commanding officer, perhaps suspicious of the whole enterprise, declined to attack the miners. After several days of stalemate, he asked and received permission to withdraw his men. Harpending's title to the mine ultimately was confirmed. In the meantime he had taken out nearly a million dollars in gold.

Our general western course has been drawing us closer to the immense urban complex of the Los Angeles region. It was stated earlier that there was gold-mining country within daily commuting distance of downtown Los Angeles. It should be introduced with some words of caution.

The locale is the San Gabriel mountain range, which rises sharply above the congested population centers. From the heights one can look down on the city of Pasadena and its Tournament of Roses and can almost, but not quite, watch the annual Rose Bowl football game. There were small but fairly rich placers worked in earlier times back in the canyons and a good deal of weekend hobby panning goes on today. It is not very productive. There have been lode mines also, most of them operating on a thin margin between profit and loss.

The San Gabriels lie almost entirely within the Angeles National Forest, with a small section extending into the adjoining San Bernardino National Forest. One has a right to prospect in these forests and from a strictly legal standpoint has the right to hold a mining claim and work a mine. Such a venture conceivably could be a success, but the venture is not recommended. These mountains have been intensively developed as a recreational area and the millions of mobile, outdoor people who live nearby swarm into them every day of the year. There are many campgrounds, many scenic attractions, many sports areas and concessions. Most of the year the danger of forest and brush fires is very great, and even in this land of little rain there can be devastating floods. The forest rangers have a job on their hands, and it may be surmised that their attitude toward major mining operations in the midst of this compound devoted to timber preservation and recreation would be cool. And before considering fighting out one's rights on a legal basis, it might be well to consider the practical outlook for making money from a San Gabriel mine. It is not promising.

The old operations, most of them spanning a few years just before and just after the turn of the century, were not notably successful. The best-known mine was doubtless the Big Horn, high on the side of 9,339-foot Mount Baden-Powell. It was discovered in 1896 by Charles Vincent, a hunter who was tracking bighorn sheep. The Big Horn produced several hundred thousand dollars in gold before shutting down in 1908. It was expensive to operate. Its reduction mill was perched on a steep cliff and building of the essential haulage road saddled the operation with a heavy initial investment. Another productive mine was the Monte Cristo, which has conferred its name on the Monte Cristo public campground on County Road N 3. The mine itself is two miles away, up a narrow canyon. This property was operated for some fifty years, but has long been idle. A state report in 1927 showed output the previous year of $70,000 from a series of parallel veins.

The little settlement of Acton is a center of old mining scenes and the Governor Mine there was considered one of the best in the San Gabriels. Nearby Soledad and Aliso canyons yielded good placer returns in the 1890s. One can now travel through both these canyons on paved roads, and in Aliso can find several abandoned lode mines. Several less accessible canyons, Bouquet, Cave, Dry, Castaic, Haskell, Palomas, and a few others all were auriferous. Russ Leadabrand, who has written extensively on California byways, says that the last activity here of any scope was during the depression of the 1930s.

The San Gabriels offer mass recreation but it does not follow that they are a land of gentle slopes and easy strolls. Away from the roads and developed level stretches they are extremely rugged. Any serious exploration involves descents into deep canyons and steep climbs out of them, mostly over large rock formations and somewhat dangerous slopes of loose rocks. The road network is quite extensive. This is largely for fire-fighting purposes, but the multiplicity of paved, well-engineered routes through the range is a boon to any visitor, whether he is there for recreation or is intent on finding valuable mineral.

Our examination of the California deserts started in the southeastern corner of the state with routes leading westward from the Colorado River crossing at Yuma. There is one route that was not explored, the most southerly one. That is the one taken in 1829 by William "Pegleg" Smith, and no account of California gold would be complete without the story of the Lost Pegleg Mine.

It is told here out of geographical context because the Pegleg country, on the borderline of San Diego and Imperial counties, is something apart, an enclave that cannot be appended to other gold-bearing districts without confusion.

A search for Pegleg's gold as a specific objective can scarcely be advised. Men close to the mystery in both time and personal associations have hunted long and tenaciously without success. The count of those known to have perished in the search stands at seventeen and at least that many more have vanished in the Pegleg country without a trace. But to prospect there with a wider aim, to look for gold in general in the Oriflammes, the Volcans, the Vallecitos, even in the secretive Santa Rosas far to the north, this would make good prospecting sense. The record is unchallengeable as to the wealth that nature has scattered through these ranges and only a fraction of it has been found.

Pegleg's story, as far as it concerns gold, dates from 1829, twenty years before the rush of the forty-niners. Smith had been engaged in trapping beaver. Among the eastern gentry of that time the beaver-skin hat was very much in vogue. With a shift in taste to the tall silk hat, so conspicuous in photographs of President Lincoln, the demand for beaver pelts slackened. In Smith's day the furs were a stock of great value.

Smith had been trapping in Utah. In a fight with Indians, the bone of his left leg had been shattered by an arrow, and it is related that Smith himself had the will power to amputate the limb. He then fashioned the wooden support that gave him his nickname. In 1829 he was camped on the Colorado River at the present site of Yuma. He and his companions had accumulated a heavy and valuable stock of furs. They now planned to cross desert and mountains to the Pacific Coast, which was the nearest market. Exports of various animal pelts, along with hides and a few other western products, were shipped around the Horn to the eastern states. The cargo of pelts was loaded on burros and the party started out.

Their journey was made long before California's now productive Imperial Valley became irrigated farmland. Their first stage therefore was over a flat, alkaline plain some sixty miles wide and utterly dry. When the men had crossed this expanse and entered the broken country of present-day San Diego County they already were weary and dangerously short of water. They had only vague information on a course which a few predecessors had followed via water holes at spots now bearing the names of Carrizo and Vallecito. The latter place, because of its historical associations, has been made a county park. Carrizo also appears on modern maps as a historical site. After pushing through a blinding desert sand storm, Pegleg's party feared they were going astray and might be in a blind canyon. Accordingly they came to a halt, to assess the situation and to wait for the wind to abate.

Smith was extremely agile in spite of his wooden leg. He now decided to do some climbing, partly to get directions, chiefly to try to sight some green spot that indicated water. The prominence that he climbed appeared to be one of three small hills. Smith saw no sign of water but he was able to determine that his party was indeed off course and must backtrack.

As he rested, idly picking up rocks, he found one piece to be exceptionally heavy, doubtless metal rather than rock. At one time in his wandering he had been able to fashion bullets out of native lead, so he gathered up a few more of the heavy pieces. They were of irregular round shape, black in color, and they lay in profusion over the entire surface of the hill. "Looked like black walnuts," he said later.

The party managed to get to water, possibly at Carrizo, and from there to complete the long journey to the Spanish settlements on the Pacific. In Los Angeles, Smith learned that the heavy black nodules were solid gold. The black coating was manganese, which adheres to countless small rocks when they are exposed on the desert for many years. The coating is colloquially known as "desert varnish" and such black rocks are a familiar sight to every prospector. Usually they are merely rocks.

Smith was not immediately impressed by his good news, because the beaver pelts had sold for a high price and he had gone on one of his long, quarrelsome drunks. It was only as he emerged from this befuddlement that he realized his good luck and tried to recall the circumstances of his discovery.

The matter of the three hills is crucial in the Pegleg puzzle, and their probable nature and appearance have been argued over and over. As often as not they were called "buttes." The term "butte," by definition, refers to "an isolated hill or small mountain with steep or precipitous sides." Buttes within this meaning they were not. Even the conception of three conspicuous hills is misleading. No one can estimate how many weary days' journeys were made by early searchers who hoped to find a trio of sharp, unmistakable signals of sudden wealth jutting up in their line of vision.

It is now well established that there is no such distinctive set of three hills anywhere in the Pegleg country. Searchers agree that certain light conditions must have given Smith the impression of three separate little summits. He had made his climb in the late afternoon, when the low slant of the sun's rays can bring even slight irregularities of a landscape into sharp relief. When Smith climbed down, the anxious party had moved on without delay, out of this canyon and back to the correct trail. The treasure site was never seen in the revealing light of midday.

We have a concensus on the most promising area for a present-day search, but the career of Pegleg himself must first be followed to its close.

The drunken trapper appears to have made himself obnoxious in the quiet Pueblo de Nuestro Señora de Los Angeles, and the *alcalde* ordered him to leave. The historian Bancroft indicates that with the aid of some of his barroom companions he took with him more than a hundred head of horses rustled from an outlying rancho.

For the next seventeen years Smith engaged in trapping, horse trading, possibly in horse stealing, and also in guiding emigrant parties through the Southwest. Much of the time he worked out of Santa Fe, New Mexico.

When California came under the American flag, Smith organized a party to go in search of his nugget-covered hill. A well-equipped group departed from Warner's ranch, which was the first permanent settlement in this remote desert country and even today is probably the supply point nearest the search area.

The hunt was not a success. Pegleg's original observations had been vague, his memory was doubtless more so. The party found nothing encouraging and Smith, possibly afraid of his disappointed companions, slipped away from them and crossed into Arizona.

When the California gold rush got under way, Smith, always the opportunist, operated a lucrative supply station along one of the central California travel routes. His later years were spent for the most part in San Francisco. He had judiciously saved a few of the nuggets. The city by the Golden Gate was dominated by the subject of gold throughout this period and we hear of Smith showing his pocket-worn specimens to patrons of this and that saloon or gambling house with a view to a possible investment or, at the minimum, a few free drinks. He seems to have been enough of a character to be written up in *Hutchings' California Magazine*. An accompanying steel engraving portrays him as well dressed and of dignified posture, though the artist Vance had given him a sly, sidewise look.

Smith lived until 1866. This means that for thirty-seven years he carried about the knowledge, sketchy though it was, of a great gold deposit. He used it for a polite form of begging and probably for a few minor swindles. But except for the expedition from Warner's ranch we learn of no serious effort on his part to return to the discovery site. It is not surprising that some regard the whole thing as a hoax, yet the men who most closely investigated the story have generally considered it to be true. His discovery is mentioned in a factual way by Bancroft and other early historians.

Among those who have pieced together the original Smith statements and the results of many futile hunts made for the nugget-laden prominence, the favored areas appear to lie somewhere along the line now followed by San Diego County Road S 2. This is the route of the old Butterfield stage line and its contours would seem the natural choice also of individual travelers. The route is a rambling one through southeastern San Diego County. It passes Vallecito, where we can say with some assurance that "Pegleg was here." At what point along the way his party wandered off course remains a sheer guess.

The distance along San Diego County Road S 2, from where it leaves Warner's Ranch to where it terminates at the hamlet of Ocotillo on Interstate 8, is a trifle over sixty-three miles. Some interesting place names are encountered along the route, among them Earthquake Valley, Agua Caliente Springs, the Well of Eight Echoes, and Egg Mountain.

Rugged, rocky, utterly arid country encloses the road on both sides. For the prospective searcher the little mountain ranges along the route may be listed: Volcan, Sawtooth, Pinyon, Jacumba, Coyote, and Tierra Blanca, as well as the San Felipe Hills and the Carrizo Badlands.

To desert mining men and Western history buffs all this is known as Pegleg Country, although to date cartographers have avoided enshrining the old scamp's name on their maps. Under any name it is tough, dangerous country and the ultimate discoverer of Smith's nugget field will doubtless have earned his reward.

The Pegleg experience is by no means the only instance of gold being found on or near the surface in this region. Local Indians knew of spots where nuggets could be picked up and also of extremely rich ore in place from which gold could be pounded out or gouged out without any heavy mining. Such gold often showed up in the purchase of food and other supplies. By persuasion, promises, or threats white men occasionally got a grudging agreement to reveal one of these places, but it appears that such a promise was not made good in a single instance. Superstition, or fear of vengeance by the gods or his fellow tribesmen, always turned the Indian back.

One instance of the Indians' ability to safeguard their mining secrets concerned the Oriflamme Mountains, a rugged, forbidding range in the general Warner's Ranch-Pegleg region. A herd of cattle was being driven to the distant Colorado River and the cowboys made camp in or near Earthquake Valley at the western base of the Oriflammes. Among the men was a young Indian of the local Digueno tribe named Julian Cabrias and a white youth named Harry Yarnell. The two had become good friends.

From this camp the Indian slipped away about dusk and was gone little more than an hour, certainly not as much as two hours. This from Yarnell some years later. When Julian Cabrias returned he had a handful of gold nuggets, which he silently presented to his friend. As if anticipating questions, he warned Yarnell that it would be quite useless for the latter to ask information or to make any search for the source of the gold. The mine was a tribal secret.

As might be anticipated, a somewhat older Harry Yarnell, now separated from his Indian friend, went back in search of the riches. The cattle camp was located without difficulty, but an industrious search over a period of days produced not the slightest indication of gold.

Several high-grade ledges were discovered and mined independent of any Indian assistance. They all were secretive, one-man operations and not a great deal is known about them. In desert tradition, four such success stories stand out. They concern a Frenchman, a Negro, and two Germans.

The first of these was known only as Frenchy and his story is brief indeed. It is confined to short visits to the neighboring mining camps of Julian and Banner, both in the rising terrain just west of the Pegleg country. Frenchy paid for his supplies, drinks, and gambling out of an ample supply of native gold. Its character showed that it came from vein matter which had been pounded up in a mortar and panned out. In other words, he had ore in place, rich ore, and possibly the makings of a great mine.

One night in Julian the Frenchman was found lying dead in the street, shot in the back.

Not long after this, one of the town's Negro residents, whose name also escapes the record, began cashing in gold similar to the Frenchman's. There may have been gossip, but neither then nor later did any evidence come to light to connect the two men or to identify the source of gold. The newly prosperous miner lived on in the area for years, occasionally helping a needy friend and at one time lending a cowboy

$5,000 to set himself up in the cattle business. Local tradition places his rich mine in either the Laguna or Volcan mountains.

Many searchers for gold in the Pegleg country have looked speculatively toward the Santa Rosa Mountains, rather far to the north of the region we have been discussing. They are an extensive range and can best be approached from the modern resort and residential developments on the north shore of the Salton Sea. State Route 86 serves these communities but runs at some distance from the mountains themselves.

Several stories of rich pockets have come out of the Santa Rosas, notably the one regarding a German named Nicholas Schwartz. In the early part of the century Schwartz spent years prospecting in the Santa Rosas and eventually took a tidy fortune out of a shallow pocket mine. He then retired to Chicago. Schwartz told friends that there was plenty of gold left in his diggings and that a crossed pick and shovel, buried a little below the surface of the ground, identified the spot. Though his mine was secret, his place of residence was known. It was a stone cabin, and the gulch where it was located went down on maps as Rockhouse Canyon, though the rock house itself has long since collapsed into a heap of rubble. A branch canyon is known as Old Nicholas.

Although Rabbit Peak, the high point of the Santa Rosas, is barely eight miles from Desert Shores and other Salton Sea communities, the range remains roadless, empty, and largely unexplored. Since the southern reaches of these mountains are now incorporated in Anza Borrego Desert State Park, the hopeful miner should try for a strike in the northern section. Keep above an east-west line drawn through Rabbit Peak and you are in public domain where your mining location should be legal.

Anyone sifting clues to Pegleg's gold or other bonanzas in the Pegleg country will find repeated references to Warner's ranch. It was the only outpost of civilization in those early days. Today, the location, with its hot springs and a few venerable buildings, is a modest resort and the logical base for prospecting in the area. It is now known as Warner's Springs and is reached by State Route 79.

"Long John" Warner, so called because of his height and slender build, came to California from Connecticut for reasons of health. In 1837 we find him in Los Angeles, a partner in a small store, speaking Spanish, and a friend of Pio Pico, last governor of Mexican California. That year he married Anita Gale, an English girl and ward of the governor. With Pico's help he obtained a grant of land covering the surprisingly fertile and well-watered little valley in the midst of the desert.

Over the years Warner's was a boon to miners and Indians alike as a watering place and source of supplies. It also was the most attractive stop on the twenty-two-day journey between the Middle West and California via the Butterfield stage.

There were at least two periods of danger in the quiet ranch life of Warner and his English wife. The rancher had become a Mexican citizen and when, during the war with Mexico, American troops entered California he was arrested. He narrowly escaped summary execution at the guard house in San Diego, the charge being that he was a Mexican spy. Cleared of all suspicion, he returned to his ranch.

During the next few years Warner had as a trusted employee and second in charge a fellow New Englander named William Marshall. In 1850, Indians in the back country were resentful of a tax levied on their cattle by the now Americanized county government. Marshall, who had married an Indian girl, plotted with the more violent tribesmen for a full-scale uprising against all white settlers. One result would be to place him in ownership of the ranch and its herds of cattle.

Informed of the danger by friendly Indians, Warner sent his family away but himself remained to defend his place. He was soon under siege. Long John's long rifle killed four attackers who imprudently showed themselves. Then, under cover of night, he stole away to get help. The next day the enraged siege party took over the house and store, wreaked as much damage as possible, and drove off the cattle. Even at this early date people were coming to the hot springs seeking relief from various ailments. Four such persons, virtually invalids, were at Warner's at this time. All were slaughtered by the raiders.

Responsible Indian leaders had not joined in the outbreak and the warlike faction was soon rounded up. Marshall was found guilty in the murder of the four innocent health seekers and was hanged.

Warner lived on through the gold rush and through nearly fifty years of California statehood. The ranch was where many prospectors outfitted and was a place to which they gratefully returned after scorching days in the desert. Today a chain-link fence protects the old ranch house as a historical relic and as noted there is a modest resort in operation. The vast, empty area surrounding it is little changed from the days of Pegleg Smith and his hopeful followers.

Westward in San Diego County, one finds less and less to invite the prospector, although the gold mines of Julian and Banner, the two old camps on the edge of the eastern desert country, may well be worth another look. Their heyday came soon after the Civil War and their mines were developed largely by young Southerners who came West.

About twelve miles south of Julian there once was considerable mining activity on and near Monument Peak. Still farther south, on Interstate 8 and about midway on its course through San Diego County, is Pine Valley. The name applies to a small town and to an adjacent mining district. There are proved gold-bearing mineral formations here but the values per ton so far encountered have not been encouraging.

The city of San Diego, as a seaport, received occasional gold and silver shipments and has received an even heavier cargo of alluring stories of rich lodes, found and lost. These emanate from the rugged mountains to the south. The localities, however, are in Mexican territory, the peninsula of Baja California, and lie outside the scope of this report. True, on the American side there is a persistent legend of silver, to be dug out in chunks of the native metal, in San Diego's back country. The favored location is in the heights overlooking the west shore of El Capitan Reservoir. Another version would have us try San Miguel Mountain, which is sixteen miles to the southwest. From this unsatisfactory prospect, we turn to the scene of California's major gold production–the Sierran Gold Belt and its principal feature, the Mother Lode.

3
California's Mother Lode

The great gold rush of 1849 and succeeding years was directed toward the central part of California. The southern desert we have been examining was unknown, or at any rate of no interest, to the Argonauts who were heading west across the Sierra Nevada. To those who swung south over the old Spanish Trail it was a region to be endured and forgotten as they hurried toward the proved riches of the placer streams and quartz mines far to the north.

Among the many auriferous formations that awaited the gold-seekers, the one which deservedly struck the world's imagination was the phenomenal concentration of gold known as the Mother Lode. The name is a translation of *Veta Madre,* a term used by Mexican miners to define an impressive zone of mineralization in their own country. In California it applies to a well-defined fault structure extending for 120 miles along the lower reaches of the Sierra, and which is laced with gold-bearing veins ranging in width from knife-blade thinness to a massive 150 feet.

The first readily identifiable person to tap the wealth of the Mother Lode was the celebrated scout and frontiersman Kit Carson. His strike was made in 1849. It was near what is now the town of Mariposa, a picturesque little place which vacation parties pass on their way to Yosemite National Park. Carson came upon an outcrop of rusty quartz which contained visible flecks of gold. With two associates he worked the crumbling surface rock with pick and shovel and recovered the gold by use of the ancient, slow-motion arrastra, a primitive ore-grinding device powered by horses or mules.

The restless Carson apparently was not connected with the operation for long. In 1859 the mine passed into the hands of another noted

Western explorer, being situated on a 44,000-acre tract of land which an appreciative government granted to John C. Frémont. Carson's original outcrop proved to be only the tip of a large and consistently rich ore body. Developed into the great Mariposa Mine, the deposit was worked with success until 1915.

By chance, this first great mine on the Mother Lode also marked the lode's southern extremity. For near Mariposa two giant cracks in the earth's rocky crust come together—the San Andreas and the Garlock earthquake faults—and the structural conditions which encouraged the deposition of gold come to an end.

From Mariposa the Mother Lode can be traced northward to Georgetown, originally named Growlersburg, an isolated village not a great distance from where James Marshall made his electrifying discovery of yellow particles in Sutter's mill race.

There is an "East Belt," somewhat removed from the main Mother Lode formation. One of the features of the East Belt which has lured prospectors since the earliest days is the profusion of "seam" deposits. As the name indicates, they are narrow veinlets which would be dismissed as of no value except that in numerous places the gold is highly concentrated. Many of the dazzling little chunks of specimen ore seen in museums and in private collections have come from the seam mines.

There also is a branch of the Mother Lode running off to the west called the "Gold Thrust." Its values were so consistent that successful mines were crowded together along its entire length.

The major mines along the Mother Lode, virtually all of them inactive today, will be of little interest to the individual gold hunter. They were giant operations and it would be a giant operation to rehabilitate one. Perhaps the most celebrated of these old producers is the Kennedy, near the town of Jackson. State mine officials soberly report that it contains 150 *miles* of underground workings. From these involved subterranean galleries came ore that yielded $45 million in gold bullion. Miners toiled a full mile below the surface of the earth. The elevators carrying them to their working level made a much longer vertical trip than those of any modern skyscraper. Few of the Mother Lode mines were developed on the heroic scale of the Kennedy, but many of them went several thousand feet in depth, with numerous working levels branching off. The workings were necessarily equipped with ponderous and costly machinery.

To dismiss the large, historic mines from current prospecting interest is not to dismiss the Mother Lode itself. There are many shallower, less-developed workings where it would be much simpler to resume mining. One may also reasonably investigate mere prospect holes, at least to the extent of seeing whether any promising vein matter has been exposed.

With regard to such lesser remnants of early-day mining, snap judgment understandably is to the effect that they must be worthless. In most cases this probably is true. Yet there are other reasons why some were deserted, why a small miner or small partnership may have pulled up stakes. One reason was the lure of the Comstock Lode. In 1859 this tremendous treasure chest of silver and gold had been opened up just across the Nevada state line. The ensuing excitement, which was fully justified by the size and richness of the great bonanza, took hundreds, perhaps thousands, of miners away from the California diggings. Borderline prospects and even better than borderline were in many cases abandoned.

There were additional factors to account for the abandonment of claims. Among them were family ties, discouragement with early results, offers of employment at steady wages, and plain lack of money to continue operations. In some cases a new locator, more persistent in his exploration, proved to be a lucky man. Mining history shows that many successful properties proved their wealth only when being worked for a second or third time, commonly under new ownership. The pattern is repeated time and again in federal and state reports.

Some of these abandoned diggings may classify as "pocket mines." There were many of these along the great lode. They represent the same gamble as the seam mines, the finding of concentrations of gold in rock not workable by standard methods. A single miner could delve away in a pocket mine with no help and little expense. Some of these early-day free-lancers took out fortunes. Their boring around in twisting, nondescript excavations in search of visible gold was known as "gophering," and doubtless was the despair of mining engineers. But the pocket miners knew what they were after and had an axiom, "Follow your ore even if it climbs a tree."

Is there any remaining chance of coming upon one of these rich pocket veins? Certainly not at Carson Hill, scene of the most intensive mining of this type, yet a brief look at those diggings will be instructive. They furnish an example of the lavish mineralization with which nature, in rare and widely separated locations, has surprised and enriched the prospector.

The unimpressive little peak, near the present town of Carson Hill, was so rich that virtually every square foot of it was spaded up and sometimes spaded up again. There were many spectacular discoveries. At the head of the list stands California's, and perhaps the world's, largest gold nugget. This single mass of pure metal, dug up almost at grass roots, weighed in at $43,534 at the old $20 price of gold. A replica now exhibited in San Francisco is somewhat larger than two clenched fists held together.

It was at Carson Hill that an elderly Negro, vaguely hopeful of getting into mining, was directed by a group of barroom roisterers to a distant spread of barren earth which they considered worthless and upon which no one had filed. In his innocence the old man took up a claim and kept digging, and at last, in this unlikely ground, he struck it rich. A modest fortune rewarded the amateur miner and the recognized limits of the Carson Hill gold district were suddenly extended.

If Carson Hill is mined out, as mature mining judgment indicates, and similar locations already prospected more or less thoroughly, where can one look for a comparable discovery today?

The best opportunities would seem to be amid some of the dense growths of brush that are found in many parts of the lower Sierra. The red-branched shrub manzanita can be a formidable barrier to prospecting. Because its branches interlock at heights of waist level and up, a manzanita thicket can be penetrated only on one's hands and knees. Equally frustrating are masses of hardy, fast-spreading Scotch broom. Various other wild growths, unfortunately including poison oak, seem equally intent on hiding any mineralized outcrops that may exist. One can find little in mining records or anywhere else on such a commonplace matter as to whether miners have crawled or hacked their way into this or that batch of undergrowth. How many of these thickets have been prospected? Certainly many of the stubborn patches show no sign of ever having been penetrated by man.

As late as 1970 there was a discovery and an inconclusive search at such a spot. A prospector roaming government land south of Interstate 80, near the town of Colfax, picked up a piece of black and white rock with conspicuous incrustations of gold. This was obviously rich ore. It seemed quite evident that successive rains and erosion had carried the fragment down from a nearby rise.

This slight elevation covered only an acre or two, but its entire surface and the possible bonanza that it sheltered were obscured by a dense growth of manzanita. The prospector's exuberant invasion of this tangle was not a success. After hours on his hands and knees, scratched and bruised and dismally aware of how little of the terrain he had been able to examine, the tired searcher gave up. He could only resolve that at some later date, with ample time at his disposal and with a camp set up nearby, he would resume the battle.

For a close-up look at Mother Lode ore, an actual gold-bearing vein in place, the California Division of Mines can be helpful. Its geologists tell us of an exposure on State Highway 49, near the now inactive Mary Harrison Mine and 1.8 miles south of the drowsy gold rush town of Coulterville. Immediately north of the surface ruins of the Harrison Mine there is a highway cut that slashes through a typical part of the Mother Lode vein system.

"This locality is easily accessible," says the state mineralogist, "and is an excellent place to observe gold ore in place and the minerals which are associated with it. The ore itself consists of gold-bearing iron pyrites, usually somewhat oxidized."

Highway 49 comes to another excellent exposure near the Maxwell Creek bridge only two tenths of a mile south of Coulterville. Here the vein matter "is accentuated in many places by rust-colored zones which mark the positions of sulphide gold horizons."

At these and other places where highway work has exposed mineralization, the traveler who will pull his car aside and make an inspection can add considerably to his knowledge. In earlier days he might also have added to his wealth. One of the chapters on Nevada will tell how a family traveling in a horse-drawn wagon paused to look at an unusual rock formation and soon had a mine for which they were offered $2 million.

There will be much more to say about lode mining in California, especially about certain neglected areas outside the Sierran Gold Belt, but it is time to put the important matter of placer mining into the picture. This has been the first, obvious type of mining in almost every gold field, the drilling and blasting of ore in place being a later development. In some fields, including the Klondike and the beach at Nome, placering has been the only type of mining. The hidden and perhaps distant ledges from which the gold has washed down are never discovered, or at any rate never exploited.

Until a gold-bearing gravel deposit is exhausted it can yield very rich returns. As one of the more extravagant examples, let us take the operations at the old town of Columbia, California, now a state historical preserve and once lauded as "Queen of the Southern Mines."

Columbia is in the foothills of the Sierra. Around it is an expanse of rough limestone bedrock with a surface of holes, crevices, ridges, and other irregularities. Through the centuries this limestone contrived to trap and hold much of the gold—nuggets, flakes, and dust—washed down from the huge and bountiful segments of the Mother Lode that lay above. When some lucky forty-niners started to pan the overlying gravel they discovered one of the richest, most accessible accumulations in all mining history. The official estimate of recovery here is $87 million. A church still standing on an undisturbed little plot undoubtedly shelters a fortune beneath its foundations.

Lesser but similar deposits were found along all the streams flowing from the mountains into California's great central valley. The methods of recovering placer gold, ranging from the most primitive up to the sophistication and great capacity of the bucket-line dredge, have been described succinctly by Charles V. Averill, a mining engineer with the California state government. In a publication commemorating the one hundredth anniversary of Marshall's discovery, Mr. Averill wrote:

"Methods for the recovery of gold depend largely on the high specific gravity of that metal. A particle of gold of a certain size is so much heavier than a particle of associated rock of the same or even larger size, that a current of water washes the particles of dirt and rock away and leaves the gold behind.

"The earliest efforts to recover gold after Marshall's discovery were directed toward those places where flakes and nuggets of gold had been concentrated on rough bedrock and in crevices by the natural flow of the streams. Small tools were used to dig in the crevices and to pick up the flakes of gold. Soon the miner's pan came into use. An ordinary frying pan can be used to concentrate gold providing it is free of grease, but the familiar miner's pan with its gently sloping sides is more efficient. The capacity of the pan is so limited that gravel very rich in gold is needed to pay by this method. A skilled operator is able to pan less than one cubic yard of gravel during a day's work.

"Rockers or cradles and sluice boxes were the next devices used by early day miners. The rocker is a crude concentrating machine made of wood that combines the shaking motion of the pan with some features of the sluice. [It is rocked back and forth sideways, like the traditional baby's cradle.] Riffles or obstructions are placed across the bottom to catch the heavy flakes of gold. With the rocker, both the gravel and the water are introduced by hand.

"With the sluices, the water flows by gravity but the gravel is shoveled in by hand. The miner is able to wash several cubic yards a day—still a small amount. By 1860, gravel that would pay to work by these methods was getting hard to find, and production of gold declined.

"In the meantime, the hydraulic method of mining was developed. Great jets of water were directed against banks of gravel hundreds of feet in height. [These deposits were ancient river beds uplifted by geologic action.] Some of the jets of water were 9 inches in diameter as they left the nozzles. . . . The gold was recovered in riffle-sluices, and quicksilver was usually added to aid in its recovery. The debris deposited in the streams by hydraulic miners became so objectionable to agricultural interests that the method was stopped by injunction in 1884. . . .

"About 1900 connected bucket dredges were introduced. These great machines contain a continuous chain of buckets for digging gravel and taking it aboard a floating barge. [Such barges float on ponds of their own making, scooped out in front and filled in behind.] . . . The dragline type of dredge is a later development. The heavy digging equipment is an ordinary dragline excavator that travels on the bank. The floating barge carries only the screen, equipment for concentrating the gold, and the tailing stacker. Placer gravel was handled so cheaply by these dredges that gravel running 10 to 20 cents per cubic yard could be mined at a profit."

The day of the gold dredge, as far as the United States is concerned, has clearly come to an end. One reason, among others, is the increased concern for the environment. Dredging leaves massive rows of heavy boulders on the surface of the land. Near Leadville, Colorado, such man-made ridges can be seen stretching into the distance like the Great Wall of China. There are piles higher than a house on the outskirts of California's growing capital city of Sacramento. These stones cover land that would be of great value for home or business development, but an economically feasible way to remove and dispose of them apparently has not been devised.

Where the gold-impregnated land had been adequately sampled and found to carry sufficient values, the dredge operation was an assured success. In the 1930s the dredges were cutting up the landscape without hindrance. One operating company could afford to buy up the town of Hammondton at sellers' prices in order to dredge the area. The same corporation, based in San Francisco, acquired another proved expanse of gravel at Bulolo, deep in the New Guinea jungle. The profits in sight justified the construction of an air strip in this remote spot, the breaking down of a multiton dredge into thousands of components, and flying the monster machine to faraway Bulolo piece by piece.

With regard to California placers, an item of passing interest is that diamonds occasionally have been recovered along with the gold. The most numerous finds were at Smith's Flat near Placerville, the Hangtown of early California mining days. A judge of the superior court became interested in the matter and compiled a list of some two dozen diamonds discovered there, together with their weights. He told of several residents who wore the local stones set in rings or brooches. Diamonds also were taken from the sluices at Cherokee, an old hydraulic mining camp near Oroville, in Butte County. The sources of the stones were not discovered, but the possibilities were taken seriously enough for the state mineralogist to publish a substantial report on diamond mining in general and its slender history in California. A story persists that a pipe of blue clay, a sort of core reaching from the earth's fiery interior to the surface, and typical of South Africa's deposits, was found in Butte County and that emissaries of the great De Beers diamond syndicate managed to get the exploration closed down. There is a record of a company formed to explore for the Butte diamonds. The enterprise was dissolved without any official report of production, but any machinations of the South African cartel are quite unsubstantiated.

Nearly all the Mother Lode settlements and several that lie in additional gold-bearing zones to the north are served by a state highway appropriately designated No. 49. This route rambled around from one mushrooming camp to another in the 1850s and it still connects the old sites today. Dipping down into canyon after canyon, it runs for 162 miles along the Mother Lode proper, although the lode's extent by airline measurement is only 120 miles. For an additional 115 miles northward, Route 49 serves a country also rich in gold-mining history, but which geologists differentiate from the Mother Lode. For the entire auriferous zone along Route 49 they prefer the term "Sierran Gold Belt."

From several viewpoints, scenic, antiquarian, and recreational, a drive along 49 is a worthwhile trip. It is especially important to anyone contemplating involvement in the resurgent gold-mining industry. Between the towns of Jackson and Plymouth he will be on the ten-mile stretch of the Mother Lode where mining was most intensive. This segment accounted for more than one half of the total production. Far to the north he will be in the Grass Valley district, which is much smaller in compass than the Mother Lode but where a concentration of high-grade vein systems produced more gold than the great lode itself.

Beyond Nevada City the forty-niner road climbs to an old and interesting settlement quite conscious of its historical heritage, Downieville. Its stone, iron-shuttered buildings were constructed largely with the takings from treasure-laden Goodyear's Bar and similar rich gravel deposits along the Yuba River. So numerous were placer miners along the Yuba in bonanza days that it was said a message could be started on its way at Marysville, in the valley lands downstream, and be passed by word of mouth almost to the river's source in the High Sierra.

On each side of the Yuba, rugged, craggy mountains stretch out into what seems an illimitable distance. Many rumors of rich strikes and many pieces of rich ore have come out of this difficult country. The one spot where successful mines have materialized is the high, hidden camp of Alleghany. The ore here was often of very high grade, especially in the Sixteen-to-One Mine. This unusual property managed to continue operations long after economic factors and government fiat had brought California gold mining to a standstill.

Near the terminus of Route 49 at Sattley but at a higher elevation we encounter one of the great puzzles of western mining, a situation which will be examined later in this chapter. This concerns the Meadow Lake district, where massive bodies of gold ore unquestionably exist but which, for reasons still debated in the engineering fraternity, have brought only disappointment and financial loss to those who have tried to mine them.

One of the towns along Route 49 which has received much attention is Angels Camp, the locale of Mark Twain's story *The Jumping Frog of Calaveras County*. An annual Jumping Frog Derby commemorates that fictional creature and crowds the little town with more visitors than it can easily handle. Our interest is not in Angels Camp itself but in an area six, perhaps as much as ten, miles to the southeast where a certain tunnel exposes a "wagonload" of gold-studded quartz, said to equal any ever taken from the Mother Lode. This is the lost mine of one of the most tragic losers of them all, the boardinghouse hanger-on William Moyle.

Seven miles down Route 49 from Angels Camp is the ghost of a once riotous camp called Slumgullion, a name later changed to

Melones. Ruins of the old Vignoli trading post are about all that remain, but Melones still appears on many maps. Mrs. Bill Moyle, the dominant member of the family, operated a boardinghouse there in the 1850s, while her husband spent his days prospecting the hills. It was a long and discouraging hunt for Bill and apparently it went on for years. There had been several high-grade discoveries in the district. The name Melones, Spanish for melons, is traced to the fact that the ore from the local mines often contained gold of the shape and size of melon seeds.

When Bill Moyle returned from what was to be his last trip, the melon-seed designation was not adequate for what he had found. In Baldy's saloon he threw a heap of ore on the bar containing nuggets as large as peanuts. Moyle bought and continued to buy drinks all around. He boasted that there were tons of such ore in the face of his tunnel, "enough right in sight to load a wagon." The night and the celebration wore on. At length it was noted that the host was missing. They found him lying on the floor, not drunk but victim of a massive paralytic stroke.

Unable to go back to his great strike, unable even to give adequate directions because of a clouded memory, Moyle lived on for several years, never enjoying the wealth that lay somewhere back in the hills.

A mining man named Milo Bird was a teen-age boy living in Melones at the time of Moyle's discovery and its tragic aftermath. Like everyone else in the little camp, he knew everything that Moyle had divulged about the rich vein—in fact, he knew a little more than the others. In later years he passed on his clues to readers of *Desert* magazine. They ran like this:

The Stanislaus River flows through this area and south of Melones describes an almost circular meander known as Horse Shoe Bend. The land between the two points of the shoe, is pierced by an old tunnel. This, said Bird, is at least a reasonable place to begin any search, for we know that it was in the area several miles north-northeast of this river bend that Moyle was doing the prospecting that led to his great discovery. Up in that area there is a man-made shortcut between the higher back country and the town of Melones. Moyle was seen leaving this shortcut and following an animal trail up one of the many deep gulches that cut through the hills. It was Milo Bird, on his way home from a day of fishing, who saw him there. As he grew older, Bird realized the potential value of this information, but try as he would he was unable to remember that chance encounter with any accuracy or identify any particular gulch. So Moyle's secluded tunnel, which cuts into a gold vein of indisputable value, remains lost.

As one continues north on State Route 49, he comes to California's earliest and most productive lode mining district. It was at Grass Valley that the Argonauts first turned from the comparatively easy mining of placer deposits and began to wrest gold from veins of solid rock. The town is situated 55 miles by road northwest of the state capital of Sacramento, and the present-day urban area includes picturesque Nevada City. Grass Valley gold mining had a long history. It started in 1851 and several mines, among the largest in the state, continued operation until 1959.

"Grass Valley is probably the most beautiful active mining camp in California," wrote geologists Oliver E. Bowen and Richard Crippen in 1949. "It is set in a well-watered coniferous forest of great beauty. Many stands of large trees have been spared the loggers' ax and tower in dark green borders about the open land. . . . Even the mine dumps and buildings are more or less masked by trees so that the scenery suffers little by their presence."

Writing for the state centennial report on California mining, the geologists note that "the careers of such famous personalities as Lola Montez and Lotta Crabtree are closely associated with the history of Grass Valley." These were two charmers of the West's early-day music halls, and la Montez even had something of a European reputation, for she was a consort of the so-called Mad King Ludwig of Bavaria. As for Lotta, she is remembered for a fountain she had erected in downtown San Francisco, one which did more credit to her heart than to her sophistication in the arts.

It was noted earlier in this report that the lode mines of the circumscribed Grass Valley district produced more gold than those along the entire 120-mile extent of the Mother Lode. Greatest of the Grass Valley operations was the Empire-North Star, whose 200 miles of underground workings and $120 million output outdid the Mother Lode's Kennedy Mine. Since the former is a consolidation of several large producers, the two are not strictly comparable. The original Empire Mine dates back to 1851.

Grass Valley's second largest mining enterprise, also a consolidation, was the Idaho-Maryland. In the late 1940s, employees at the San Francisco airport became accustomed to seeing a private plane sweep down from the direction of the mountains and land on a little-used runway at the edge of the field. Always awaiting the plane was a black limousine, and there was a quick handling of a mysterious cargo. The big car would then speed away. This was the manner in which gold bars, poured at the Grass Valley plant of the Idaho-Maryland Mine, were conveyed to the United States Mint in San Francisco.

Besides these mammoth operations there were other Grass Valley mines which poured out millions. Some encouragement to latter-day prospectors is provided by the case of the Lava Cap Mine, whose rich lode remained a "sleeper" through decades of active mining all around it. The geology of the Grass Valley district was believed to be fully understood and explored when some hopeful miners took to the field during the depression days of the 1930s. Almost at the edge of the Nevada City residential district they discovered good surface indications and went on to develop a new mine. In nine years the Lava Cap yielded $12 million in gold. Their success encouraged others to investigate a nearby claim where the surface had barely been scratched. This became the Murchie Mine and it was only slightly less valuable than the Lava Cap.

Like all other gold mines, the Grass Valley properties were closed down early in World War II. Miners throughout the West still grimly remember the War Production Board order of 1942, which classified gold mining as a non-essential industry. The manpower and materials it used were needed for more war-related industries.

The enforced closing meant the abandonment of many mines, for during the war years their workings became hopelessly flooded. The principal Grass Valley mines, however, were valuable enough to be rehabilitated and they soon were in operation. Postwar inflation, especially the general rise in wages, gradually narrowed the margin between operating outlay and the value of the gold recovered. The miners admittedly needed higher pay to keep pace with rising living costs and increased wages in other industries. The mines could not pay more. There was no labor dispute in the ordinary meaning of the word, for both management and labor understood the economic squeeze. In 1959 gold mining in Grass Valley quietly came to an end.

In considering the mining prospects of the Southern California desert, the principal highways were used as guides. In this chapter, we have seen that State Route 49, sometimes referred to as the Golden Chain Highway, serves most of the Sierran Gold Belt. It will be profitable now to look at possibilities along two other routes in central California. One is Interstate 80, a major transcontinental highway which carries the bulk of truck and tourist traffic to and from San Francisco.

The other is U.S. 395, which runs from Canada to Mexico. We shall be concerned with 395 from where it enters California south of Reno down to where it connects with roads to the Death Valley region. It is a scenic road, running largely along the east side of the Sierra Nevada. It climbs to more than 8,000 feet at Conway Summit and descends almost to sea level in the sunken wastes that are the outliers of Death Valley. This rather lonely road runs through some highly mineralized country.

Turning first to Interstate 80, we will make an approach from the west, that is, going toward the mountains, which will bring us into the gold country gradually. As the highway leaves the wide Sacramento Valley and slants into low foothills there is some evidence of old-time mining to be seen. A gaunt headframe or a mound of tailings stand among what are now prosperous orchards. The mines here were working on a giant batholith, a massive intrusion of granitic rock that contained just enough gold to be tantalizing. The little trade center of this district was optimistically named Ophir. The ores were low grade and both this area and the main bulk of the batholith, which extends east and is buried thousands of feet beneath later rock formations of the Sierra Nevada, may be dismissed for any immediate practical purpose.

From an elevation of 1,000 feet to say 4,000, miners found considerably better values. Lode mines often had concentrations of rich, free-milling ores close to the surface. Pike Bell, a small rancher, hacked into a little outcrop on his land and took $35,000 in gold out of a hole which he described as "about twice the size of a washtub." George Wingfield, who became Nevada's leading capitalist of the early twentieth century, got his start with a $50,000 strike almost at grass roots in the Three Queens Mine near Forest Hill. The Black Oak Mine, three miles off Interstate 80 at 2,000 feet altitude, yielded $60,000 in its initial stage. It was shut down when the deeper ore, though still of value, carried gold in the form of sulphides, which called for a more costly treatment process.

This is our first acquaintance with sulphides, also known as pyrites and white iron. They can constitute a highly valuable ore body if they carry high enough values, but their commonly yellow luster has aroused false hopes in many a beginning prospector or unwary investor. For this reason they are known as "fool's gold." It may be desirable to go into the subject at this point.

The fool's gold term also has been applied to sparkling yellow particles observed, through the water, near the shore of a lake or stream.

Such shiny spots usually are minute pieces of mica reflecting the sunlight with a deceptive yellowish tinge. As soon as the gravel containing these mica flakes is scooped up for panning, their nature becomes apparent. One might as well go on and pan the gravel out—there might be real gold there also.

No bright light or reflection is necessary to give color to iron sulphides, a truly yellow compound which is frequently encountered in highly mineralized regions. The appearance is much like brass, the shine being brighter but of lower color saturation than one would observe in native gold. Even the beginning prospector will not be handling minerals very long before he can recognize these sulphides at a glance.

The slightly more subdued luster of gold does not mean that its brightness is unimportant in identifying it. The old prospector's rule of thumb is that free gold will continue to look like gold no matter how one turns the specimen around or how the light strikes it. The temporary illusion due to light effects will be lost if the substance is not gold. One of the chief deceivers in solid rock specimens takes us back to mica. The form known as muscovite often occurs in ore of real or imagined value and in certain lights can look distinctly solid and yellow.

The yellow sulphide fool's gold occurs in liberal masses and once was something of an item in the tourist trade. When certain communities in Colorado were enjoying the benefits of mining and tourism simultaneously, children would gather at the railroad stations and industriously sell "specimens" to the sight-seeing visitors on the train. The rocks with their flashy yellow elements may have thrilled the more innocent purchasers.

It should be noted that even high-grade gold or silver ore does not necessarily exhibit any visible metal, even under magnification. The rich ore of the Comstock Lode was tossed aside by the first prospectors on the scene until a Mexican from the older Sierra Madre mines recognized its true value.

Interstate 80 for many miles describes a center line through scenic Placer County. As one climbs higher into the mountains he can turn aside to inspect abandoned mines here and there, but they are not as numerous as might be expected. "For a region that has produced so much placer gold," wrote C. A. Logan of the California Division of Mines, "the quartz mines of this county have not been extensively worked."

Placer County also is lacking in circumstantial accounts of lost bonanzas. One story has a party of three Scandinavians returning from the North Fork of the American River with bags of extremely rich gold ore which they exhibited in Auburn, the county seat. They never were seen in the county again. The location of their strike was known only as being somewhere in the main or side canyons of the North Fork, all of them extremely precipitous and to be explored only by exhausting climbing. There is no record of any search for this lost bonanza. Encouraging float ore, always white quartz with gold liberally embedded in it, has been picked up in Jefferson, Brushy, and Shirt Tail canyons, all feeding into the North Fork of the parent American River. In Bunche Canyon a bed of auriferous gossan was mined. By definition, gossan is decomposed, mineralized rock of rusty color derived from oxidized pyrites. At this mine, the Annie Laurie, the deposit was so soft that it could be shoveled or scraped up and the gold recovered by placer methods. At a depth of twenty-five feet, reported the state mineralogist, a great dyke laced with gold-bearing seams was encountered. Attempts to work this formation at the old price of gold were not profitable and it presumably still awaits development.

By continuing in Bunche Canyon or by taking a better access road out of Auburn one may reach Forest Hill. This small logging community is perched on the Forest Hill Divide, which separates the North and Middle forks of the American River. Both streams were liberal sources of placer gold for the lucky first comers of the California gold rush, and such names as Mammoth Bar, Rattlesnake Bar, Horseshoe Bar, and dozens of others recall the days when these gold-laden sand spits were populated from dawn to dusk and were yielding rewards ranging from handsome day wages to sudden fortunes.

Such streams were all but worked out by the forty-niners and we are not concerned at this point with the meager replacement of gold that has been possible in the last hundred years. For high on the Forest Hill Divide are gold-bearing streams of another kind. They are ancient river channels that have been lifted high and dry by geologic convulsions, carrying with them, cemented into place, the dust and nuggets deposited when they were flowing streams.

Millions of dollars have been recovered from these old channels. On the authority of geologists and mining engineers who have examined the deposits, it may be categorically stated that uncounted millions in gold remain. The locations are known, the ancient channels accurately mapped.

Such maps go back to Waldemar Lindgren and other federal geologists of an early day. They show that the "tertiary channels"—a name assigning them to the tertiary period of the Cenozoic era, in geologic terms—are found at intervals along the lower parts of the Sierra Nevada for about two hundred miles. The ancient Jura River has been traced across all of Plumas County in the north, and two very extensive tertiary lake-type gravel deposits have been outlined in addition to the channel itself. Other ancient river beds are found as far south as Mariposa.

For intelligent prospecting and appraisal of the old channels it will be helpful to know how they were created. Sometime in the extremely dim past gold was eroded out of exposed quartz veins in the mountains, washed into the rivers of that ancient time, and trapped here and there in the stream beds. Subterranean pressures, often resulting in cataclysmic volcanic action, would lift the stream beds to new locations in an entirely altered landscape. A new series of rivers would pour down the mountains. As they gouged out their canyons through the centuries they would cut through the old river courses. At these places they would carry off the gravel and gold of the earlier deposits, adding it to the gold which their own waters held. At favorable spots this double load would be trapped and form an extra-rich placer bed. This double cycle of gold deposition is known as reconcentration. Such a cycle could be repeated again and still again. The period of upheavals in the now tranquil Sierra, say geologists, covered at least 200 million years. At places old channels may lie one above another, but the typical deposit is a single old stream bed, ordinarily covered by a capping of lava.

What about the practical matter of recovering some of this known wealth? For reasons which will become apparent, mining today probably will have to be conducted much as it was in the past, despite great advances in earth-moving machinery. A description of the little-known process of drift mining at Forest Hill is therefore in order. Drift mining is somewhat different from operations in a quartz mine in that the workings are in gravel. While the material has become more or less cemented through the ages, it is a long way from being solid rock. Timbering of a drift mining tunnel nearly always has been necessary. Then, since the deposit corresponds to the size of the ancient river, the tunnel may be very wide, with the face or breast of the excavation one hundred or two hundred feet or more across. The unstable roof of this great underground room had to be supported by strong posts, often tree trunks cut to the right length. Each one was capped by a heavy, flat board typically thirty inches in length. These supports were needed about every eight feet. Where the gravel was strongly cemented, it was possible to leave supporting pillars standing instead of placing timbers.

With some exceptions, the tunneling would be on bedrock, for that is where placer values are concentrated. How much of the gravel lying on the bedrock would be removed and milled varried with the nature of the deposit. Working everything five, six, or even seven feet above bedrock would be typical. In some mines, so much gold had permeated the rocky floor that this would be blasted out to a depth of two or three feet. Rails were laid into the long adits of a drift mine, sometimes to the length of a mile, and the broken-down material of the ancient channel hauled out in cars holding from 1,600 pounds to a ton. Predominance of white quartz in the deposit was considered a good sign, but there also was a darker formation known as "blue lead" which was encountered over a wide area and which usually was productive. Where the river gravel rested on a base of black slate this bedrock usually could be mined at a profit, being more permeable than other rock and thus absorbing more gold.

The most prosperous days at Forest Hill were in the latter part of the nineteenth century. Gold was worth $20 an ounce and miners were paid $2.50 to $3.50 a day. There also were highly competent Chinese miners who ordinarily worked for $1.75 a day. Mines with gravel averaging $4.00 a ton operated for years with an unbroken record of dividends. Some were able to pay out to their stockholders one third or one half of the gross value of the gold recovered.

There also were failures. Some of the old river channels which looked as promising as any others proved quite deficient in values. An "adit," a term mining engineers prefer to "tunnel," might be run into the earth a thousand feet with the intention of reaching auriferous gravel. Such gravel might prove rich and profitable or it might prove too lean even to pay operating expenses.

Besides the known, mapped channels which have not yet been mined at all, there are some old workings where operations came to a stop with much demonstrably gold-bearing material still in place. The California Division of Mines notes several such cases in a report covering Forest Hill, Iowa Hill, and Michigan Bluff and their numerous inactive mines.

The Bake Oven Placer, six miles by trail northeast of Michigan Bluff, is believed to have an auriferous gravel deposit thirty-five feet thick and 1,200 feet long with only fifteen feet of overburden, or soil lying above the pay gravel. The Copper Bottom Mine, three and a half miles north-northeast of Iowa Hill, was abandoned with much gravel unworked, and also with a promising bench deposit fifty feet above the main channel. The Home Ticket is a mile and a quarter from the old mining camp of Last Chance. It was rich enough for later-day miners to work the cemented gravel left in the old supporting pillars. Left uncompleted was an inclined shaft directed toward a supposedly good deposit lying below. In the canyon below Michigan Bluff lies the Big Gun Mine, which has a recorded production exceeding $1 million and in which considerable gravel is known to remain. In one brief revival there, two partners struck a rich concentration that yielded 1,200 ounces of gold in a single week's work, worth about $24,000 at the old gold price.

The question arises as to whether bulldozers and other modern machinery might be used in reopening these old mines. There are few instances of the ancient channels occurring where they could be worked from the surface. They normally lie so deep as to be reached only by underground mining. Even if a deposit could be established as favorable for bulldozing there could be insurmountable environmental problems.

It is time now to turn to the second great deposit of known but unmined gold that was mentioned along with the old river channels, the massive ledges of the Meadow Lake district, and to describe the colossal losses in money and manpower incurred in the spectacular rise and fall of the Meadow Lake metropolis of Summit City. Why a coterie of new millionaires did not emerge from Summit City is one of the enigmas of California mining. The evidence is confusing and the entire episode is akin to those historical and literary mysteries that from time to time have intrigued the academic world. Later we will consider one engineer's conclusions as to how Meadow Lake still can be made to pay off, as to how, in fact, a new generation of miners might well produce those expected millionaires.

Where the Sierra Nevada attains an elevation of some 5,000 to 7,000 feet above sea level there are a number of sparkling natural lakes. Meadow Lake, however, was artificially created by a massive stone dam constructed in 1857. The location is in eastern Nevada County. The nearest point on Interstate 80 is the ski resort of Soda Springs, but the lake can be reached only by twisting, difficult roads which do not connect directly with Interstate 80. The lake was created at a time when the great mines of Grass Valley were coming into production, and a reservoir at this site would feed easily into the ditches and water lines of the South Yuba Water Company. To the south of the site, Old Man Mountain towers to 7,789 feet. The surrounding area for many square miles is laced with clear streams and dotted with lakes and ponds. Except when under a deep blanket of snow, it was and is as delightful a recreation spot as one is likely to find.

The men employed to build the dam did not go prospecting in the vicinity. Such indifference to new territory was unusual in those days. It has been explained, rightly or wrongly, that the granitic character of the exposed country rock was so unfamiliar to them that they wrote it off as to any possible mineralization. The discovery of gold at the lake was made six years later by a trapper, Henry Hartley.

Hartley was a Pennsylvanian who joined in the great westward migration of the 1850s. A young single man, he drifted a bit, working here and there, and arrived in California in 1860. For several years he supported himself by trapping for furs, living principally in a cabin he built at Meadow Lake but coming into Dutch Flat, Nevada City, and other gold rush towns for business and occasional companionship. He was essentially a loner and more or less remained one as the Summit City activity later seethed about him.

It was in the summer of 1863 that Hartley observed free gold in a surface vein of dark brown rock. This same rock could be observed at many places and in liberal quantities along the gentle slopes running up from the lake and also on the rugged sides of a prominence that today appears on maps as Hartley Butte. The young trapper was leisurely about following up his discovery, but by September he and two friends had staked mining claims. These covered two strong outcroppings running parallel to each other. Following the custom of the time, they declared the area a mining district and named the mine, their company, and the district Excelsior. "Excelsior," the state motto of New York, is defined by the dictionary as denoting "an aspiration to attain greater heights." Hartley's discovery was indeed much higher than any proved mine of the period—as will be seen, perhaps too high.

The partners' remote location and the impassable snows of winter guarded their secret for longer than was usual. With the summer thaw of 1864, one other group of miners arrived, found four separate ledges of gold-bearing ore clearly exposed on the surface, and staked additional claims.

By the next year, 1865, the news was out. From the still booming Comstock Lode at Virginia City, from the Reese River mines in central Nevada, from the whole string of gold-mining towns along the Sierra foothills, miners, promoters, tradesmen, and professional men and not a few prospective investors swarmed to the shores of Meadow Lake. During the summer months of 1865 arrivals were counted at fifty a day, then a hundred and more each day. The townsite of Summit City was laid out that summer. As time went on and the radius of gold discoveries extended farther and farther, other camps and towns were established. They included Enterprise City, Paris, Ossaville, Richport, Mendoza, Rocklin, Carlisle, Baltimore City, and Atlanta.

The flowering of such gold rush settlements is a familiar story and to recount this part of Meadow Lake's history would not contribute substantially to the consideration of the gold deposits themselves. It may be said, however, that in its best days Summit City had more than six hundred houses and a commensurate business district. During the summers of 1866 and 1867 the population, including the unhoused element that camped out or slept on the floors of hotels, saloons, gambling houses, and mine and mill buildings, has been estimated at 10,000. In the winters of those years, when the snow lay up to thirty feet deep, the people who had stayed on, huddled in their houses and often tunneling from one dwelling to another through the snow, numbered scarcely two hundred.

Let us look now at the ore deposits that attracted such a throng at Meadow Lake. Exaggeration of values was to be expected in contemporary accounts, but here is what appears in the record of those first years:

The original Excelsior vein was stated to have extended, in clearly identifiable outcrops, for more than five miles.

Similar veins, many of them paralleling each other, were of such size and quality as to encourage the erection of five reduction plants the first year of the boom.

The area of proved mineralization was extended farther and farther from the original strike. The number of mining claims filed, each one purporting to be based on valuable ore in place, was just under 11,000.

Assays of ore samples were showing extremely high values. The Excelsior vein showed a gold content in various samples of $9,000 to $61,000 a ton, and it was noted in the press that not a single piece of ore from this property assayed less than $65. Somewhat lesser but quite high assays were run on vein matter from the Enterprise, Mohawk & Montreal, U. S. Grant, and other mines. The values of a considerable number of ore samples from various early claims fell within the range of $600 to $700 per ton.

These reports all were for that first active summer of 1865. Later, assays soared up to $93,000 a ton on selected samples and reports in the $600-plus range noted above continued to be common. A collection of specimens displayed in Virginia City and represented as typical Meadow Lake ore had per ton values ranging from $250 to $50,000.

Ample allowance must be made here for the temptation to pick out superior ore for assay instead of getting a truly average sample of the entire vein, of the rock that is actually going to be mined and milled. Rich selected specimen ore has been employed in many mining projects to sell stock or to impress individual investors. In view of the eventual failure of the mines, it was widely believed that this exploitation of occasional rich pockets explained the whole boom-and-bust episode, that there never were any substantial deposits of commercial grade ore.

This conclusion can hardly be accepted. During the three most active years of Summit City and its environs we have a history of reputable assayers finding high gold and silver values, chiefly gold, in innumerable samples submitted by miners and coming from veins over a considerable area. That all these men were deceiving the public or themselves with selected, unrepresentative ore from their properties is hard to believe. Moreover in those days most of California's population was highly sophisticated in mining matters. Such spectacular ore as that running $61,000 a ton would be shrugged off as hand-picked and meaningless, but it is not so easy to ignore the many assays that ran between $60 and $600 per ton. Such ore would be of commercial grade by a wide margin.

We have evidence that deposits within this range were worked at several places and with modest success. Scattered printed items relate that men crushing the weathered surface rock and panning it by hand were recovering about $10 in gold daily. At the Mohawk claim, the greater capacity of the rocker brought returns of $24 a day from similar material. An arrastra was operated at the Wisconsin claim and its crew reported takings of $25 a day per man. In 1865 this represented a fine income.

Much of the vein matter exposed at or near the surface contained free gold visible to the naked eye. Too much of this was seen by too many people for the fact to be denied.

We will soon suggest a basic reason for Meadow Lake's failure, but first some contributing factors may be noted. For one thing the working season was extremely short. The heavy winter snowpack melted slowly and at that elevation the snows of a new winter came early. Haulage of machinery was slow and costly. At one ore mill, transportation made up one quarter of the gross cost of its equipment. And, although a great deal of money was expended at Meadow Lake, no single property appears to have attracted financing on the scale that developed the great mines of the Comstock or the Mother Lode. As for the small miner, once the weathered surface of his ledge had been excavated he found the vein matter to be exceptionally hard to mine. Accounts agree as to the excessive labor required to put in drill holes and the time lost in sharpening and resharpening drills. Also, as has been noted, there was a tendency among the small miners of that time merely to open up a promising ledge and sit back in hope of selling their "coyote holes" to a promoter or investor. At Meadow Lake, where even well-equipped and working properties were failing to show any profit, the market for mere prospects must have been slim indeed.

What was the actual reason that demonstrably good ores did not give up their values?

The facts about those active days and their frustrations are scattered about in fragmentary mining reports and even more fragmentary accounts in contemporary newspapers. Recently a Purdue University professor, Paul Fatout, has pulled much of this information together in his *Meadow Lake Gold Town*. The book is an over-all history of the place and its people rather than a technical mining paper, but Fatout has investigated the mines' problems with some thoroughness.

In brief, the trouble appeared to be that at Meadow Lake nature had locked her treasure into a rock that was complex in composition, troublesome to mine and crush, and enormously defiant of man's attempt to extract its gold. True, there were decomposed surface areas where free-milling gold was found. There also were numerous rich pockets. In the main body of the deposits, however, gold and silver occurred in combination with iron, manganese, copper, cobalt, lead, nickel, arsenic, and sulphur. It is possible that even this array of unwanted complications is not complete. The art of spectographic analysis of ores was years away, and if any comprehensive qualitative report on the Meadow Lake ore was made by current chemical procedures it has escaped the record.

Now the occurrence of gold in combination with iron sulphides is common as mines get below an upper oxidized zone. To work these lower ores calls for an expensive changeover from free-milling operations. Some mines are shut down at this stage. Where gold values promise continuing profits, the changeover is made. The new treatment is a standard one and there is every promise of success.

At Meadow Lake the ores were so "refractory" that all recovery methods failed. In mining, refractory has much the same meaning as when we speak of a refractory child, that is, unmanageable. When the original Excelsior Mine bravely started up its new mill, ore that showed $200 a ton when assayed yielded only $10 to $11. The rest of the metal, still locked into the native rock, was carried off in the tailings. This was the story everywhere. Ore that displayed bonanza values in the laboratory refused to give those values up.

Many treatment processes were tried out, a few of them fanciful but most of them following sound metallurgical principles. There was the Hazen process, the Burns process, the Deethken process, the Churchill, the Maltman, the Crosby, the Gleason, and on and on. Some were devised by practical mining men, some by inventors who were intrigued by the problem and its prospect of reward. One which received a great deal of attention and at last attracted some eastern capital to Meadow Lake was that of young Robert M. Fryer. In his laboratory he definitely could crush and treat the complex ore and get out the gold, although he remained quite secretive about how this was accomplished. A well-financed and heavily publicized plant was built to employ the Fryer process. The fanfare of a Fourth of July opening was followed by many weeks of silence. Eventually it was announced that the young inventor-promoter had suffered severe burns at the plant and had left for the East. The $10 million corporation went under new management. Fryer did not return, little or no bullion was produced, and the plant soon stood idle.

The successive failures of this and that process finally wore out even the most optimistic of mining men. By 1890 the district was virtually dead. Summit City itself was obliterated. Many of its structures collapsed under the heavy snows. Doors, windows, hardware, finally the weathered lumber itself, were carted off to more permanent settlements. A visitor at the turn of the century and a visitor today would find the country much as it was when the trapper Hartley first struck his pick into the gold-bearing outcrop of the giant Excelsior formation.

Later, when California's excellent mining and geological reports were being published, the state mineralogist commented with regret on the paucity of information about Meadow Lake and its problems. This lack was partly made up in 1936, when a San Francisco mining engineer, A. L. Wisker, published a report based on extensive research of documents and a personal inspection of the old workings.

Wisker uncovered records of the actual gold recovery in more than a dozen brief mill operations. The yield varied from $13 to $98 a ton. He confirmed that the upper, oxidized zone, that is, of free gold amenable to simple, efficient recovery, was quite shallow. Below, the gold was held in complex ores that consisted of 90 per cent pyrites. The great smelter at Selby, on San Francisco Bay, boasting the tallest smokestack in the world, had not been built in the 1860s. It could have handled this ore, but at that time the only comparable plant was at distant Swansea, Wales. Wisker reviewed in technical detail the reduction processes available to miners at Meadow Lake and as a mining engineer he found them all quite inadequate. This, he contended, explained the discrepancy between the excellent, often sensational, gold content shown by assays and the sorry returns when the ore was milled.

Even should average values prove lower than indicated, Wisker argued, the mines would be profitable because of the massive size of the deposits. Noting that California's premier gold producers, the mines of Grass Valley, were working veins averaging only two feet in thickness, the engineer cited Meadow Lake formations that range from twelve to fifty feet across.

The lake itself is now owned by the Pacific Gas and Electric Company, which took over many of the old dams and ditches of early mining days for the hydroelectric generation of power. The mineralized area, however, extends far beyond the shores of the lake.

The one man who appears to have mined at Meadow Lake with some success was the original discoverer. Henry Hartley, as a principal owner of the Excelsior, shared in the repeated disappointments experienced in trying to make an efficient recovery of gold. He stayed on beyond the boom, a lone man amid the network of ledges whose potentialities had attracted so many thousands. And he prospered. Crushing and sluicing the oxidized surface ores, occasionally chancing upon a rich pocket, he built up a comfortable little fortune.

More than a decade after Summit City had vanished, Hartley, now fifty-three years old, met and married a young Englishwoman of twenty-three. She was Alice Marion, a lady of considerable talent as a painter, and as a pianist as well. In the fifth year of their marriage Alice was given a trip to London, where she made some tentative efforts to sell the Excelsior. While she was away Hartley died. It was determined that someone had introduced a fatal dose of opium, then readily obtainable, into Hartley's food. A disappointed but still hopeful suitor of the young Englishwoman was suspected but no arrest was made.

It is ironic that the surviving Mrs. Hartley became a better-known figure throughout the West than her husband. On her return from abroad she settled in Reno. She was soon giving art lessons and also was soon involved with Nevada's most prominent banker, M. D. Foley. When a child was expected, there was a bitter quarrel over "legalizing her position," which culminated in her drawing a pistol from somewhere in the voluminous attire of the day and shooting the banker dead. Alice's trial was a *cause célèbre* throughout Nevada and California. Public opinion was strongly in the widow's favor, but a jury brought in a verdict of guilty of murder in the second degree. She went to prison, but after serving two years was pardoned by Nevada's governor. As for the Excelsior, attempts to make it pay were made by several successive leasers. Like their predecessors, they all gave up, baffled by what seemed to be not only a stubborn but a truly malignant ore.

4
The Lonely Trans-Sierra

The mining fraternity was surprised to find gold deposits at as high an elevation as Meadow Lake. It was the last surprise of that sort. There have been a few other, although minor, workings at such an altitude but in general the geologic conditions become less and less favorable as one approaches the crest of the Sierra Nevada. At Donner Pass, named for the ill-fated emigrant party of pre-gold rush days, Interstate 80 begins a descent and we will soon be looking at the rather abrupt eastern face of the Sierra Nevada range.

On this eastern side we are in a mineral region completely divorced from the Sierran Gold Belt and in country that is commonly described as semidesert. We also are near U.S. 395, a highway which will take us on a long trail southward. It is a trail through mineral districts which many mining men consider neglected. Certainly it has its share of lost mines and of mines which are not lost at all but have been abandoned under mystifying circumstances.

It was through this country that the buoyant small-town physician Dr. Darwin French hunted that archetype of hidden bonanzas, the Lost Gunsight Mine. Here another medical man, Dr. John Randall of San Francisco, kept up a costly search for the Lost Cement Mine, only to be betrayed by his trusted lieutenants. At the base of this eastern escarpment of the Sierra Nevada is the Dead Sea of the western world, Mono Lake, which some authorities associate with $15 billion—yes billion—in unrecovered gold. Death Valley also may be conveniently dealt with here, for it lies farther north than most people picture it and is only a slight jog, as desert distances go, of eighty-one miles off our course down U.S. 395.

The territory to be examined takes in three counties, Alpine, Mono, and Inyo. The first may be quickly disposed of. It is a high, sparsely populated county consisting mostly of federal forest land. The road connecting the county seat, Markleeville, with U.S. 395 passes a number of old gold mines. The problem in these operations seems to have been a heavy content of zinc in the ores. The existence of strong veins and attractive values is substantiated by official mining reports, and from time to time there has been tentative work done in hope of being able to ship ore to a smelter at a profit. Higher gold prices might well attract the considerable capital required to construct an efficient treatment plant right at the mines. The principal routes into Alpine County from central California are not kept open throughout the winter, but it usually can be approached from the Nevada side.

The long stretch of the U.S. 395 we are to follow starts at the towns of Coleville and Walker, near the Nevada state line. At this point the highway follows the canyon of the Walker River, named for one of the West's early mountain men. Gold has been found at various spots along the Walker, but to date no rich or pervasive mineralization is of record. Not far from its headwaters, however, there were two great bonanza camps, Bodie and Masonic. High, difficult, unpopulated country encloses the Walker's course, and one may speculate as to whether these uplands ever have been thoroughly prospected. Most of the region is part of the Toiyabe National Forest and is thus open to mining location.

One can spare himself any prospecting at the once booming gold camp of Bodie, for it is now a state historical preserve. Many old buildings still stand on its bleak, wind-swept location at an 8,368-foot altitude. Its first citizen, Bodie himself, froze to death in an attempt to guard his fabulously rich discovery through the winter. Isolated from any law enforcement agency, the camp became one of the toughest of them all and the term "bad man from Bodie" joined the vocabulary of western folklore.

During this lawless period a family at another gold town, Aurora, was moving to Bodie for reasons of employment. The Aurora weekly newspaper told of how the little daughter, aged nine, had left a piteous note reading, "Good-by, God. We're moving to Bodie." There also was a newspaper in Bodie, and the next issue carried the editor's riposte. Conceding that the note was accurately quoted, he suggested that his fellow editor had used the wrong punctuation. What the child had written, he contended, was: "Good! By God, we're moving to Bodie!"

U.S. 395, after attaining an altitude of more than 8,000 feet, dips down to the town of Lee Vining. This point is of interest to the gold hunter. It looks out upon the vast, dreary expanse of Mono Lake, about which cluster various legends of lost gold. It also is the takeoff point for the Tioga Pass road, now a seasonal route into Yosemite National Park but originally built to tap the riches of the great Tioga, or Sheepherder, Lode, the origin of one of the great mysteries of California mining.

Mono Lake is a very large, almost circular body of water. It has no outlet and is so salty that its only aquatic life consists of a minute, shrimplike creature found near the shore. There are two treeless, uninhabited islands, Paoha and Negit, and the sixty-odd-mile circumference of the lake is uninhabited too except where the federal highway touches it. Pale gray, statuelike formations of tufa stand here and there where receding waters have left them, and to the south rise a series of dark volcanic cones known as the Mono Craters.

A short distance from the north shore of Mono Lake are several gulches which have been worked for placer. One area, at least, is reputed to be quite rich and to have supported a miner named Shephard for a long span of years. R. W. McAllister, a collector of old mining data, talked with Shephard in 1953 and related that the old man's policy was to draw on his gold in the ground only as he needed it, just as he would draw money from a bank. On a mountain behind his cabin, and about 1,000 feet higher, an interesting contact was shown McAllister. A formation of red rock lay against the dull brown rock that constituted the rest of the visible surface. This contact was understood to be the site of a high-grade gold-bearing ledge and a well-concealed shaft. The discoverer had disguised the workings, McAllister learned, when he left for Carson City, Nevada, to get medical treatment. The miner had died there, leaving, so far as a disappointed Shephard and his friends could learn, no map or clue to his bonanza.

The approximate location of Shephard's cabin is the start of one of the longest arrow-straight highways in the United States, an undeviating 27 miles. This road runs northeast across the Nevada state line, where it becomes State Route 31, then curves north to the town of Hawthorne. The country along this stretch is flat, but a digression to any of the distant hills will put one in potential gold country that has not been too thoroughly prospected. An exception would be the Mono Craters, or more precisely that part of the crater country which is clearly lava. Yet, either hidden among or somewhere south of these barren volcanic cones lies a strange red formation which, on the basis of ore carried out and displayed to scores of astonished people, is one of the richest gold deposits ever found. The exact location of this deposit is no longer known. It is the Lost Cement Mine.

We should not leave Mono Lake for the craters, however, before reporting, with some skepticism, on the supposed treasure of the lake itself.

In 1923 a rumor of unknown origin spread along the West Coast to the effect that the quiet, salty waters of this desert sink were rich in gold. The report aroused little public excitement and seems to have been ignored by practical mining men. Yet to some romantic souls and men of speculative bent it had great appeal—especially in the state of Oregon. At least eight groups of investors took shape there, intent on getting a share of such great wealth. Several of the groups incorporated and offered stock for sale. Hasty investigations were made into the technology of recovering metals held in suspension. Equipment was designed, and its components bought or manufactured for shipment to the novel bonanza.

There was nothing implausible in the basic idea. Sea water is known to hold a small amount of gold and attempts to recover it have been made. San Francisco Bay and several other bodies of water on the globe have been identified as holding somewhat more than the minuscule average amount. The element boron has been recovered from ocean water on a commercial basis.

Among the individuals who joined the rush to Mono was A. Fred Eads of San Gabriel, California. He found many mining claims staked along the shoreline and seven plants pumping out the lake's water in small quantities. The recovery systems were described as rather crude, but we have no record of how they operated. Eads had some association with a former Columbia University physicist who was on the scene, Dr. Herschel Parker. The professor's laboratory tests, Eads related, showed gold of approximately $1.00 in value to each ton of water. It was estimated at the time that Mono's water totaled about 15 billion tons, which could, with sufficient optimism, be translated into $15 billion.

Dr. Parker appears to have favored an electrolytic process for recovering the gold, but he experimented also with mechanical and chemical means. The professor finally gave it up as a practical matter but, according to Eads, maintained his faith that the gold was there.

The various Oregon companies also failed and one by one withdrew from the scene.

Today, in the scanty population near Mono Lake, it is hard to find anyone who has heard of the excitement of half a century ago, or who knows of the supposed wealth in the desolate sheet of water. One may speculate as to whether current work on the desalting of sea water may develop a technique that is applicable to Mono's reputed gold.

The riches of the Lost Cement Mine, as has been indicated, are believed to be among or near the Mono Craters. These barren cones are encircled by straggling, unimproved roads and trails. To drive inside the group, one finds only a single, two-mile road that leads to Pumice Valley. Even the thoroughgoing auto club scouts have been unable to place a dotted line on the map to show any route that penetrates very far into this weird, dark country. Exploration among the craters clearly must be on foot. The favored location for the lost ledge, however, is south of the principal craters and in that less rugged terrain quite a network of sketchy roads exists.

The name most reliably associated with the Lost Cement Mine is that of the Dr. Randall of San Francisco, previously mentioned. This physician saw samples of the rich ore and possessed a map which attempted to show the mine's location. He acquired 160 acres of land which he believed included the site. He kept as many as a dozen men on a payroll for two years to search for the deposit. Nevertheless he had to return to his San Francisco practice without a glimpse of the elusive lode itself.

The original find was made by two brothers, German immigrants, who were members of one of the covered-wagon parties which, in 1857, were still making their way to the California gold mines. This group was attacked by Paiute Indians near where the hamlet of Benton, California, now stands. The pair in question escaped during the confusion. Heading on west, the two Germans found water in a small stream, probably McLaughlin Creek. Later it appears that they camped near the headwaters of the Owens River in what is now known as Clark Canyon. This would place them about six miles southeast of the southernmost of the Mono Craters.

It was about this time, perhaps a day or two later in their trek, when they were resting within sight of a dark gray cone, that one of them chipped off a piece of protruding rock and found it studded with coarse gold. From the first time the story was told, and always thereafter, the gold-bearing deposit has been described as "red cement." Just what the wayfarers meant by "cement" in those days is uncertain. The word comes from the Latin and historically it seems to have applied chiefly to the mortar used between building stones. One can speculate, therefore, that the material was fine-grained in whole or in part, possibly a conglomerate with its constituents "cemented" together. Its red color always was accented. The deposit could scarcely be identified as a typical vein of gold-bearing quartz.

Survival was the primary concern of the two men, for they still had the Sierra to cross. In this region the range is high, steep, and forbidding. The discoverers made a hasty attempt to cover the outcrop and then pushed on.

One of the pair now disappears from the picture. In 1860 we find the other in San Francisco, in ill health and receiving treatment from Dr. Randall. The man failed to recover. Hopeful of returning to his strike, he had kept some of the amazingly rich rock and had drawn a map while the events of his journey were fresh in his mind. These items he turned over to the physician. The dying man's story must have been really persuasive and circumstantial, for it led Dr. Randall to commit a great deal of time and money to a search.

The quarter section of land the doctor acquired, by homestead or through some other Land Office transaction, was described at the time as being in Pumice Flats. Is this the same locality as today's Pumice Valley? If so, it is a depression three miles south of the south shore of Mono Lake and barely a mile from California State Route 120, which is a secondary paved road branching off U.S. 395 just five miles south of Lee Vining. Pumice Valley's elevation is 7,150 feet. Its location fits in very well with the course of the Germans, who were trending north in the hope of encountering other westward travelers before crossing the Sierra.

Dr. Randall's search began in 1861. The names of some of his employees have been preserved. There was a Gideon F. Whiteman, who was a foreman of sorts; a Van Horn, whose name had bobbed up in several mining rushes, and a German who, like so many of his nationality in early-day America, was merely dubbed "the Dutchman." Later, men named Carpenter, Colt, and Kirkpatrick were involved, and in addition there was the menacing figure of Joaquin Jim, the leader of the warlike Paiute Indians. The modern treasure hunter may make some wry acknowledgment to Joaquin Jim for the fact that the Lost Cement Mine still awaits a lucky discoverer. We shall soon see why.

Dr. Randall appears to have been double-crossed by two of his employees, the German and Van Horn. These two found the red cement. They crushed a quantity and panned out a substantial stake in nuggets and gold dust. Salving their consciences, if any, with the theory that the deposit lay outside their employer's 160 acres, they went to Virginia City, Nevada, and prepared for real mining. Some crony of Van Horn's was taken into partnership. Like so many of the actors in pioneer mining adventures, his name has been lost to the record.

Arriving at the rich ledge with tools, powder, food, and other supplies, the three partners first cut trees and laid the floor of a cabin. Winter comes early at that high elevation, far removed from the mild marine climate of the California coast. But there was to be no winter for the miners. Joaquin Jim and a party of braves had been quietly watching the intruders and the Indians now moved in. The prospective millionaires were allowed to keep the clothes on their backs and, as the story goes, were given one of their own horses. This was perhaps to allow them to pack enough supplies to get well out of Paiute territory. They were warned never to come back.

Van Horn necessarily heeded the threat, for he suffered some unspecified illness that soon rendered him virtually an invalid. We know that he made his way to Sacramento and there boarded one of the passenger boats that up to the mid-twentieth century plied between Sacramento and San Francisco. Van Horn realized he was near death and, encountering a friend named Carpenter on the river craft, felt impelled, like the ailing first discoverer, to give an account of his bonanza. The other two partners, Van Horn said, were at that time in Virginia City. He was convinced that they hoped to evade the Indians and make an early return to the mine. Once again a dying man was giving directions for finding the "red cement."

Carpenter's information was passed on to some adventurous friends, Colt and Kirkpatrick. The waybill, written or oral, proved quite accurate. The two young men were able to follow it to the stark, unfinished log cabin, surrounded by the stumps of trees. Beside it were two skeletons. The impatient gold miners from Virginia City had found that Joaquin Jim meant what he said.

All we know about the other results of the Colt-Kirkpatrick expedition is that they failed to discover the valuable ledge. Friends said they had found unusual red rock in several places but none of it contained gold. The original rich outcrop was doubtless well concealed by the

Paiutes. This was something that Indians made a practice of doing at various places where they had driven out white miners.

Dr. Randall had failed. The men who may be said to have defrauded him of the discovery were dead. The latecomers to the cabin site found nothing of value and passed from view. But the mysterious seepage of news that characterized western mining camps had carried word of the Lost Cement Mine to many ears. In 1879, more than twenty years after the original strike, a search was still going on. The San Francisco *Post* found it important enough to send a correspondent, J. W. A. Wright, to the scene to write the history of the mine and describe the current activity. But a headline telling of its rediscovery has yet to be written.

In this general area but to the west of U.S. 395 lies a massive silver deposit that once brought $163,000 at a sheriff's sale but from which hardly a dollar in silver ever has been recovered. The giant deposit received little notice from western mining men but it was acutely noticed by a small, private group of New England investors. It drained their bank accounts of $350,000.

For their money and their sustained confidence in the great lode the distant backers got a long, well-engineered tunnel, but it stopped tantalizingly short of what some geologists believe is a great bonanza. And if the New Englanders were unable to bequeath a vast mining fortune to their heirs, they did bequeath to western mining one of its major puzzles.

The property is the Tioga Mine and the operating group was the Great Sierra Mining Company. The veins are the Sheepherder, held by two claims totaling 3,000 feet in length, and the Great Sierra, covered by four claims extending 6,000 feet. They are found on the rather flat, plateau-like summit of Tioga Hill, which is situated on the very crest of the Sierra Nevada, surrounded by peaks towering to 11,000 and 12,000 feet above sea level. The elevation of the outcrops themselves ranges from 10,600 to 10,800 feet.

"On the surface the Tioga veins are large ones," wrote H. A. Whiting, a state mining engineer. "In the instances observed they were 10 to 15 feet wide on the average, though in some places not more than 6 feet, and in others as much as 40 feet between walls." These would be phenomenally large ore bodies.

Someone unnamed had spotted the great silver lodes in 1860 and left a location notice. In 1874 a sheepherder observed the veins and found the old document. Four years later, the outcrops still lying untouched, the sheepherder located them for himself. He apparently sold out, for skipping to the early 1880s, we find the New England company organized and a well-financed development under way. There was one other mine in this high and rugged country. Five miles away by air line but much more by tortuous mountain trail was the May Lundy, which ultimately recorded a production of $1.5 million.

At the Tioga it was decided to tap the big deposit at a depth of about 750 feet by means of a crosscut tunnel. This is standard policy when contours of the land permit tunneling, for in the long run it reduces operating expense.

Getting machinery high into this jagged country was an exasperating, even tragic, undertaking. Many mules were killed and some men lost their lives. The dragging of an immense boiler to the tunnel site was narrated later as if it rivaled the labors of Hercules, and the account was scarcely exaggerated. Even maintaining a work crew was a logistical problem. At length the investors approved the building of a road in a westerly direction, down into the hills bordering the San Joaquin Valley. This construction cost $64,000, an immensely larger outlay in the 1880s than it would seem today. This later became the state's Tioga Pass road. Even now it can be kept passable only for a brief summer season and many a traveler finds the steep, twisting route a bit unnerving.

While management struggled with supply problems, the driving of the tunnel continued. It was a workmanlike job, six feet wide and seven feet high, accommodating two of the narrow tracks used for ore cars. The ambitious plan was moving on schedule.

In 1884 money ran out. At this time the tunnel was 1,784 feet long. Engineer Whiting, reporting to the state mineralogist, estimated that at this time the face of the long bore was directly under where the Sheepherder vein appeared on the surface. However, the vein had a "dip"—a technical term for slant or deviation from the vertical—of 70 degrees. It was going to take 270 more feet of tunneling to reach the ore body. Beyond that, at a distance of nine hundred feet more, the Great Sierra vein also should be encountered.

The personal problems of the New England investors have not come to light. Some may have been impoverished, others merely discouraged. The six claims had been patented and had the legal status of any other privately owned real estate. They came up at a sheriff's sale and a corporation with the slightly altered title of Sierra Consolidated Silver Company appeared as the new owner. It was very much the old group, with some defections and some additions. The $163,000 sale price probably represented transfer of very little cash, merely a readjustment of credits and liabilities among the owners.

The operations under this new regime are obscure. In a brief passage on the Tioga Mine in the state mineralogist's report for 1894 no recent developments are noted and the report merely rehashes earlier information on the deposit. All we know is that work stopped and that the costly burrow into Tioga Hill remains unfinished.

The portal of the tunnel and the great dump of excavated rock can be seen today from the ruins of the old mine headquarters. The little settlement that grew up there was given the name of Bennetville. Even at its elevation of 9,800 feet, early optimism had pictured it as a coming little city. Still higher, on the lofty flats where the veins show on the surface, there was another settlement called Dana. It can be reached by a foot trail starting a little north of the tunnel, the climb taking about two hours. A massive stone building, with walls two feet thick, still stands at the summit.

Continuing south on U.S. 395 one comes to Mammoth Lakes and their well-known recreation area. There are known gold deposits here but it can scarcely be recommended for prospecting. Winter skiing and summer camping pre-empt much of the area and snow exposes the ground to prospecting only during brief seasonal interludes. Any search for gold-bearing veins in this region could reasonably start at the Mammoth and Headlight mines. Both are on strong mineralized formations and lie high above the recreational development.

The next point of interest as one travels south on U.S. 395 is Bishop, the mini-metropolis of this Trans-Sierran region. It is a logical base for exploration of several districts that have seen considerable gold mining in the past.

One such area is reached by U.S. 6, which runs north from Bishop along the base of the White Mountains. This range shows much scattered mineralization, but appears to have been somewhat neglected because of more active mining that developed to the east and south. A thirty-four-mile stretch between Bishop on the south and Benton Station on the north is of most interest. There are gold mines, nearly all inactive, on both sides of the highway, and just east of Benton Station is the ghost town of Montgomery. Clearly visible in the White Mountains are sixteen major canyons, a few with dubious roads but most of them traversible only on foot.

Along much of U.S. 395 the traveler is in the valley of the Owens

River, and from Bishop southward the flat expanses of that stream's narrow flood plain are always close at hand. In an early-day coup, the Metropolitan Water District of Los Angeles acquired title to Owens River water and it is now piped from a series of reservoirs to the teeming population of Southern California.

On the east side of the river valley the Inyo Mountains rise gradually. This entire range is of interest from the gold-mining standpoint.

In the low mountains immediately east of Bishop mining has been extensive. A reconnaissance of these diggings may be made by driving up Silver Canyon, then up some steep, twisting grades to a radio relay station, thence south at a rather high elevation to connect in due course with the pavement of State Route 168 just below Westgard Pass. A great many mines and prospect holes will be observed on this trip. Many more exist on the ridges and in the canyons that slope down from the road. It is up to the individual as to which he wishes to examine and try to evaluate. Most of the workings are within the Inyo National Forest. Any new discovery is open to mining location and any abandoned claim delinquent in annual assessment work may be relocated by a newcomer. Some of the mines were patented in the early days and during the long doldrums of gold mining went tax delinquent. Whether any such property is available through a tax sale may be determined at the county seat, Independence, forty-three miles farther south.

At Independence we are near the Green Monster Mine, a rich discovery of early times, and near the string of additional operations it inspired in now depopulated Mazourka Canyon. For reasons which it is hard to track down, Mazourka Canyon always has been well regarded by mining men. The quiet little town of Independence, however, no longer holds its breath in expectation of a great bonanza. Along the twisting, fifteen-mile canyon are numerous old workings, including the Brown Monster, not quite the equal of its Green counterpart but for a time a profitable operation. The road ends at the base of Waucoba Mountain and the now inactive Bluebell Mine.

There has been intermittent placer mining all along Mazourka Canyon and large nuggets have been found. Some have been unearthed high above the dry stream bed, a circumstance which has puzzled the old-timers. A miner named John Amic, prospecting on a hilltop, picked up a nugget of obvious placer origin that weighed in at $300.

There is a secret concentration of gold in or near Mazourka Canyon once known to the local Indians and possibly still known to some of their descendants. John W. Dixon quotes an old resident of the area who witnessed two deliveries of placer gold made by a Paiute Indian to a bar in the town of Lone Pine. The brave had damaged the establishment's furniture and fixtures in a drunken outburst and was making restitution under threat of arrest. Dixon's informant, John Gorman, said the Paiute made brief trips into the hills for each installment and handed over coarse gold in little buckskin bags. As with various other Indian mines, it appears that this deposit could be tapped by a tribesman in case of real need, but strong taboos forbade its being worked on a scale which might reveal it to white men.

On the other, eastern, side of the Inyo Mountains, where they slope into one of the most barren, untraveled sections of all the southwestern desert, there is much to interest the modern gold hunter. Of exceptional interest to the more adventurous is the startling outcrop of gold-studded quartz found by the old Foreign Legionnaire Alec Ramy. One says "startling" because it was found by an exhausted, desperate man in no condition to pick or dig for mineral but was looking only for water. To come to Ramy's attention in such circumstances, it must have been prominently exposed, striking the eye at once.

Ramy was no stranger to desert conditions. During his Foreign Legion enlistment he had been stationed at a remote Algerian outpost, well into the Sahara. Emigrating to America and drifting west, he worked at Virginia City and was lucky enough to find and lease a gold and silver claim on the edge of the Comstock Lode from which he recovered $30,000. A volatile, gregarious man, he soon dissipated this stake in high living among a group of hangers-on and fair-weather friends. Further ups and downs in the Comstock area followed. The district itself went into a decline. Ramy, now a mature man with several years of day labor behind him, decided to take his savings and strike off into the California desert.

There was winter after winter of fruitless prospecting in and near Death Valley, each followed by a necessary respite during the intense summer heat. The autumn of 1904 found Ramy in the Last Chance Mountains. This is an extremely isolated range that can best be described as lying northwest of Death Valley. The much publicized "castle" built by Death Valley Scotty is the best landmark in this area, for it is shown on all maps and is a tourist attraction. A ten-mile tramp from the castle would take you to the southern tip of the Last Chance range. From there twenty-five miles of completely desolate peaks and canyons extend north. An old sulphur mine on the western flank, long abandoned, is about the only sign of human penetration.

Somewhere in or near the Last Chance Mountains, or possibly in the neighboring Saline range, Ramy suffered the bad fortune to have his burros escape. This was during a night camp, so his food and water had been unloaded from the animals and were still beside him. The old prospector was at a great distance from any known habitation. He had two canteens of water and he had food, but nevertheless was in a really dangerous plight. The rest of his story is one of desperate wandering, much of it thirst-crazed and beyond future recollection.

One thing he did remember was hacking away at an outcrop of rusty quartz, full of coarse gold, the thing he had been dreaming about in all his years of prospecting. It was located, he said, between a *tinaja,* which he had found completely dry, and a tiny seepage of water from under a great boulder, the fluid that had saved his life. These obscure desert features were hardly notable landmarks, even to Ramy himself. He believed that at that time he was getting close to Saline Valley, a wide, flat depression south of the two mountain ranges.

His next recollection was of being among friendly Indians, with a young squaw bathing his swollen feet. The Frenchman, in the course of years, had become known to many of the Death Valley tribesmen. Two of his rescuers, Grapevine Dick and Coldmore Jack, knew who he was, and told with some relish how Ramy had been encountered staggering across the desert singing, as best he could with his parched throat, something loco, a song, Ramy later assumed, remembered from his days in the Foreign Legion. The desert's annals are full of the manias resulting from thirst and their strange manifestations.

It was nine days before Ramy felt able to travel on his own. With the usual miner's instinct, he had held onto a bundle of his rich ore. Apparently he gave his red friends a specimen, for such a piece turned up in the possession of Crisante Santavinas, a Death Valley character who attempted without success to find its original location. The next person to see a sample was Alfred Giraud, at that time a young sheepherder and a fellow countryman of Ramy. Giraud was with his flock along the path of Ramy's return to civilization and was able to assist the again ailing and discouraged wanderer. We know little about the rest of the Legionnaire's life, except that he and Giraud met again after several years and became fast friends. As far as Giraud could tell, when he discussed the matter with Ken Worley of Indian Wells Valley in the 1950s, Ramy never tried to find his way back to his bonanza.

Giraud himself had Ramy's story at firsthand and had vague intentions of making a search. Meanwhile he acquired some land, soon was herding sheep of his own. He was pretty well tied down by this operation and in the course of years found himself comparatively well off. The young man's dream of the treasure hunt eventually was only an old man's reverie. It was too late for the search.

Giraud had a few clues. He believes that Ramy lost his burros near Dry Mountain, a landmark rising 8,726 feet just outside the western boundary of Death Valley. He also revealed that some unspecified foot ailment was making walking quite painful for the lone prospector. Water would be Ramy's chief objective. He evidently missed both Upper Warm Spring and Lower Warm Spring, the only known water sources in this great expanse of desert. The lower spring may be reached today by a dirt road which starts at the mouth of Hunter Canyon near the old Vega Mine. It probably would be to the west of this road and north of both springs that Ramy found his life-saving water seepage and his golden outcrop. He would not be climbing mountains in his wearied and painful condition. The best advice therefore would be: Don't climb, just look along the base of the Saline Mountain range.

So much for the lost ledge, the quick fortune, that lurks east of the Inyos. For the prospective developer of an old mine, there is what might be termed a lost mining district along the eastern flank of the range. It is lost in the sense that isolation, a poor road running to great distances between watering places, and excessive cost of getting supplies in and getting ore out have caused many promising mines to be abandoned. The district might be said to have been awaiting the motor age, in particular the age of the four-wheel drive.

The twisting dirt road that runs along this neglected eastern side of the Inyos is all of fifty miles long. It is best entered from the north. Big Pine on U.S. 395 is the nearest town. Recent construction has pushed a paved road across the Inyos from this point, and mileage should be taken from Big Pine. There are turnoffs onto dirt roads leading south at 14.1 and 15.3 miles, and these two routes soon converge into the single unnamed and unnumbered road along the base of the mountains. If one fails to make one of the turns, he is off on a long, dry, and lonely trip to the equally lonely northwest corner of Death Valley.

After making the turn and heading south, he quickly comes to one of the most extensive and best-proved gold deposits of the Inyos. This is the ancient river channel of Marble Canyon. In a report published in

1938, W. B. Tucker and R. J. Sampson of the California Division of Mines stated that the channel was two hundred feet wide and could be traced for nine miles. This would make it a deposit of extraordinary size. It runs in an east-west direction at an altitude of 7,000 feet.

"The gold occurs on bedrock and is fairly coarse, ranging from the size of wheat grains to nuggets," Tucker and Sampson wrote. "Development consists of several shafts sunk to bedrock, with depths ranging from 70 to 115 feet. The principal drawback to working these claims is lack of water."

The Marble Canyon miners used the dry washing methods that have been described earlier. In one operation, the Anderson placers, some records were kept. The pay gravel, that is the excavated material valuable enough to wash out, was from three to six feet thick and fifteen feet wide. Regarding other workings, it appears that pay dirt was struck almost anywhere on the great formation where a shaft was sunk to bedrock. Some reports had the gravel running $5.00 a cubic yard, a suspiciously round figure not vouched for by the state field men.

This is the only known placer ground of significance along the east side of the Inyos, or at least the only one figuring in available official reports. Additional deposits, possibly of substantial value, cannot be ruled out. There quite definitely are lode formations of remarkable size and continuity. Most are located in the Beveridge Mining District, which takes its name from a short, rugged cleft known as Beveridge Canyon in the heart of the productive area.

In official mining reports, "district" is not used in a general sense synonymous with "area" or "territory." A mining district, often abbreviated M.D., is a legal entity. Miners in remote areas, far from any seat of government, had the privilege of organizing such a district, and its regulations, so far as they did not conflict with state or federal statutes, had the force of law. Among the first acts of such a district would be appointment of an officer to record mining claims and the creation of some kind of tribunal to adjudicate disputes over claims. Such organization was very much to be desired at the isolated Beveridge diggings.

Paralleling Beveridge Canyon is Hunter Canyon, where the massive veins of the Gold Standard Mine can be traced for 2,100 feet on the surface. The general tenor of the vein matter is not recorded, but the portion that was laboriously shipped out of this remote location contained three ounces of gold and fifty-eight ounces of silver to the ton.

In addition it ran 11 per cent copper. There was even better ore in the Custer Mine, valuable enough to be packed across the mountains on burros and shipped to a smelter at Midvale, Utah. These shipments ran 35 per cent lead and 5 per cent copper. Gold ran from two to three ounces a ton and silver thirty to eighty ounces.

The Casey Mine included five parallel gold-bearing veins. One of the most highly regarded mines was the Burgess, whose nineteen claims took in a series of outcrops at a 9,200-foot elevation near the top of the Inyo range. The Big Horn Mine made shipments running $155 a ton in gold and silver. Silver-lead mines, as noted earlier, have been minor but in individual cases very profitable operations in California. In the Beveridge district the Bunker Hill was such a mine, its ore running 33 to 60 per cent lead, with thirty-three ounces of silver and some minor gold content.

The Golden Eagle Mine is situated in Hahn's Canyon, and the name of Hahn recalls a rare incident in which local citizens were at odds with U. S. Army troops stationed among them in the early pathfinding days of the West. According to Russ Leadabrand, previously mentioned as a writer on western exploration and something of an explorer himself, the individual was C. F. R. Hahn, who had hired out as a civilian guide to a survey party of the Corps of Engineers under command of Second Lieutenant D. A. Lyle. The party's assignment was to find whether there was any practicable route between the Owens River and the distant Amargosa, the chiefly underground stream in Death Valley. The year was 1871, and the uninformed, some would say stupid, order from headquarters called for making the survey in the midsummer desert heat.

With the military survey party a few days out in the parched Saline Valley, the guide Hahn went ahead to search for water. He never returned. The soldiers were left on their own. After great hardship, emaciated, and with their shoes worn out, the army men made contact with a standby detachment awaiting them on an established trail.

Back in Owens Valley there were murmurs among the settlers that no adequate search had been made for Hahn and that the guide had been left to perish in the desert. The corps' operations continued and on a second exploration trip a well-known local man, William Egan, was engaged as guide. He disappeared under the same circumstances as Hahn. When the survey party returned without Egan, local feeling ran high against Lieutenant Lyle and his commanding officer, First Lieutenant George M. Wheeler. No legal case could be made against the officers, but the disappearances and the blame were debated for many years. A local editor, at a later date, insisted that the episodes had seriously retarded exploration and road building east of the Inyos. If so, the feud contributed to the isolation and mining difficulties of the Beveridge district.

Entrance to these mining areas from the north has been recommended. If the long road past the Marble Canyon and Beveridge districts is followed on south to its end, the traveler will come out on State Route 190. This good highway diverges from U.S. 395 near the town of Lone Pine. It is the route we will now follow to examine the known and the lost mines of the Death Valley region.

In its course eastward Route 190 passes scattered mine workings surrounding the old town of Darwin. The Coso and Argus mountains are of interest in this region, but curiously enough the Nelson range to the north has seen little mining activity. Whether the Nelsons are barren of surface indications or are insufficiently prospected is hard to determine. They deserve at least a question mark on the modern gold hunter's list.

Route 190 next dips into Panamint Valley, a smaller version of Death Valley, and then climbs over Towne's Pass and into the great sink itself. The name Towne preserves the name of a ragged, half-starved emigrant who found and lost the Gunsight ledge of all-but-solid silver. The Lost Gunsight is the first, and one of the most celebrated, of all California's lost mines.

The story begins in 1849, when one of the largest wagon trains of the entire western gold rush was assembled at Salt Lake City. The travelers were waiting out the summer heat. The group swelled to at least several hundred ox-drawn vehicles and some reports put the number at a thousand. Their occupants represented parties from Illinois, Missouri, Kansas, and several other states. All had now banded together at Salt Lake City and had employed as guide a Mormon explorer and frontiersman, Captain Jefferson Hunt. Their route was to be along what was euphemistically termed the Old Spanish Trail, a mere trace left by early wayfarers through the southwestern desert. It was a dry and dangerous course, but it had the advantage of skirting around the great, roadless barrier of the Sierra Nevada.

With cooler weather, the ponderous migration got under way, but at a stop known as Little Salt Lake there was dissension. It was caused by a document produced by one party of emigrants, something of unknown origin referred to as the "Williams map." This map purported to show a shortcut to California which would save about five hundred miles of travel. Captain Hunt refused to swerve from the original course. Nearly all the emigrants stayed with the guide, and although suffering some hardship they arrived safely in California.

A large splinter party turned west, however, hoping for a much shorter trip. After sober second thoughts and vistas of unknown mountains ahead, there were many defections from this group. There remained twenty-seven wagons whose occupants continued to struggle onward under the now questionable guidance of the Williams map.

Quite a literature exists dealing with the hardships, heroism, and tragedies of the early desert crossings, and it is beside our purpose to go into such history in detail. Of interest here is the fact that the Williams map led the dissidents straight into Death Valley.

By this time the hardships in rainless southern Utah and Nevada, and the crossing of the sear Amargosa Desert, had taken some toll of life and a heavy toll of provisions, of reserves of water, and of mutual tolerance. The travelers split again into even smaller parties, each following its own desperate counsel as to how to survive. The names of these units, the Jayhawkers, the Bennett-Arcane party, the Georgians, the Martin party, and the Reverend J. W. Brier party are preserved in western history.

It is uncertain which of these segments included the traveler Towne, probably the Jayhawkers, but accounts agree on the circumstances of his discovery of the great silver deposit. Trailing behind his group in the hope of shooting some game—an unlikely stroke of luck in Death Valley—he discovered that the forward sight of his rifle was missing. By a startling coincidence, there was some metallic substance in a rocky outcrop close at hand. He dislodged a small piece and, if the original version of the incident is accepted, managed to pound it into the small slit in his gun and cut it to a serviceable shape.

He then looked at the metal more closely. He decided it was silver and he saw that it occurred in the rocky slope in lavish amounts. Towne returned tight-lipped to the rest of the party, but before they were much farther along the trail he could not resist talking. For what he had seen, he insisted, was a whole hillside of silver.

Several of the younger men in the group, including a Dr. McCormick and one Willie Webster, went back to see for themselves. Towne's statement, while somewhat extravagant, was essentially true. Before the investigators lay a great deposit of what was incontestably native silver.

We hear no more of the man Webster, but Dr. McCormick, only a year later, definitely confirmed the discovery. Impressions were fresh in his mind as he talked with a good friend, P. A. Chalfant, about the matter. The conversation took place at Rough and Ready, an active mining camp in Nevada County, California, where the physician presumably was in practice.

McCormick said that where the ore was exposed, which was on a considerable part of the mountainside, it had "strings or wires" of silver running all through it. The rock was extremely hard, but he and his friends had managed to collect a number of pieces. In camp they pounded them up and extracted a quantity of wire silver. The doctor said the rock was practically bound together by silver strands. He showed Chalfant a handful of the metal in this stringy shape.

These facts were included routinely in an account of Chalfant's own gold rush experiences published in 1872, and they later were amplified from family sources by W. A. Chalfant, a pioneer Inyo County editor. The story in less detailed form had circulated much earlier. Among the hopeful treasure hunts it inspired were those of Dr. Darwin French of Oroville. He led two ambitious expeditions, and while his activities produced some valuable geographic knowledge, they contributed little knowledge about the Lost Gunsight. They did make it evident that the bonanza was not easy to find.

Among those who have sifted the scanty clues, the concensus is that the silver ledges lie somewhere in the lower slopes of the Panamint Mountains. This is the range that forms the western boundary of Death Valley. The deposit would be on the inner, valley side of the Panamints, for Towne's party is known to have been trekking south on the valley floor. This means that they would be at the base of, rather than in, the mountains. Towne himself, in his search for game, might have climbed higher. The Dr. McCormick account unfortunately does not help in this matter of elevation.

But does the Gunsight story represent something real? A generation of early California miners believed implicitly that the deposit existed and that it was as rich as represented. One finds it hard to reject the circumstantial story of Dr. McCormick, told within a year of his being at the very spot. He had the native silver to back him up.

Failure of successive searchers to discover a spot so casually come upon by one of the first travelers in that area may seem hard to understand. The problem will become more apparent when one has tried the search himself, or at least has had a personal look at Death Valley. The likely search area extends along the Panamints for forty-five miles, from Tucki Mountain in the north to Sugar Loaf in the south. The floor of the valley is at elevation 0—sea level—and the mountains rise almost a mile into the sky. Famed Telescope Peak soars still another mile higher. The whole range is steep, rocky, and waterless. As late as 1925 two tough, experienced mountain men found $15,000-a-ton ore in the Panamints, gold in this instance, and were unable to locate the rich spot again. Here is the story of Asa M. Russell and Ernie Huhn, known in the desert country as Panamint Russ and Siberia Red:

The two were staying in an old stone cabin in the southwest corner of Death Valley National Monument. The cabin was and is near an anvil-shaped crag known as Striped Butte. This landmark rises from the center of Butte Valley and is shown on maps of the monument. The Panamints tower almost straight up behind the cabin, culminating in Manly Peak, 7,196 feet in altitude. From a one-legged pioneer in the region named Carl Mengel, who gave his name to Mengel Pass, Russell and Huhn had learned of favorable indications of gold on Manly Peak.

It testifies to the difficulty of climbing in the Panamints that the two prospectors would spend one day in exploration and then two days resting up and examining their samples. Panamint Russ later warned any searchers on Manly Peak to go in pairs and keep close together. With loose boulders everywhere, he cautioned, a misstep could mean a broken leg or a broken back. It is no place to be alone.

Russell and Huhn did not follow this precaution. The two separated on each trip, for they were intent on covering as much ground as possible. They found numerous stringers with promising mineralization but too narrow to warrant further investigation. Some larger outcrops panned out to indicate values in free gold of $25 to $40 a ton at the $20 gold price. But it was not ore such as this that the partners were after. Old Mengel had been talking of something more spectacular.

There came a Sunday when the partners were at their cabin, with a dozen or more rocks lined up on a shelf ready to be crushed, panned out, and evaluated. The key to a fortune lay among them.

Literary historians deplore the fact that Coleridge managed to write only a fragment of *Kubla Khan* because, as the poet was setting the lines down on paper, a well-meaning "person from Porlock" called at his house and interrupted his train of inspiration. The program of Russell and Huhn was similarly interrupted by two well-meaning persons from Panamint. They were friendly miners from that old silver camp and the four men had much in common. There was much talk and reminiscing around the outdoor fire. Next day there was quail hunting. There was more exchange of prospecting experiences. Then there was need to drive across and out of Death Valley to get supplies at Shoshone. At last the visitors moved on. The chore of panning out the samples was not accomplished until three weeks after they had been gathered.

One of these rocks was an extremely heavy piece, of medium gray color. It had been brought in by Huhn, the older and more experienced of the partners. He remembered, or thought he remembered, that it came from a ledge three feet wide and exposed for a linear distance of about twenty feet. On the steep side of a draw, Huhn said, and near the contact of two differing bedrock formations. When the pulverized material had been washed down to the usual residue, the bottom of Huhn's pan was half covered with gold. It was estimated that the vein would run $15,000 to the ton.

That was the only piece of the bonanza ore that the partners ever saw. The discovery place had not been marked. The jumbled, broken country played its usual tricks of confusion and deceit. The two searched for days and days and Huhn himself remained in Butte Valley for three months, all without result. There was one explanation, aside from human forgetfulness and error, that might explain what, to the two men, seemed an almost mystical disappearance of solid matter, of a gray rock formation that simply should be there. During their costly three-week interval of visiting and travel there had been rain. Along the high ridge of the Panamints it had been unusually heavy for desert country. They conjectured that the runoff might have started a landslide down the steep slopes and that all surface indications of the ledge were now covered up. It was not very convincing, even to themselves.

Twenty years later Huhn died, reasonably prosperous from talc mining in the Coso Mountains. Russell had developed a small business in Los Angeles. With no wish to try another search himself, and now relieved of any obligation to keep the story secret for Huhn's sake, Panamint Russ passed on his information to a new generation of gold hunters by publishing it in the September 1955 issue of *Desert* magazine. His chief adjuration: Keep on the east side of Manly Peak and work in pairs.

Some distance from Manly Peak but in the same Panamint Mountains were two rip-roaring mining camps that achieved a considerable production of gold or silver in their brief existence. These were Skidoo and Panamint City. Both are reached not from Death Valley but from the great trough to the west, Panamint Valley. Mining operations at both places were well financed; hot, arid Skidoo, for example, being served by an eight-inch water main extending twenty-three miles over difficult mountain terrain. Between 1905 and 1918 Skidoo shipped more than $3 million in gold. A paved road winds up Emigrant Canyon to within seven miles of the old ruins there. Modern accommodations are less than an hour's drive away at Stovepipe Wells.

Panamint City was a silver camp that had a brief productive life in 1874–75, punctuated by a great deal of gunplay. Money was poured into its mines by Senators John P. Jones and William M. Stewart, two of Nevada's early-day moguls in mining, politics, and the law. The place could be reached only by a steep, narrow road up Surprise Canyon, which was termed, doubtless with good reason, the worst stretch of road in the nation. When a cloudburst ripped much of the roadbed away and clogged several narrow passages with boulders, no one possessed the money and confidence to restore access and keep the camp alive. Although an impressive amount of silver was mined, the operations as a whole appear to have been losing ones. In the belief that Panamint still held much valuable ore, a number of Hollywood entertainment people organized the American Silver Corporation in 1947 to resume operations. There was said to be a million dollars in capital available for deepening the mines, but no actual production of silver materialized.

No deepening of a mine is necessary to get to the $41,000-a-ton gold ore somewhere east of Panamint City if one can discover the rich ledge of a vanished prospector known as Alkali Jones. One of the true desert characters of the model of Shorty Harris, Seldom Seen Slim, and Badwater Bill, Jones was on a tramp from Skidoo to distant Searchlight, Nevada, when he found his bonanza. The outcrop, of white quartz in a pink country rock, was on a small butte rising out of the flatness of Death Valley. It appears to have been on a line, or nearly on a line, between Skidoo and Coffin Peak. The latter is a high point in the Black Mountains near the present scenic attraction known as Dante's View.

Old Alkali could not pause to do much work. He was short of water and also was contending with a fierce desert sandstorm. When he reached Searchlight, he had packed enough of the ore with him to pound out $180 in gold, and one of the pieces submitted to assay gave the $41,000-a-ton figure. Jones told his story in a letter to his sister in the East. He wrote that he was buying three burros and would proceed across the valley well equipped to work his great strike. Alkali never was seen again. Years later two Shoshone Indians found an old pack outfit under a mesquite tree, on what could reasonably have been the prospector's route. Some items in it were definitely identified as old Alkali's.

We know the deposit was on a small butte. If you find the right one and see a typical location monument of that time—a pyramid of rocks about three or four feet high—look among its stones for the flat tobacco can in which miners commonly enclosed their claim notices. If the paper states that the claim is the Golden Eagle and is signed Jones, you are on a ledge of fabulous value.

Any mention of Skidoo and Panamint City also involves mention of Ballarat, the southernmost camp in Panamint Valley. Named after the great Australian gold-mining center, Ballarat functioned chiefly as a supply and rest stop for miners. One feature was its Post Office Spring, where roving prospectors left and received written messages. There have been mines, several of some importance, in Ballarat's backdrop of mountains.

Remnants of early mining structures are encountered south of Ballarat on both sides of the highway, up to the boundary of the Naval Ordnance Test Station. This military withdrawal covers a very large area. Outside the restricted area, to the east, are the Owl's Head Mountains, scene of several promising strikes and still partly unexplored. The Owl's Heads can be approached by automobile only by a long detour via State Route 127, then by a dirt road ending at the Black Magic Mine. Beyond are 140 square miles of roadless and all but trackless desert mountains. A Lost Dry Lake and an Owl Dry Lake lie far in the interior. For one who can work out a water, transport, and general maintenance program, this region would seem to offer a real frontier of mineral exploration.

Thus far we have been examining possibilities on the west side of Death Valley or in nearby Panamint Valley and its environs. On driving east across Death Valley, perhaps with an eye out for Alkali Jones's lonely butte, one is approaching the state of Nevada. The state line may be crossed via Daylight Pass in the north or via Death Valley Junction in the south. Twenty-five miles south of the junction on State Route 127 is tiny Shoshone, a historic gathering place for miners who had business in the nearby Black or Greenwater mountains.

There are two other mountain ranges along the eastern boundary of the national monument. They are the Grapevines, north of Daylight Pass, and the Funerals, to the south. The Funeral Mountains are believed to shelter another great gold deposit, the Lost Breyfogle Mine. The range itself is in California, but Breyfogle's adventures, and those of other men inspired by his sensational discovery, are more properly told as part of Nevada's mining story.

5
Around Nevada's Rim

Gold or silver or both have been mined all over Nevada. There have been richer areas in the world than the so-called Silver State, but few which have shown such a wide dissemination of mineral deposits or have witnessed the creation of such a large number of transitory mining camps.

On a map of old ghost towns and mining settlements prepared by an industrious investigator and historian, Stanley W. Paher, one can run his eye around the whole perimeter of the state or trace almost any random route through the interior and find the names of these old locations everywhere—nearly six hundred of them.

There are only two blank spots of considerable size. One is the huge desert territory taken over by the government for its nuclear bomb tests. This is in the southern part of the state just north of Las Vegas. The other is a particularly arid and forbidding area in the northwest corner of the state, stretching westward from King Lear Peak. We shall be looking at present-day mining possibilities around the edges of the test site and also try to shed some light on this mysterious northwest corner, an area of some 3,600 square miles which, except for very small colonies at widely separated lakes, has been shunned even by the native Indians.

The multiplicity of old mining camps in Nevada does not mean that all the promising mineralized zones have been pre-empted or worked out. On the contrary, they have much to offer the modern prospector and miner. Many of them were deserted under conditions that raise serious doubt as to whether their potential was ever fully realized. There are scantily worked prospects in the public domain that stand ready for the taking under the mining location laws. Others, privately owned, are available for lease to anyone willing to expend the labor and a little money for further development.

To understand this situation, and to understand why some of these workings may offer a genuine opportunity today, it is necessary to look briefly at Nevada's checkered economic history.

The state's first period of exploration and development began in the late 1850s and was in part a backwash from the California gold rush. It culminated in the great days of the Comstock Lode at Virginia City. By 1878 the Comstock mines, after pouring out some $300 million in silver and gold, had become all but unworkable. Speaking generally, Nevada mining went into a decline.

The second boom may be dated from the discovery of silver at Tonopah in 1900. Frenzied prospecting and rich discoveries followed. Even more frenzied were the stock promotions associated with every new mining camp. The money poured into these schemes by credulous investors the country over was enormous. The boom town of Rawhide, whose population never exceeded 8,000, is said to have had 125 brokerage houses, all engaged primarily in mail-order stock sales.

During this chaotic speculation, the development of the genuinely valuable ore deposit and the worthless excavation that was merely a basis for stock promotion became hopelessly confused. Among those confused and misled were many of the practical miners themselves. Even when a miner uncovered a vein that showed good promise there was a tendency to put down only a "coyote hole" and then try to sell the property for cash. The discoverer would hope, usually in vain, for an eastern capitalist to take over. Failing that, he would hope for a profitable deal with some blue-sky promoter.

In 1907 there occurred a financial panic of nationwide extent. Even mines of demonstrable value had to close down as working capital dried up. The general hard times also affected the small, individual operator who was trying to mine and mill commercial grade ore. A result of this boom-and-bust phenomenon, one may reasonably assume, was that many of the ore discoveries of the early twentieth century were never definitely established to be either good or bad.

Another explanation of Nevada's abandoned mines lies in the tendency of miners the world over, and notably in Nevada, to rush from one strike to another. Rumors of new and richer discoveries, many of them well justified by the facts, would quickly depopulate one settlement and call another one into being. These dramatic shifts of population make interesting history, but our only concern with them is to note the reason that certain mineralized areas were abandoned and also to note that they may well deserve renewed exploration today.

As we turn to descriptions of specific opportunities for today's prospector, we are faced with a diffuse, statewide, crazy-quilt pattern of old camps and the ore deposits that gave rise to those camps. To approach the matter with some order, let us first follow the main transcontinental highway through Nevada, Interstate 80, which runs almost due west from Salt Lake City to Reno. With its connecting roads, federal, state, and county, Interstate 80 serves most of northern Nevada.

Driving into Nevada from the east on this route, one passes through country devoted primarily to farming and cattle raising. North and south of the highway are remains of a few old mining settlements that flourished briefly on copper or silver-lead ores. The first approach to a real gold area is made by turning north from Interstate 80 at the little junction town of Deeth. The road, mostly dirt but graveled in places, leads eventually to Twin Falls, Idaho, but our interest is in the gold-bearing formations of Jarbridge Canyon, near the Idaho state line. This is a district of proved gold deposits. It is also the general location of the rich Lost Sheepherder Mine.

It is a difficult sixty-nine-mile trip from Deeth to Jarbridge. The elevation of town and canyon is 6,000 feet and snow lies in the deeper hollows for eleven months a year. There are barely two months for prospecting, somewhat variable but generally considered to be July and August. In the mountains rising immediately east of Jarbridge men have been trapped and have perished in heavy snows occurring as early as September.

It was to be a lucky trip through Jarbridge Canyon for the family of D. A. Bourne when, in 1908, the little group set out in a horse-drawn wagon from Steptoe, in White Pine County, Nevada, where they had been visiting, to return to their home in Boise, Idaho. The Jarbridge River flows north and the road to Idaho runs along the stream. This is in the depth of a canyon whose sides rise steeply to a skyline 2,000 feet above. Along here Bourne's practiced eye rested on a rocky intrusion in the canyon wall that called for investigation. He stopped his team, recognized the formation as gold-bearing quartz, and chipped off a sample. Then he hurried the family along under a threatening sky. In Boise the rock from the canyon wall assayed $73,000 a ton in gold.

This was the start of the Jarbridge mining boom. Other valuable deposits were discovered and became the profitable Bluster Mine, the

Rex Mine, the Success Mine, and so on. Despite the short working season and the remote location, the district produced gold and silver of a value variously reported to total from $10 million up to $50 or $60 million. The Guggenheims and other large mining interests were involved and for milling operations a power line was run in from a source seventy-five miles away. Bourne, the original discoverer, was offered $2 million for his property but preferred to retain and develop it himself.

At one time there was a seasonal population of about 1,500 at Jarbridge, but isolation, long winters, and snows ranging up to eighteen feet in depth militated against any normal civic development. Moreover, all the habitable land along the river was in continual danger of avalanches, and in one stormy twenty-four-hour period landslides crashed down the canyon walls in nearly a hundred places.

As always, production eventually tapered off, but some work continued until gold mines were closed by the War Production Board order of 1942. A handful of old-timers have remained in Jarbridge. Among these, and among the members of a younger generation who retain the old log structures as summer cabins, there is a subject that usually will bring a gleam to the eye—a lost gold ledge and its guardian skeleton up in the highest of the high country, namely, the aforementioned Lost Sheepherder Mine.

The facts about the Sheepherder come to us in most detail from Mrs. John Pence. It was her husband, a local rancher, who year after year financed searches for the mine during the brief summer interludes. Mrs. Pence told her story in 1910, and it has been corroborated in some details and repeated by the western writer Nell Murbarger in her *Ghosts of the Glory Trail*. The incidents begin in 1890 with a prospector named Ross, who found a piece of extremely rich float—one is reminded of Bourne's $73,000-a-ton specimen—and undertook to trace it to its source. His trail led him out of the canyon and then higher and higher into the mountains on the canyon's eastern side. He would appear to have been on either Jack or Jenny Creek and later to have reached the area between Jarbridge Peak, towering to nearly 11,000 feet, and a small plateau known as Sawtooth Ridge. Local people say these creeks, Jack and Jenny, are the only feasible routes into the higher elevations. They would doubtless be the natural avenues for anyone seeking the lost mine today.

Further specimens of broken, gold-bearing rock were found by Ross. They convinced him that he was on his way to a rich outcrop. A major winter storm was gathering over this high country, however, and he was in real danger. Ross divested himself of his pick and short-handled shovel for faster traveling, and, as he explained later, these abandoned tools would serve as markers when he again took up the trail. The snowstorm was a heavy one and it closed the high mountains for that winter. Ross had kept ahead of the storm and he came at length to a camp at a much lower elevation to which sheep had just been herded from their summer range above. The flock belonged to John Pence and the name of the sheepherder in charge has come down to us as Ishman.

Ross was fed and made comfortable. While at the crude sheep camp, he confided his experiences and hopes to Ishman. That is the last we hear of the prospector except perhaps as an unidentifiable skeleton. We know that next summer the sheepherder again took the Pence flock high up in the Jarbridge range and that he was looking for a pick and shovel, for stray pieces of gold-studded quartz, and for the outcropping vein from which they came. The brief summer went by.

When the season was over, Ishman brooded over his situation and at length decided he needed the help of his employer. In a session with John Pence, he told of Ross's visit to the sheep camp and of Ross's story. Then he went on to tell of his own long searches day after day and reluctantly revealed that he had found the rich lode. He had broken off some specimens, which he now exhibited. They were extremely rich, shot through and through with coarse gold. He also told of finding a skeleton near the discovery site.

Was it Ross himself who had come to an end? In that night at the sheep camp he had in effect provided his new friend with a waybill to the mine. Did Ishman then kill the prospector? Or did Ross himself originally have a partner whom he had eliminated in order to keep the prospective bonanza all to himself? These are questions which Jarbridge miners argued through the years.

In any case, this remote, storm-bound ledge appears to carry some inherent bad luck. After a winter of impatience, John Pence and his man Ishman set out in the earliest spring weather. Filled with confidence and excitement, they climbed too rapidly up the steep slopes and through the rarefied air. Ishman collapsed. He died without regaining consciousness and his death left Pence with little information on which to act. The sheepherder had concealed the surface of the vein with brush and earth and had suspiciously withheld any detailed directions from his employer-partner.

Pence prudently continued to devote himself to sheep raising, but each summer he would grubstake one or more prospectors, equip them with the meager information available, and send them out on the quest. The search went on for years. During the decades of wet weather, with summer rains and runoffs from winter snows pouring down the mountain slopes, one may believe that Ishman's hastily contrived cover-up of the lode has been washed away. Again, the outcrop could be so situated that slides from above could bury it still deeper. But open to the eye or hidden under the soil, the rich deposit is still there.

The immediate Jarbridge area can scarcely be recommended to the modern prospector. The phenomenal values in the Bourne and other workings attracted hundreds of competent miners and exploration in those early days must have been quite thorough. However, there is an East Fork of the Jarbridge River from which little has been heard. This stream lies on the eastern side of the lofty tangle of peaks and canyons roughly designated as the Jarbridge Mountains. It is shown on detailed maps of the area. The eastern slopes might well share the mineralization that created the great bonanzas of the western side. Excessively rough, cold and roadless, the East Fork country was all but inaccessible during the original stampede. Today much of it lies within the Humboldt National Forest. This means a certain number of roads and trails for fire-protection purposes. Is there some real pioneering available here?

In the other direction, to the west of Jarbridge and lying between Cornwall Mountain and Twin Buttes, is the ghost of Bruno City. It was the scene of gold placer workings and of good silver ore in place, and the abandonment of Bruno City has not been satisfactorily explained. Its hundred or more miners are believed to have been lured away by the White Pine discoveries of 1870, which depopulated various other camps. Discoveries at Bruno City were extensive enough to give the name Silver Mountain to a nearby elevation, and the Nevada press briefly recorded good ore and good prospects there. In driving north to Jarbridge, the turnoff point for Bruno City is Charleston, also an old mining camp and for a time one of the toughest and most lawless in the state. The ruins of Charleston stand on what is now a privately owned ranch.

Returning to Interstate 80 from this northern detour, one continues west and soon is at Elko, the trading center for a very large part of northeast Nevada, both now and during the hyperactive mineral exploration of the early 1900s. It was from Elko that countless prospecting expeditions set out and to Elko that the members returned for supplies. As often as not there were ore specimens to exhibit and stories of success, sometimes achieved, sometimes anticipated, to tell the local newspaper editor.

Leading north from Elko is State Highway 11, and the territory it serves is inviting to the modern prospector. One reason is the established fact of valuable silver deposits over a wide area. Mines with good ore in them were abandoned before the turn of the century owing to the collapse of silver prices. Another reason is that the old ghost towns–Tuscarora, Midas, White Rock, and others–are the jumping-off places for large expanses that were imperfectly explored in earlier days. The latter might yield profitable secrets to the prospector with a well-stocked, four-wheel-drive station wagon, which would make it possible to spend considerable time in a promising spot. The plodding burro prospector too often had to keep moving in search of water and forage.

The north-south axis of this territory extends a good hundred miles through all but empty country, from Interstate 80 up to the Idaho state line. On the eastern fringe is the site of Tuscarora, one of the best-known of the nineteenth-century camps. Because both its gold placers and the high-grade silver lodes of Beard Hill and Mount Blitzen were pretty thoroughly worked, even through the depressed years of silver, Tuscarora may have little left for today's miner. The richest mine to contribute to Tuscarora's reputed $40 million production was the Grand Prize. One valuable mine, the Dexter, was shut down because of corporate financial trouble and its true potential remains unknown. The nearby camp of Falcon was abandoned because the silver occurred in extraordinarily hard rock. Here, as in many other inactive mines, one speculates as to what modern technology can do about formerly insoluble problems.

The intermittence of all-gold and all-silver deposits throughout Nevada is illustrated by the fact that Tuscarora's neighbor to the west, Midas, was strictly a gold camp. Veins were strong and values consistent but the general content of the ore was too low to warrant shipping to a reduction plant. One or more small mills eventually were erected, then closed after the stock market crash of 1929. That some life and some gold remained at Midas was proved by a later mining and milling operation which continued with apparent success until closed by the wartime order of 1942.

A turnoff north of Chicken Creek Summit leads to several old towns now abandoned along with their surrounding mines. These are Cornucopia, Good Hope, Edgemont, and White Rock. The Leopard and Panther mines at Cornucopia had ore running $1,000 a ton in silver. Edgemont's production was in gold. The improved road approaching these localities ends near Chicken Creek Summit but wheel tracks may be followed on north into the Owyhee Indian Reservation and a junction with State Route 43.

From almost any point on Route 11 and its branches one may look westward into a featureless and, as far as official maps go, an almost roadless landscape. It is a territory of more than 7,000 square miles. What do we know about this desolate area and its possible mineralization?

The basis for optimism is that it is almost completely surrounded by valuable deposits of historic record. On the east, Tuscarora and its neighbors have been noted. On the southern edge is the current mining at Carlin, a huge operation in low-grade gold ore, plus the cluster of old producers near Battle Mountain and Golconda, and the highly successful Getchell Mine. The last is one of Nevada's great mines. Originally a gold property, the Getchell was allowed to operate during the World War II shutdown because its by-product, arsenic, was needed. Later the primary output was tungsten, some $30 million worth of that metal being recovered.

To the west of this 7,000 square mile expanse is the old National Mine, a fabulously rich gold property which shipped one carload of ore that ran $100,000 to the ton. The National was discovered in 1907 by one of the earliest motorized prospecting expeditions. It took its name from the discoverers' automobile, the National being an early-day luxury car contemporary with the better-known Pierce-Arrow. A number of other successful mines, lode and placer, were worked in the National area. Among the once flourishing camps were Buckskin, Spring City, Vicksburg, Rebel Creek, and Dutch Flat. Both sides of the Santa Rosa Mountains, a highly mineralized range trending north-south, have continued to interest mining men and have attracted sporadic operations up to the present time.

So much for the perimeter of the semiexplored region.

From any of the locations named, plus points in southern Idaho, one may take off into the interior. Some local information is desirable, as always, but as to new gold and silver discoveries the newcomer probably will have to follow his own hunches. A few lakes of fair size exist, or to be more precise, they are shown on some maps. Three small tributaries of the Owyhee River have their sources in this region and, when they flow at all, run north across the Idaho line. There are a few isolated cattle ranches. These would be the only possible sources of gasoline and supplies.

Our main course has been along Interstate 80. At the town of Winnemucca, named after an early Paiute chief, Interstate 80 swerves sharply to the southwest. It passes the sizable trade center of Lovelock and continues on to Reno.

A glance at the map will show that the seventy-two-mile stretch of highway between Winnemucca and Lovelock forms a rough boundary for another desert area of immense size. It occupies the entire northwest corner of Nevada. Two items of special interest in this territory are the rich Jumbo gold discovery of the 1930s and the Lost Hardin Mine, with its slabs of solid silver.

The Jumbo affords evidence that not all the rich lodes were discovered during Nevada's boom days, that such sleepers can exist even in the midst of old mining camps.

The Slumbering Hills, some five miles distant from a Western Pacific whistle stop known as Jungo, were the scene of some gold mining dating from 1910. There was enough activity to create three ephemeral settlements, Jumbo, Daveytown, and Awakening. For forty years a G. B. Austin maintained a small frame hotel at Jungo and in his spare time did some prospecting. It was in 1935, while poking around near the earlier workings in the Slumbering Hills, that Austin chanced upon exceptionally rich gold ore. The mine paid for itself from the start, including the cost of a thirty-ton mill. He named it for the nearby ghost camp of Jumbo. It was the first big gold strike the country had seen in years, and coming in the depth of the great depression Austin's good fortune received a great deal of publicity. Former President Herbert Hoover, who was a mining engineer by vocation, jolted over many miles of washboard desert road to inspect the new bonanza. The mine afterward had a checkered history, at one time being owned by Harrelson L. Hunt, the Texas oil magnate. Austin himself seems to have been amply recompensed for his discovery. Despite a modest rush to the discovery area and some attempts at promotion, the Slumbering Hills, Jungo and its little hotel, and the encompassing desert are slumbering again. Will a later George Austin find a later Jumbo Mine?

For the optimist who hopes to gather up the slabs of pure silver found and lost by the late James Hardin, the record indicates a location about forty-five miles northeast of Gerlach. This is a small desert community reached by State Route 34. That silver can occur in the form described by Hardin has been established by the dazzling Planchas de Plata Mine near the Arizona-Sonora border. Both occurrences were in a gray "ashlike" formation. In both places, large pieces of native silver could be picked up by hand or easily pried out of their matrix.

The scene of the Hardin strike was a particularly forbidding section of Nevada's northwestern corner called the Black Rock Desert. Hardin was with an emigrant train bound for the Pacific Coast, and although they were traveling in the gold rush year of 1849 it appears that this party was organized primarily to settle on agricultural homesteads. Such migrations already had entered the fertile, well-watered valleys of Oregon, and the group was far north of the known routes to the California gold fields.

With the fourteen wagons halted for a rest, Hardin and others scattered out to hunt game, a few species of which can exist amid the sagebrush and dry lake beds of this arid country. Hardin came upon the silver slabs and summoned several companions to the site. They agreed that this was silver and that there was enough in sight, as they expressed it, to fill all their fourteen wagons. Filling the wagons with silver, however, was just what the company could not do. The food, water, and other supplies carried in the wagons were much more precious in their present situation. Moreover, in the unsettled country ahead there was no place to sell silver, only a fur-trading post at faraway Astoria. The silver haul could be of no immediate value to them. Another consideration was fear of the hostile, far-ranging Bannock Indians, who might sweep down at any time. The wagon train moved on.

The fact that Hardin and his friends waited nine years to go back to the silver is perhaps understandable. Survival in a new land was the first concern, and clearing, plowing, planting, and the many other labors required on a self-sustaining farm took time. Also, in any part of the West that lacked an army outpost the danger from Indians remained acute.

Eventually Hardin went back over the wagon trail. He was quite unable to find the silver, and we last hear of him making a try in 1859. He then disappears from the picture. There followed, however, much confused and expensive activity by others who had heard his story. Newcomers identified a soft, dark rock in the Black Rock Mountains as silver ore. It in no way resembled Hardin's description but the discoverers were convinced of its value. For several years they kept a well-paid chemist named Charles Isenbeck busy working out a method to extract the supposed silver from the strange rock. At one time Isenbeck announced that the material was running $70 to $565 a ton. This set off something of a rush. More claims were filed, several ore mills were constructed, and the facilities grouped around them took the name of Hardin City.

The construction of the new mills brought a showdown. When the so-called silver ore was subjected to standard recovery methods it proved worthless. Isenbeck, whether guilty of deliberate deception or not, disappeared. Investors in Isenbeck's operations or in the mines themselves were losers to the extent of several hundred thousand dollars.

From the admittedly hazy record, it appears that Hardin's original silver discovery and the fiasco of Hardin City are two separate matters. The black rock of the Black Rock Mountains obviously was not the silver deposit discovered during the covered-wagon trek of a decade before. Hardin himself never said it was—in fact, his whereabouts during all the excitement is unknown.

The Black Rock range appears on most maps of Nevada. It rises just to the west of the extensive Black Rock Desert. On some maps Hardin City is designated by the symbol for a ghost town. There may remain some foundations and a few adobe walls unless they have weathered into the desert soil. This, the location of the fictitious Hardin silver mine, is a reasonable starting point in a search for the real one, and it is Hardin City that is referred to in the directive "forty-five miles northeast of Gerlach."

Beyond the Black Rock Mountains there is a very large amount of vacant territory extending to the California state line on the west and that of Oregon on the north. It can be traversed by a graded earth road extending more than 100 miles from Gerlach, in Nevada, to Alturas, California. The geologic prognosis for valuable discoveries in this region is not favorable, although in ore-spangled Nevada the true prospector does not rule out anything. There have been lead-silver finds that called the substantial camp of Leadville into being, while over in Alturas they will tell you that somewhere in here the lost gold mine of Captain Dick awaits a discoverer.

A Modoc Indian, Captain Dick appeared in Alturas off and on over a period of years with gold that clearly had been pounded out of very rich ore. Attempts to follow him to his mine always failed, and the wily Indian kept his secret until his death. His course always was eastward from Alturas, which would bring him into the Warner Mountains of northeastern California. Mining men have found little encouragement in this range, and local opinion is that the Captain Dick ledge is farther on, somewhere in northwestern Nevada.

As Interstate 80 continues southwesterly toward Reno we come to the Humboldt Mountains, once described as "the richest mineral region on God's footstool . . . gorged with precious ore yielding $4,000 to $7,000 a ton." The mines in this area may be grouped together geographically, although the discoveries and development were widely separated as to date.

The belt was extensively worked for both gold and silver, and there are about a dozen old camps lying along an axis of barely twenty-five miles. The gold placers are all but worked out. In the lode mines of the Humboldt range the yield was almost entirely in silver. Today's miner will have to decide whether current prices of that metal will make existing ore deposits profitable. One may also note that since the Humboldts were mined there has been a century of erosion as the summer snow melt coursed down the steep sides of the range, raising the prospect of newly exposed outcrops.

To reach the old camps, there is a turnoff from Interstate 80 at Mill City. One has seventeen miles of pavement on State Route 50, then finds himself on a dirt road extending twenty-six miles farther. The route rejoins Interstate 80 at Oreana. The district also can be entered from the Oreana (southern) end.

The principal towns of the Humboldt district were Unionville, in attractive, well-watered Buena Vista Canyon, and the twin settlements of Rochester and Lower Rochester. The widespread mineralization also gave rise to a number of small camps, including Prince Royal, Santa Clara, Star City, Lima, Spring Valley, Fitting, Panama, American Canyon, Packard, and Humboldt City. Little or nothing of them remains.

Mill City, the turnoff point on Interstate 80, is also served by a railroad and in 1862 was laid out as something of a metropolis. More than five hundred city blocks were platted. It was to be the site of reduction plants for the Humboldt mines, powered by water brought in by a capacious ditch ninety miles long. The water project was abandoned after some $100,000 had been spent in excavation, and Mill City never developed beyond the village stage.

A few of the old Humboldt ghost towns are shown on large-scale highway maps. The lesser ones, possibly more promising for present-day exploration, may be spotted in library reference material at nearby Lovelock or at Reno. The principal canyons that will lead one into the recesses of the Humboldt range are, from north to south, Star, Coyote, Buena Vista, and American. The northern two thirds of these rugged mountains offered little possibility of road building, and exploration must be on foot. In the south a skein of old dirt roads will take one well into the range.

There are some promising tungsten occurrences in much of this area. The Arizona Mine near Unionville, which yielded $3 million in silver, is now considered to be of potential value for its tungsten.

Some thirty miles due west of the Humboldt mines are the abandoned gold camps of the Seven Troughs mining district. This area is reached by driving south to Lovelock, then heading northwest on State Route 48. The mines at Seven Troughs, Mazuma, Vernon, and Tunnel boasted of high-grade ore running up to $100,000 a ton in gold.

North of Seven Troughs by about twenty-five miles of untended road is Scossa. The discoveries here date from 1930. There were profitable mining operations for six years, until the original veins were worked out. Nearby is the site of an earlier camp, Rosebud. Overpromotion and high-pressure sale of townsite lots brought a small community into being, but a discouraged original population soon moved away. Values here were in silver and the true value of the deposits remains uncertain.

Our course through northern Nevada ends at Reno. There has been mining around Reno, right up to the city limits, but this area can well be by-passed by the modern prospector. Development, residential and industrial, has spread out and turned the few old diggings around the city into valuable real estate. This expansion has not reached the once promising Olinghouse gold district thirty miles eastward, and its possibilities will be examined later in connection with the mines and camps of central Nevada.

To continue an orderly course, we leave Reno and head south to connect with U.S. 95, which runs through mineralized country for a distance of five hundred miles. It serves the once celebrated mining centers of Tonopah and Goldfield, the nonmining city of Las Vegas, and finally Nevada's southernmost town, mine-encircled Searchlight.

Carson City, the Nevada state capital, which we pass thirty miles south of Reno, offers little in its immediate area. Of interest, however, is the state museum, where one may view a remarkable reconstruction of underground mine workings. Various operations are shown with life-size figures of miners and rock formations accurately reproduced. Carson City also is a point of departure for Virginia City, location of the Comstock Lode. The great mines there, corporation-owned and all idle at this writing, have nothing to offer the individual prospector. It also may be assumed that the nearby terrain was intensively prospected time and again in the city's heyday. Two mineralized areas that lie a little outside this zone are the Florence district and another lying around the ghost town of Jumbo.

Almost due east of Carson City there is one spot that holds out some present-day promise. This is the lofty old gold camp of Como. It is reached by a steep dirt road starting at Dayton, an interesting little hamlet on the Carson River where ponderous mills once handled much of the Comstock ore.

Owners of several old mines and prospects at Como have stubbornly held onto them through the years with confidence that they would eventually pay off. Excessive stock promotions during the early boom probably impeded businesslike development of the mines, and certainly clouded the picture as to the value and continuity of the gold-bearing veins. Quite close to Como is the site of a slightly earlier settlement, Palmyra. The two camps are surrounded by government land which is open to mining location.

Not a great distance by airline from Como but reached by car only by detouring around to Silver Springs are the abandoned camps of Ramsey and Talapoosa. Gold was found at Ramsey in 1906. The too frequent pattern of townsite promotion instead of honest mining followed, and Ramsey was all but deserted by the following year. Nearby Talapoosa was a much earlier settlement than Ramsey, dating back to 1863. Values there were in silver. Leasers worked the deposits off and on over a long period of years.

South of the Como-Ramsey area there are mountains of varied mineralization and some promise, the Pine Nut range. They lie just east of the twin towns of Minden and Gardnerville and rise to an elevation of 9,450 feet at the summit of Mount Siegel. These mountains are primitive country, with only a scattering of Indian dwellings and a few isolated ranches on their west side and virtually no inhabitants on their east. Environmentalists may take some heart from the fact that this unspoiled territory is barely thirty airline miles from the gambling casinos and $40-a-day hotel rooms of Lake Tahoe.

There have been promising gold and silver leads at many places near the 5,000-foot level of the Pine Nut range and showings of turquoise and of cinnabar, an ore of quicksilver. The higher slopes to the east appear to have been imperfectly explored, possibly because of the Indians' extreme hostility to anyone who threatened their important harvest of pine nuts. Mining had too often meant the ruthless cutting of trees for firewood. The largest concentration of early mining activity was a mile north of Pine Nut Creek on the western side of the mountains.

Investigation of the now accessible eastern slope seems to offer the best chance of virgin discoveries. It would be reached by a detour to the town of Wabuska on Alternate U.S. 95. A dirt road branches off here and runs south for twenty miles along the base of the Pine Nuts, terminating at State Route 3 near Wellington. There are no ghost camps along this stretch, and areas of known mineralization can best be identified by the few old mine workings. Numerous canyons extend back into the mountains and any of them might be worth exploring for float ore.

In dealing with Como, Ramsey, and the Pine Nut Mountains we have been on several different highways and local roads. It is time now to get on our long route southward, U.S. 95. On joining it east of the Pine Nuts we run into two large restricted areas, the Walker River Indian Reservation and the Naval Ammunition Depot. The latter is a sprawling bunker-filled storage area, very much off limits to civilians, with its own headquarters town, Babbitt. The adjoining civilian town is Hawthorne, county seat of Mineral County.

Indians on the Walker Reservation always have done a little prospecting. Occasionally one of them has some partly developed diggings for sale or lease or is looking for financial backing. Approach through the United States agent on the reservation is advisable. Mining on Indian lands, as will be made clear in connection with other locations, is possible but is subject to special contractual arrangements.

No contract is necessary to look for an extremely rich and extremely visible outcrop of gold ore east of the towns of Luning and Mina, a little south of Hawthorne. This would be Tim Cody's lost ledge.

A piece of Tim's dazzling, gold-studded rock may still exist at Luning. For years it was in possession of a county commissioner there, who obligingly showed it to hopeful lost-mine hunters. Cody himself died in the early 1930s.

Tim's adventurous wanderings were among the little desert mountain ranges to the east of the Luning-Mina area. There are two parallel ranges running north and south. The nearest, or western, are the Pilots, then come the Cedars. North of the Pilots is the Gabbs Valley range, and even farther north is Pilot Cone, which is one of the landmarks to be noted in any search. The country is so parched that some map-makers, in indicating the ghost towns of Golddyke, Pactolus, and Simon, add the warning "No Water." This applies also to a curious rock formation known as the Petrified Spring.

The designation Stewart Spring marks a dependable water source. This remote, unpopulated oasis has saved the lives of many prospectors, among them Tim Cody himself. It is also the last definitely known spot where we can place Cody on the trail to the lost mine.

Cody was an Irishman, tall, gangling, and good-natured. He would work in some mine long enough to make a stake and then equip himself with food, water, and a good deal of whiskey for an indefinite stay in the hills. His wanderings at one time took him to the area that later became the prosperous Simon Mining District, but somehow Cody failed to recognize the impressive mineralization there.

In 1908, Cody was on one of his periodic tramps and was replenishing his water at Stewart Springs. He decided to go on north to Golddyke, then an active mining camp. The distance was fifteen miles. On most of the journey, but not at all points, Paradise Peak was a visible landmark, towering just behind Golddyke itself. It still can guide one to the scanty ruins that remain at the old Golddyke workings.

Somewhere, while crossing a dry lake bed as he afterward told it, Cody lost not only sight of the peak but lost his entire orientation. Daylight was fading. Resigned to an overnight camp, the prospector climbed up a nearby knoll to get a view and re-establish his sense of direction. We know from Cody's later accounts that this knoll was fairly well covered with juniper trees and that it was of no very great height. We do not know, however, whether Cody was referring to an isolated rise, as one would assume, or whether he used "knoll" in a sense rather common in the far West, that is, to describe a spur of land jutting out from a larger ridge or mountain. At any rate, it is on the summit of a knoll that the extravagantly rich ledge of gold ore was found. Found and lost.

After Cody had got his bearings, his practiced miner's eye rested on a band of white, iron-stained quartz near his feet. Breaking apart a few of the loose, weathered fragments, he found the fractured surfaces heavily impregnated with gold. The vein was well defined, even in the dimming light. A substantial body of ore was clearly visible.

Thanks to his climb, Cody had sighted Paradise Peak to the north of him and also Pilot Cone, still farther north and on a line a few degrees to the west of Paradise Peak. Now he knew pretty well where he was. He used his prospecting pick to cut several gashes in a nearby tree, put several pieces of ore in his pockets, and then descended the knoll to pick out a comfortable campsite. His experiences from that point on explain why his discovery joined the list of lost bonanzas.

At times the bachelor miner was a heavy drinker. Whether or not he had any firewater left in his pack at this stage is not known. But perhaps due to premature celebration of his good luck, or because of heat, thirst, and exhaustion, Cody was out of his mind for the next three days. All he remembered of this period was that he eventually found himself at the abandoned mining camp of Pactolus. Someone had left a cache of water and flour there. Cody was able to revive himself and get back to civilization.

In more than twenty years of search by Cody himself or Cody and friends, it proved impossible to find the blazed tree or that streak of mottled white quartz so lavishly laced with gold. Near Stewart Springs a piece of ore exactly matching Cody's specimens was picked up in 1915. Tim himself was not with this party. The excited searchers were unable to locate any outcrop remotely matching their encouraging find. It was speculated that this piece did not come from nearby but had been dropped by the Irishman himself during his three days of wandering.

There are still people in Mina and Luning who can describe the general area in which Tim Cody's lost gold must lie. Maps of the Geological Survey will pinpoint Stewart Springs, where Cody started, and Pactolus, a significant spot in his ramblings. They also will show Golddyke, Paradise Mountain, Pilot Cone, and all the contours of the landscape in between. One of the elevations shown must surely be what Cody called his knoll. It is all desert country, still unpopulated, still inviting a search.

Continuing south from the Mina-Luning section on U.S. 95, one gets into a flat country of dry lakes and salt marshes where a prosperous borax industry once flourished. The road then swings eastward to Tonopah, the once great silver-mining center and still an interesting and active little city. It is the principal population center on the long route between Reno and Las Vegas.

Well before reaching Tonopah, one will sight the desolate Monte Cristo Mountains north of the highway. The Monte Cristos were the scene of a comparatively modern gold rush. This occurred in 1924 and gave rise to the town of Gilbert. The three Gilbert brothers, miners and prospectors all their lives, discovered ore in the Monte Cristos running $96,000 a ton in gold. It was so valuable that it was sacked and stored in the vault of a Tonopah bank. When the word got out, a new rush was on. So many people swarmed in that a townsite was laid out and within a year there was a population of eight hundred. The daytime population was swelled by many persons who drove out from Tonopah, only twenty-eight miles away. As in other places with good but limited mineralization, a few of the earlier, choice locations paid off handsomely. Most of the hopeful gold hunters necessarily went unrewarded. Little remains at the site of Gilbert and it has doubtless had its day. What may remain to be discovered farther back in the Monte Cristos is another and entirely speculative matter.

Opposite this range, to the south of U.S. 95, is the Silver Peak Mining District, long the scene of highly profitable silver mining and the location of some very large ore-reduction plants. There always has been some activity around Silver Peak, often by leasers working the older mines. Present-day possibilities would appear to be in reactivation of known producers rather than in new discoveries.

Beyond Tonopah is the equally famous mining center of Goldfield. We will look briefly at the history of these two places, but as in the case of the Comstock Lode, this report will not go into their massive operations. The principal mines are corporation-owned and in some cases consolidated into very large properties. In their busiest periods conditions were favorable for intensive exploration and development. Ample skilled labor, easy financing, a fair cost-price ratio, and general mining enthusiasm were among these factors. One wonders what underground values can be left after so much activity. There were a multitude of claims filed, however, and not all of them were adequately explored. Their status as to availability for relocation can be learned by examining the courthouse records in Tonopah, for Nye County, in Goldfield for Esmeralda County.

In historical sequence the discovery of Tonopah comes first. The story often has been told of how Jim Butler, hay farmer and part-time prospector, discovered his multimillion-dollar Mizpah outcrop at the future site of Tonopah. When he picked up a piece of rock to throw at his wandering burros it seemed unusually heavy. It was silver ore and it came from a vein exposed almost at his feet. Jim Butler suddenly had a mine. The first shipment, a single ton of ore from this outcrop, returned $800, a substantial sum in those days. Butler took a partner, Tasker L. Oddie, the $50-a-month district attorney of Nye County, later governor of Nevada. They staked a number of contiguous claims and went on to become rich men, along with various friends, leasers, and other miners who rushed in while there were still rich veins to be claimed. Butler leased out sections of his rich deposits for limited periods with only oral agreements. It is said that these were never violated by either side. For some leasers, this meant carrying their tools away for good at high noon on a stipulated date, with thousands of dollars of embedded silver still staring them in the face.

Happy days were here again for Nevada mining, and it received an even greater stimulus from the fabulous discoveries at Goldfield. Tonopah, its treasure discovered in 1900, was a silver camp. At Goldfield, where the first hint of mineral wealth came in 1902, the values were in gold. The ore right at grass roots was of extremely high grade. In such mines as the Mohawk, Florence, Jumbo, Red Top, and Combination the yellow metal occurred in such concentrations that fears were expressed for the stability of our then gold-based monetary system.

In these and comparable vein systems, "high-grading," the theft of rich ore by the $4.00-a-day miners, possibly set a world record. The estimated production at Goldfield up to World War I was $125 million. Of this, there was $45 million that authorities consider to have been stolen. The output passing through legitimate channels was $80 million.

What profit, if any, accrued to Goldfield's first known miner has escaped the record. He was Tom Fisherman, an Indian, who first exhibited gold ore from the Goldfield outcrops while visiting in Tonopah. Two young prospectors who chanced to see Tom's samples took action at once. They hurried to the site and soon were working a bonanza. This pair, Billy Marsh and Harry Stimler, are commonly listed as the founders of Goldfield.

The modern gold hunter may wish to pause in both Tonopah and Goldfield. Neither is a ghost town, although Goldfield exhibits a number of sepulchral buildings which stand vacant owing to the dwindling mining activity and the dwindling population. Its massive, empty hotel has been used as an eerie background in several motion pictures. Either of the old towns makes a suitable supply base for the prospector who wishes to try his luck in certain more remote and less exploited fields a little farther south.

The first of such suggested areas is the Gold Point district. It is reached by turning west off U.S. 95 at a point fifteen miles south of Goldfield. The road is paved for a distance, then tapers off to an improved road, and finally to faint wheel tracks across the desert. The principal old camps along the route, all of them ghosts or near ghosts, are Lida, Gold Point, Gold Mountain, Palmetto, Oriental, Pigeon Point, Tokop, and Sylvania.

Gold, both lode and placer, and silver too have been taken out of this region at many places. Accounts of the early operations and the circumstances of the closing of the mines are rather disappointing as to detail. Inefficient milling operations, some of which failed to recover even 50 per cent of the values, accounted for some closings. Litigation halted work at other properties. Such reports as we have leave the conviction that the deposits never have been fully exploited, that either abandoned workings or fresh discoveries may ultimately yield good profits.

The ghost town perversely known as Gold Point was actually devoted to silver mining. Its ledges were known and worked at least twenty years before the striking discoveries to the north. Until his recent death, State Senator Harry Wiley, a well-known figure in Nevada mining and politics, made his home at this lonely outpost. There were mining rushes to Gold Point in 1880, 1905, and 1908. Intermittent working of the Great Western, the most productive mine, continued until the wartime closing order of 1942. The encouraging fact about the Gold Point ore bodies is that they persisted at depth, and in at least one case developed greater values in gold than in silver.

Much richer ore was found ten miles to the south, where a small settlement grew up and took the name Oriental. The values here were in gold. There were some astonishing specimens that were saved for public display. In a region where a natural water source is rare indeed, Oriental was fortunate to have some flowing springs. They made life easier than in other camps but the volume of water was not great enough for substantial milling operations. When the higher-grade, shipping ore was exhausted the mines shut down. Gold certainly remains at this one-time bonanza, but since 1900 the site has been deserted.

The early mining camps in this area were all in the public domain and most of the land is still so classified. Where a site was on land suitable for gazing, it sometimes passed into private ownership. Considerable areas were taken up either by homesteading or by purchase from the United States Land Office. Such a spot is Lida, which has a mining history similar to Gold Point but is now part of a large private ranch. Any prospecting or mining enterprise here would be a matter of negotiation with the property owner.

No such barriers stand in the way of the gold seeker at either Palmetto or Old Palmetto, which are over to the west near the California state line. Silver was discovered at the older site in the 1860s. More promising finds not far away led miners to establish the newer Palmetto. Off-and-on activity, much of it by individual leasers, extended well into this century. During all this period silver was bringing far below its present price.

There was similar mineralization around the now vanished settlements of Topok, Pigeon Springs, Senner, and Roosevelt Well. The latter two camps were situated in Tule Canyon, along whose length was a well-nigh inexhaustible deposit of placer gold. The gravel was not extremely rich but it furnished a living for an uncounted number of hard-working miners who labored there off and on, especially when down on their luck, for more than a hundred years. How much placer gold remains after all this time is unknown. If one can find a likely looking but undisturbed location and sink a shaft to bedrock it will help with the answer.

Traveling south from the Lida-Gold Point district, one draws near the eastern boundary of Death Valley National Monument. Part of this reserve extends into Nevada as a large triangle, and for a long distance the monument boundary runs along the California-Nevada state line. From U.S. 95 one sees successive mountain ranges to the west: the Grapevine, the Funeral, and the Black mountains. To the east and just outside the restricted atomic test site are the Bare Mountains and the Spectre range.

There are other geographical features near the highway whose colorful names have appeared in much literature dealing with the American West. There is Sarcobatus Flat and there is the Amargosa Desert, either of them bad places in which to get lost. There are the ghost towns of Bonnie Clare, Rhyolite, and Bullfrog, the latter with its celebrated Bottle House, whose walls consist of beer and whiskey bottles cemented together. There is the great baroque railroad station at

Rhyolite that has no railroad within a hundred miles of it. There is the old track of a railroad unaccountably running straight up the side of a mountain. This ends at deserted Carrara and the rails were used to ease giant slabs of marble down to transport on the plain below. Seldom visible is the Amargosa River, which for most of its ninety-mile length flows underground. And if, in this strange territory, you see an old desert rat running a speculative eye along the jagged summits of the Funeral Mountains, he may be contemplating another sanguine attempt to find the Lost Breyfogle Mine.

The Breyfogle story begins when a gaunt but powerfully built giant is encountered by a rancher named Wilson in Smoky Valley. The scene is about halfway between Death Valley and Austin, Nevada. The man is in pitiable condition. His clothes are in tatters, his bare feet torn and bleeding. His shoes are slung by their laces about his neck and contain a meager, dripping supply of water. Half-starved and emaciated, he has suffered in mind as well as body from days of wandering under the desert sun. But one fixed resolution has not weakened—he has clung to a little bag contrived from his bandanna. The bag contains possibly the richest gold ore ever seen in the Western Hemisphere.

This man was Breyfogle. According to J. Frank Dobie of *Coronado's Children* fame, his first name was Jacob. Harry Sinclair Drago, a careful researcher and writer on western subjects, has his name as Charles, with the name Jacob applying to a brother. Authorities also differ on the circumstances of the man's trek out of Death Valley. A small body of literature, some of it controversial in nature, has grown up about the subject. We must confine ourselves here to such matters as bear on the location of one of the country's most alluring undiscovered treasures.

How did the adventure start? The time was 1864, and a spring night of that year found Breyfogle and some companions camped at a water hole now known as Stovepipe Wells in the north central section of Death Valley. Today there is a comfortable hotel near the spot. Breyfogle may have had two companions from Los Angeles, McLeod and O'Bannion, or he may have been with sojourners from Austin. Such irrelevancies are part of the tangle of fact and conjecture already noted. In any case, the other men made camp at the water hole, while Breyfogle rolled up in his blanket some distance away.

Death Valley and the rugged Panamint Mountains were the home of a branch of the Paiute Indian tribe. The names of Hungry Bill and

Panamint Tom have come down to us as among their leaders. During the night a band of Indians crept up to the water hole and killed the men sleeping beside it. There was enough commotion to awaken Breyfogle. Seizing his shoes, but not taking time to put them on, he struck out into the darkness and frantically covered as much ground as he could before dawn. Having made good his escape, he hid in a depression of the flat but slightly undulating valley floor and waited for darkness before continuing his flight.

The discovery of the fantastic gold ledge is now near and we begin to get some directions toward it.

This northern part of Death Valley is hemmed in on the east by the Funeral Mountains, whose mineralization is amply attested by the productive Keane Wonder Mine and by other operations. All the early "breyfoglers," as lost-mine hunters are sometimes termed, agreed that when Breyfogle fled the Paiute attack at Stovepipe Wells his course was eastward. It was in fact east of the whole Death Valley complex that the wandering, starving man was encountered by the rancher Wilson.

Breyfogle crossed the valley floor by night and reached the mountains. For an uncertain number of days he roamed through them, his constant objective being to discover water. His first find was a salty, alkaline seepage. He drank some of this and, although it made him sick, filled his two shoes with the fluid and pushed on. His next hopeful sign was a spot of green sighted on a mountainside. As Breyfogle remembered it, this inviting spot of color proved to be adjacent to a small mountain meadow covered with bunch grass. The green turned out to be merely the sparse foliage of a single mesquite tree, and there was no surface water. But there was something else on the surface, some crumbling, chocolate-colored quartz shot through and through with gold. In one of the specimens later seen and described by reliable persons in Austin, gold made up a good one half of the total mass.

It should be noted that at this point Breyfogle was not in the extreme physical and mental distress that he experienced later. His feet were bruised and so swollen that he could not wear his shoes, he was doubtless thirsty most of the time, and his nutrition was limited to mesquite beans of debatable value. He was a physically strong man, however, and he had been on the run probably no more than three days and possibly only two. He was in good enough condition to appreciate his newly found wealth and to note its identifying surroundings. He was even careful enough to discard the first pieces of ore he had picked up and replace them with even richer specimens lying about the outcrop.

Later—was it later that same day?—he came upon a good spring. He drank and bathed his face and aching feet. The Indian menace was by no means past and this water hole was certainly known to them. Breyfogle therefore settled down for the night some distance away, more or less hidden in the scanty brush. His withdrawal from a water hole did not protect him this time. A nocturnal band of Paiutes discovered the sleeping man, clubbed him into unconsciousness, and left him for dead.

The odyssey of Breyfogle's desperate tramp out of the mountains and across the scorching flats of southwest Nevada was never adequately reconstructed by his friends and questioners. Its circumstances were unknown to Breyfogle himself. The beating he had received—he was not scalped as some reports have it—and the madness and amnesia that heat and thirst can inflict on the stoutest constitution had all but erased the experience from his memory. When, after his rescue, he had somewhat recovered his health he felt that once back in Death Valley he could get his bearings and retrace those earlier, saner steps that had carried him to the rich lode.

This was a false hope.

Breyfogle accompanied at least four search expeditions during 1865, 1866, and 1869. Two of the organizers, Jake Gooding and Pony Duncan, were believers in Breyfogle's sincerity and became his good friends. They defended him against the growing skepticism concerning his story and at times had to defend him against the physical threats of their disappointed associates. An episode in the second search gave the critics some ground for their suspicions. When Breyfogle had withdrawn from the "good" spring to sleep in the brush, he had built a low barricade of rocks around his bivouac. What protection this was supposed to give is not clear. The second set of searchers, this party led by Jake Gooding, found such a rock structure and Breyfogle readily identified it. Somewhere near, surely, they should find the ledge of chocolate-colored quartz. But it was not found, nor could they discover any spot answering the description of the little mountain meadow covered with bunch grass. Breyfogle had remained insistent on the little mountain meadow as the best clue.

The disappointment was keen, but with a later sifting of evidence an obvious explanation develops. The gold had been picked up by a mesquite tree where there was no water. The stone barricade had been built near a spring. How much time, how much mileage, how much altitude separated these two stops? It is something that nobody seems to have estimated.

The concensus of those close to Breyfogle, of those who had talked with him time and again, was that his lost mine was situated in Boundary Canyon. If so, modern roads have made the search area quite accessible, for California State Highway 190 runs right through the canyon. At the Nevada state line it becomes Nevada Route 58. From the Death Valley side, the Boundary Canyon road starts at a highway "Y" and runs through a mountain cleft known as Hell's Gate. It then climbs up to a saddle previously mentioned, Daylight Pass. A little to the west of here is Daylight Spring, which conceivably could have been one of Breyfogle's water holes. The road then descends to almost flat country and takes one past old Rhyolite and Bullfrog and into Beatty, Nevada. After crossing Daylight Pass you are definitely out of Boundary Canyon.

As for Breyfogle himself, after the last of his futile searches, in 1869, he went to Eureka, in eastern Nevada, and obtained an office job with one of the mining companies. His health had been failing and in the winter of 1869–70 death put an end to his personal appearance in the growing Breyfogle legend. The puzzled man tried hard to make maps of his wanderings and to inscribe the all-important X in the right place. One map was given to his brother, who made a single, ineffectual search. Another, reputed to be the most promising, was in possession of a branch of the family in San Francisco. In 1938 the San Francisco Breyfogles told a reporter for the Associated Press that the treasured document had inadvertently been burned in the fireplace. A cynic might comment that this accident was a piece of good luck for the surviving Breyfogles.

The state of Nevada ends in the south in a sharp angle formed by the California state line and the Colorado River. As one proceeds down U.S. 95 past Death Valley and the scenes of Breyfogle's wanderings, he is getting near the vanishing point, yet there remain ahead of him several old mining districts of interest, conceivably offering present-day opportunities.

Eliminated from the picture are the vast expanse of the atomic test site and the recreation areas set aside by the government along the Colorado River and its chain of man-made lakes. The rest of the territory along U.S. 95 is of interest to the prospector. It shows scattered and in places impressive mineralization.

It has been noted in connection with the Breyfogle story that the Daylight Pass route out of Death Valley passes the ghost towns of Rhyolite and adjacent Bullfrog. The boom and bust at Rhyolite in the early 1900s has struck the imagination of many people. Its population soared from nothing to 6,000 in only three or four years. This was on the basis of exaggerated estimates of its mineral wealth, and it was much more rapidly depopulated after the nationwide financial panic of 1907. Few visual reminders of its former opulence are to be seen, and at this writing it numbers only six residents, none of them engaged in mining. In its best days, starting in 1905, there were some fifty gold mines being worked. Only the Montgomery-Shoshone and possibly two to five others got into ore that gave any promise of sustained, profitable operation. The Montgomery-Shoshone, although struggling under a debt of $500,000, which had been borrowed to erect an oversized mill, outlasted all the others. It closed down in 1911.

The official map of Rhyolite and its environs shows the land surface almost completely covered with the standard twenty-acre mining claims, squeezed together and in some cases overlapping. Virtually all of them were patented, which means that on abandonment they did not revert to the public domain. They remained private property. Few owners kept up their taxes, however, and this mass of twenty-acre parcels, with or without extensive mine workings, is now the legal possession of Nye County. The county is empowered by Nevada law to issue prospecting permits on them. Whether title would be conveyed to such an applicant and under what conditions is a determination to be made by the board of county commissioners.

Such reports as exist on the Rhyolite and adjacent Bullfrog areas indicate that quartz veins are numerous and that few are entirely barren. In the initial Bullfrog discovery by Shorty Harris, the surface ore was extremely rich but the general quality of the deposits was low grade. Here again we face the question of whether today's higher gold price and advanced technology, offset as they are by higher labor and material costs, can make these borderline mines attractive. Rhyolite is not a camp where mining has continued, where the ubiquitous leaser has tried his luck through the years. Rhyolite simply rose and fell.

At the junction of U.S. 95 and State Route 58 is the small town of Beatty. There have been sporadic strikes east and southeast of Beatty in the Bare and Spectre mountains. One such camp was optimistically named Transvaal, but no Rand type of gold-bearing reef ever was found.

It is 114 miles from Beatty to the metropolis of southern Nevada, Las Vegas. There is little to attract the prospector on this stretch of highway or in the immediate vicinity of Las Vegas. Farther south are two districts of important and widespread mineralization. One is centered at Goodsprings, reached via Interstate 15, the busy connection between Las Vegas and Los Angeles. The other is Searchlight, farther along on U.S. 95.

Isolated and unconnected with these or any other district is the great Potosi Mine. It is located northwest of Las Vegas, about halfway up a steep cliff that at first glance appears unscalable. The Potosi was the first operating mine in the state of Nevada. Discovered by Mormons in 1856, it poured out silver, lead, and zinc over a long period of time, seemingly inexhaustibly. Only the Empire Zinc Company, latest in its list of owners, knows what values remain in the lofty cliffside workings.

Goodsprings is reached by turning off Interstate 15 at Jean and heading west on Route 53. This road goes on to cross into California, circle around, and rejoin Interstate 15 near the ghost towns of Valley Wells and Windmill Station. The entire length is through territory of known mineralization. Goodsprings itself still contains a few residents, and the district has a history of discovery and rediscovery that takes in a surprising variety of metals—gold, silver, lead, zinc, copper, and platinum.

The Goodsprings mines have yielded an estimated $31 million. The original Yellow Pine Mine produced silver and lead from 1868 on, while the valuable Keystone Mine, discovered in 1892, had values almost exclusively in gold. When a visiting metallurgist examined rock that had been discarded as worthless at several mines, he found that it was high-grade zinc ore. There were profitable operations from then on based on zinc until commercial grade ore was exhausted. It was at the Boss Mine, near the California line, that platinum was discovered in 1914. The mild sensation was made the basis for the promotion and sale of lots in the ephemeral townsites of Platina and Boss. The platinum mining itself was not a success.

There are many old mines and prospect holes in the Goodsprings district for any who care to investigate their possible worth today. Many years of erosion, which in this country can be considerable, hold some thin hope of new surface strikes also. The gold hunter in this region will be required to widen his scope, for any values he discovers may well be in part or in whole among other metals.

There are some parts of the Goodsprings district that are better approached from the California side. Among the more interesting areas is the western slope of Little Devil Peak, as well as the western side of other mountains to both north and south of that eminence. This rugged country figures very little in official mining reports. A series of at least a dozen gulches invite exploration. They can be penetrated only on foot. If promising float is found, the rather limited size of the ridges and spurs separating these gulches will reduce the amount of ground to be covered in looking for ore in place.

The old mining center of Searchlight is south of Goodsprings. Its prospecting area is slightly restricted now because of the large federal recreation zones along the Colorado River. The scenes of the richer early-day strikes, notably the Duplex and Quartette mines, still lie outside the zone. Between U.S. 95 and the river most of the Eldorado Mountains and the Newberry Mountains remain open to prospecting.

Next to Searchlight the chief mining center in this area was Nelson, which now lies just outside the recreational reserve and is reached by State Route 60. The great mine there, the Techatticup, was known to the Spaniards of long ago. While it was in production during the American era, two strangers arrived there from Madrid. They carried a map handed down in their family with clear directions to the Techatticup. This inherited waybill gave them no standing as against the discovery and entirely legal possession by the current owners and one lost bonanza of colonial days was crossed off the list.

To the west of Searchlight, across the McCullough Mountains, is the site of now vanished Crescent, once a gold-mining center but better known for its valuable turquoise deposits.

Searchlight has accommodations and supplies for the traveler but the country around it is virtually unpopulated. Nineteen miles to the west is Nipton, on the California state line, which consists of little more than a general store and a few dwellings. Below Searchlight the southern tip of Nevada is marked by a triangular piece of public domain bounded by California on the west, the Colorado River on the east, and the federal recreation area on the north. A dirt road, designated State Route 76, runs through the region. This may well be the least explored portion of southern Nevada.

6
Through Central Nevada

Having covered the northern part of Nevada and much of the perimeter of the state, it is now time to take a course eastward from the Reno area and report on mining prospects in the central and eastern sections of the state.

There is only one road, arterial or otherwise, that crosses central Nevada on an east-west course. The multiplicity of mountain ranges running north and south make road construction difficult and there is an extremely scanty population to be served. The one through route is U.S. 50. It chooses an easy course along the *playas,* or dry lake beds, but necessarily climbs over several mountain passes. The distance from Reno to Ely, the easternmost town in the state, is 324 miles. From Ely U.S. 50 passes into Utah via the Lehman Caves National Monument.

In this considerable distance there are only three paved roads leading north off U.S. 50 and only two leading south. A great many dirt roads, however, remind us of hopeful mining enterprises in times long past. Many of the old trails lead to promising diggings that have been abandoned, occasionally under somewhat mystifying conditions. Some may be worthless, some of borderline value. There certainly are others that deserve another look.

Starting eastward from Reno, one need go only thirty miles to find some old placer and lode workings that still are of interest. These constitute the Olinghouse Mining District. It is reached by turning off the pavement at what remains of the once important railroad town of Wadsworth. A rough dirt road climbs well up into the hills to the ruins of the old camp.

Gold was first discovered in placer deposits in several tributaries to

Olinghouse Canyon, and was then traced up to outcropping quartz veins. After some reasonably successful mining of these veins around the turn of the century, a new corporation built a very large ore-reduction plant at Wadsworth and persuaded one of the boom-day Nevada railroads to lay a branch track up to the mines. Neither of these ambitious installations operated as much as a year. It is not clear whether failure was due to lack of volume or lack of value in the mines' output. Intermittent small-scale mining has managed to show some fair returns at Olinghouse in the six decades elapsing since this overblown development. A dispassionate, expert engineering study of the best of the old workings might put Olinghouse in business again.

When one reaches the town of Fallon he has left the network of roads around Reno and is entering the sparsely settled central part of Nevada. The country around Fallon is rather flattened out, the great Carson Sink lying to the north and Carson Lake to the south. The only mineral production has been salt, mined or scraped from various *playas*. It is not until we are halfway across sprawling Churchill County that a series of side roads leads to old camps that have had their day in mining history.

A turn north at a roadside settlement called Frenchman, or Frenchman's Station, takes one to La Plata's ruins and silver discoveries dating back to Civil War days. Eastern capital was attracted to La Plata and several of the best claims were consolidated. Operations were interrupted in 1867 when spectacular discoveries were made in the White Pine district to the east. Most of the miners left for the new strike and the principal operating company at La Plata moved its complete mill to one of the White Pine mines.

At the nearby camp of Wonder, mining was longer-lived and more profitable. This ghost town is noted on some highway maps along with its satellite camps of Hercules and Victor. The three places can be reached over an old, badly deteriorated road from La Plata or by turning off U.S. 50 about six miles east of Frenchman. Wonder's values were chiefly in gold and were discovered during the statewide mining revival of the early 1900s. The big producer was the Nevada Wonder Mine, which continued mining and operation of a two hundred-ton cyanide mill through 1919, consistently paying generous dividends. Farther north a few ruins mark the location of Silver Hill, where remarkably rich surface ore was discovered in 1860, and Bernice, another camp of that time.

These operations were all in either the Stillwater Mountains, which some old maps show as the Silver Mountains, or in the next range to the east, the Clan Alpine. At the southern tip of the latter is Chalk Mountain, the scene of much placer mining.

With all allowances for the energy and persistence of the early-day prospector, the matter of manpower alone makes it questionable that ranges such as the Stillwater and Clan Alpine were very minutely explored for mineral deposits. In 1860 the population of all Nevada was only 6,857. By 1870 it had risen to 42,491, but the concentration of people at Virginia City in the great days of the Comstock Lode accounted for much of this figure. Thirty years later, in 1900, the population remained at barely 42,000. It was not until the Tonopah and Goldfield bonanzas, plus other important discoveries, attracted additional outsiders that any real increase occurred. By 1910 the state's population had doubled, but because of Nevada's vast area the total of 81,875 in that year figured out to less than one individual per square mile.

The camps just described are north of U.S. 50. South of the highway the record shows similar scattered mineralization. At one such spot there grew up the flamboyant town of Rawhide. Other locations where more or less successful mining was carried on include Fairview, Eagleville, Broken Hills, Quartz Mountain, Phonolite, Carroll, and Gold Park.

Roads that lead nowhere except to abandoned mining camps do not have a high priority in the way of maintenance. Those that straggle out to the points just named are examples. When such routes are shown on maps as "unimproved," the traveler should pause and ask information locally. In Nevada, with its very limited rain and snow, the old roads stand up well. Those that are mapped are generally passable, but in case of trouble help may be a long way off.

The question for the prospector at the once rip-roaring town of Rawhide is whether any real values still exist in its overpromoted mines, which inevitably get confused with the overpromoted town itself. Rich gold ore was discovered there in 1907 by an old prospector, Zack Carson. A boom as frantic as any Nevada had ever seen was quickly under way. Amid all the wheeling and dealing that followed, Carson, who had made a legitimate and valuable discovery, was doubtless lucky to sell his claims for $50,000 in hard cash and get out of the picture.

Rawhide was the stamping ground of Tex Rickard, later known to all the sporting world as a boxing promoter; of Nat Goodwin, the much-married actor whose romantic talents were rivaled by his talent for stock manipulation; of George Graham Rice, perhaps the nation's most persuasive seller of blue-sky securities, and of silver-tongued H. W. Knickerbocker, the former preacher whose flowery oratory now sold town lots and mining shares.

At this time, the first decade of the twentieth century, Nevada was at the height of its second great mining excitement and there was a large floating population alert to any new chance. It included not only miners but merchants, publicans, gamblers, and professional men quite ready to re-establish themselves wherever prospects looked good. Many of these flocked into Rawhide. Within a year, genuine mining production plus intensive promotion had created a small city in a bleak, almost waterless wasteland. The *Rawhide Rustler* claimed a population of 22,000 for the place. The fact seems to be that the peak population of Rawhide was roughly 8,000. For what it is worth, the apostles of growth cited the fact that the red-light district, known as Stingaree Gulch, had eight hundred girls practicing their profession.

Rawhide lasted barely two years. In 1908 a fire starting in a drugstore spread to the large, flammable Brown Palace Hotel. Whipped by a strong wind, one of Nevada's fierce "Washoe zephyrs," the flames wiped out the entire business section in forty minutes.

The chief promoters did not return. A hopeful residue who started to rebuild and to work the better mines had their hopes quickly extinguished. Almost exactly a year after the fire, one of the great cloudbursts so feared in all the desert country sent a destructive flood through the town and filled many mine shafts with water. Rawhide never recovered.

Today's prospector who may wish to look over possibilities there should locate Murray Hill, where extremely high-grade ore was found. Just south of there were Carson's original discovery claims, the Bluff and Mascots No. 1 and 2. Bearings on these claims can perhaps be found in the old county records at Hawthorne. Such records also may show the holdings of the four Grutt brothers, who held onto property at Rawhide for many years and were well known in Nevada mining circles. One of the well-mineralized areas was known as Grutt Hill, others as Burro Mountain and Baloon Hill. Any of these eminences would be logical places for renewed exploration and appraisal.

Some six miles east of the Rawhide turnoff from U.S. 50 a short detour to the south takes one to the ghost town of Fairview. The rich silver ore on the side of Fairview Peak escaped notice in Nevada's early wave of prospecting, but was found and developed in the post-Tonopah boom. The area was completely devoid of water, wood for fuel, or a railroad connection within reasonable hauling distance. Only the highest-grade ore could be profitably shipped, and mines and town lasted only two years. Today heavy-duty trucks roll along U.S. 50 within a few miles of the silver ledges. Such a present-day transportation advantage might justify a fresh look at Fairview Peak and its deposits.

A little northeast of Frenchman one may view the great faults in the earth's crust left by the earthquake that jolted central Nevada in 1954. Five miles farther along U.S. 50 from the quake-zone turnoff is the junction of State Route 23. This paved road courses through many miles of desolate country and joins U.S. 95 at Luning, in the area of Tim Cody's lost mine, already described.

The one remaining point of interest is Broken Hills. Its name a bit inaccurately taken from that of a great gold strike in Australia, Broken Hills had only one important mine. This was based on a silver-lead discovery in 1913, and the mine's gaunt headframe still forms an eerie landmark, visible from miles away. In 1925 discoveries two miles southeast at Quartz Mountain seemed so promising that hundreds of miners came to the area. Broken Hills became the supply and recreation center for this new population. A considerable amount of ore was shipped from both points but the community did not last long. In 1950, the writer Nell Murbarger found a single resident, Maury Stromer, operating a one-man gold mine at Broken Hills and taking out ore good enough to provide him a living. This last holdout died in 1956.

A few miles east of Quartz Mountain there was a profitable gold and silver producer, the Penelas Mine, which operated chiefly during the depression years. The deposits in this area gave rise to two transitory settlements, Phonolite and Duluth. What remains at the Penelas Mine appears on some maps as a ghost town.

Our route along U.S. 50 now gets close to the geographical center of Nevada and the durable mining town of Austin. Neither tourism nor real estate development has had much impact on Austin, and it remains interesting and unspoiled. Fire, which has been the scourge of so many old mine camps, has spared Austin as a whole, although several individual landmarks have been destroyed. There scarcely has been a year in its century of existence that Austin has not seen some mining activity. Even today a sign or two will be noted along its one business street marking the offices of some operating, or at least hopeful, enterprises based on the demonstrable mineralization of this area.

Austin and its dozen or so surrounding camps were the outgrowth of what was known in the mining world as the Reese River excitement of 1863. It was one of the great mining rushes of western history. Central and eastern Nevada, arid, largely treeless, hot in summer and often bitterly cold in winter, had not attracted many prospectors up to that year. The few who had come encountered, sooner or later, nomadic bands of Indians and these were usually hostile. The Overland Mail stage route necessarily passed through the region, as did the short-lived Pony Express.

It was a station tender for the stage line who first chipped off some promising-looking rock in Pony Canyon along the Reese River. He thought it good enough to send to the Comstock mines at Virginia City for assay. The value of this first specimen is not recorded, but it was startling enough for the news to spread almost instantaneously through the miners' grapevine. The development came in the summer of 1862, an ideal season to pick up stakes. The rush was on. Since the migration did not swell into thousands until the following year, the date 1863 is usually assigned to the Reese River stampede. Early comers, however, already were in ore that ran up to $7,000 a ton.

While values were principally in silver, there were some worthwhile discoveries of gold. As in so many other mineralized districts in Nevada, Reese River could exhibit mines almost side by side, with one of them yielding gold and the other being worked primarily for silver. More often both metals occurred in the same vein in varying proportion. The Comstock Lode is a statistical example of this combination of the two precious metals. By weight, the metal recovered was silver by an overwhelming margin, but in dollar value the two metals virtually evened out. In the extremely rich National Mine to the north, silver and gold occurred in an alloy of the two, known as electrum.

One is necessarily skeptical of a modern prospector's chances in the Reese River district. Miners in very large numbers prowled through the entire region. The slopes of the Toyabe Mountains were covered with mining claims, each one requiring as "location work" a shaft, tunnel, or open cut ten feet in extent. It would appear that the mineralized ground was explored so thoroughly that nothing of value or promise could have escaped.

On the more hopeful side, one may suspect that this exploration was not so thorough after all. Stanley W. Paher's research for his

Nevada Ghost Towns and Mining Camps indicated that there was much frantic trading in "feet" along this or that ledge, much waiting and hoping for capital investment or cash sale, but in many—possibly in most—cases no serious work was done to determine what the deposit was really worth. That the Reese River veins did indeed constitute one of nature's treasure chests is demonstrated by the record of competent mining and milling carried on at the better properties. Aggregate production at Austin and its surrounding camps ran between $25 and $30 million.

Within a radius of about thirty miles around Austin were the fleeting settlements of Skookum, Gweenah, Ravenswood, Amador, Geneva, Yankee Blade, New Pass, Gold Park, Kingston, and Canyon City. All these arose as the result of local discoveries and there were various profitable operations, with recoveries ranging to half a million dollars or more from individual mines. Except for the unpredictable leaser, who may bob up anywhere and at any time, they are now inactive. Some of the abandonments were under circumstances that leave the real value of the deposits in doubt and hence make them candidates for renewed consideration. At Yankee Blade, for example, the mines struck water. This is a common, even an expected, hazard in deep shafts, but is unusual in such shallow workings in dry Nevada. This was in 1867. With modern pumping equipment and electric power now available, the submerged silver veins could be of interest. The best mine at Geneva shut down because the $12-a-ton ore was considered too low-grade. Fortunes have since been made on ore averaging considerably less.

These two examples would indicate that the story of Reese River mining is by no means a closed book. It is scarcely an area, however, where one can be encouraged to take the field in hope of virgin discoveries. The intensity of the early-day prospecting must be remembered. The best approach would seem to be through the federal and state reports on individual mines and prospects, plus all the information that can be obtained by local inquiry.

From Austin it is some seventy miles on an eastward course to the next center of mining activity. Eureka is now a quiet town and travelers' stop on U.S. 50, but there was a time when it was known as one of the great industrial centers of the West. Giant smelters poured black smoke and soot over its homes and buildings and the forests of distant mountains were hacked away to supply an insatiable demand for charcoal. The ores were lead-silver and the Eureka deposits constituted the first important discovery of this kind in the country. Gold also figured to some extent in the smelter returns.

Eureka is near the border of White Pine County, scene of the so-called White Pine excitement of 1869. This is the rush that called into brief existence the all but incredible settlement named Treasure City. Suggesting, except in its architecture, the lofty Machu Pichu of the Incas, Treasure City was situated near the summit of 10,000-foot-high Treasure Mountain. The site was a narrow, wind-swept shelf described as having "ten months of winter and two months of damned cold weather." Yet the lure of sudden wealth brought 6,000 persons to settle on this uncomfortable perch only a few yards below the mountain's topmost crags.

Over possibly the worst and steepest road in the West, they brought lumber to build not only houses but a business district of almost fifty buildings. There were stores, saloons, professional offices, a bank, and the inevitable honky-tonks for the recreation of the hard-working miners. "Prices," a chronicler relates, "were outrageously high." Lithographs depicting Treasure City and its setting with reasonable accuracy were scarcely believed when published in the East.

The amazing number of 13,000 mining claims, covering much of the mountain, were filed. Some of the locators were handsomely rewarded. The great Eberhardt discovery was worked as a glory hole, or open pit. From a gouge in the earth's crust only seventy feet long and about half that width, the owners took $3.2 million in silver.

"So rich is this mine," wrote Rossiter W. Raymond of the United States Treasury Department in 1869, "that its name has become almost synonymous with that of the cave entered by Aladdin." In the lower workings, penetrating the mountain below the open cut, Raymond related that "the walls were silver, the roof was silver, the very dust which filled our lungs and covered our boots was silver."

There were other great concentrations of silver besides the Eberhardt lode, and some miners were lavishly compensated for the hardships and expense of working a mine at this elevation. Treasure City was a fantastic creation and it came to a fantastic end. At the very time that the demonetization of silver had thrown the whole silver-mining industry into a panic, Treasure City caught fire. As if some power even greater than the government's money managers had decreed its end, the city in the sky burned down to the bare rocks on which it rested. It never was rebuilt.

The Eureka and White Pine districts more or less merge, and they taper off in the country north and south of U.S. 50. A number of localities besides Eureka and Treasure City deserve mention.

Small camps within a short radius of Eureka were Diamond City, Vanderbilt, Prospect, and Newark, all of them growing out of silver discoveries. At a greater distance gold-bearing veins were found. Buckhorn, thirty-eight miles north, was a gold camp that flourished off and on well into the present century. A few miles southwest of Buckhorn, on the opposite side of Mount Tenabo, is the tenacious old gold center of Cortez, and some fifteen miles across Grass Valley the lesser camps of Tenabo and Goldacres. Cortez has had a checkered history, but the existence of a number of gold-bearing veins, some of them very rich, is fully established. Silver predominated in value during the earlier operations. The principal known gold deposits are now corporation-owned and at this writing are being mined and milled on a large scale. There remains ample scope for exploration on the slopes of Mount Tenabo to the east, in the Shoshone range to the west and north, and at various elevations reached via McClusky Pass to the south.

Mineral Hill, almost due east of Buckhorn in the Sulphur Spring Mountains, had silver lodes that attracted capital from London. Hundreds of thousands of dollars in bullion went to the mint from Mineral Hill, but the income was not enough to meet the obligations of a $3 million investment. The hill from which the settlement took its name is easily identifiable. Whether the deposits are amenable to a more modest exploitation, with the help of greatly increased silver prices, remains to be seen.

About ten miles south of U.S. 50 and near the Eureka-White Pine county line lay a cluster of camps usually identified as the Hamilton district. Treasure City, already described, was one of them. Hamilton, with about the same population as its lofty neighbor, was situated near the base of Treasure Mountain and had the advantage of being on flat land with reasonably good transportation. Hamilton was the focal point of the White Pine excitement. It was first known as Cave City, because many of the improvident first arrivals settled themselves in the natural grottoes along the base of the mountain. Historians have described this stampede as the greatest since the original California gold rush. We have seen earlier that promising diggings at many places were abandoned for the lure of White Pine.

Next to Hamilton and Treasure City, the largest settlement in the compact district was Shermantown, which once had 3,000 population.

Smaller camps were Babylon, Seligman, Swansea, Picotillo, and Monte Cristo. These were all crowded into a remarkably small compass, an area scarcely larger than the standard township of six-by-six miles. On the record, the country surrounding Hamilton and its neighbors cannot be recommended for prospecting. In 1869–70 the district swarmed with miners who must have fanned out through the region for miles. Yet we find not even a small and transitory camp established outside the group already noted. The explanation appears obvious: no important discoveries were made.

From the Hamilton district it is scarcely ten miles back to U.S. 50, which is encountered near Antelope Pass. From this point one can look north into a semidesert region that, for a distance of at least one hundred miles, shows little sign of human habitation, past or present. For those with explorer instincts it should possess considerable fascination. Its scattered mountains bear the names of Butte Range, Medicine Range, and Cherry Creek Range, and there are prominent individual peaks such as Bald Mountain and Mount Taylor. There is little of record about this country either to invite or to discourage the prospector. Only on the eastern, or more settled, side of the Cherry Creek Range and the adjacent Egan Mountains do we pick up any conspicuous trails of the early miner. The town of Cherry Creek and its mines had their day and the place always has retained a few residents and witnessed some occasional mining. Across a valley is the site of Schellbourne, whose gold-bearing ledges were largely deserted in 1872 when discoveries at Cherry Creek seemed more inviting.

In the extensive, all-but-vacant territory north of Antelope Pass there are no paved roads. A tangle of dirt roads exists, but even these leave very large tracts accessible only on foot, or to some extent by trail vehicles. However, the once important town of Cherry Creek on the far eastern fringe is served by paved State Route 35. The Butte Mountains may be taken as the center of this large, vacant region. They also lie in the least approachable part. All the ground within a radius of thirty miles of their central peak may be considered as offering a chance, perhaps a rather long chance, of a virgin discovery. The nearest identifiable habitation is the Goicoechea Ranch, which some maps designate a point of historical interest.

The next point as one proceeds east on U.S. 50 is the great Ely mining center. This is copper country and a notable feature is the great open pit mine at Ruth, nearby. This highly developed copper center is hardly a place where the gold hunter will linger, although three miles northwest of Ely the old camp of Lane City may be of interest. Silver was mined here even before the White Pine rush. An eastern capitalist was directed to these mines, according to his own account, by spirit voices. The ghostly executives then directed the expenditure of more than $150,000 over a period of five years and lost all or most of their patron's money.

The spirits' mine and other workings at Lane City revealed quite a variety of minerals: silver, lead, copper, and eventually gold being discovered in workable quantities. Occasionally the gold ore was very rich, the Chainman Mine having values up to $20,000 a ton, and the Emma–not to be confused with the notorious Emma Mine to be encountered in Utah–yielding $13,000 in ore.

This multiple mineralization suggests possibilities for the modern prospector. Where nature has been so profuse with varied deposition, one may wonder whether some of the "space age metals"–beryllium, titanium, and quite a list of others–may not occur in paying quantities. Earlier generations of miners had little knowledge of and no use for these minerals, if indeed they could even have been identified. The huge deposit of molybdenum in Colorado that became the immensely valuable Climax Mine was passed by many times by gold and silver miners.

At Ely one is 556 miles from Los Angeles and 247 miles from Salt Lake City. It is at Ely that U.S. 93, an important north-south artery, forms a junction with U.S. 50. The latter soon passes into Utah and we will follow it no farther. Our course will be south from Ely on U.S. 93, for still to be considered among Nevada's possibilities is a large and desolate area, a virtual blank spot on the map, which lies between the Utah boundary and the vast atomic test site north of Las Vegas. It is one of the few primitive areas of the West that has been left unspoiled for the twentieth-century prospector. It also is the favored location for Nevada's legendary mountain of silver.

Is the silver mountain a legend or a fact?

The old-time prospecting fraternity was admittedly overoptimistic. Reports of new strikes, like many other types of news, became exaggerated as they traveled from mouth to mouth. It also was traditional for the little weekly newspapers that sprang up at most camps to present the local discoveries in somewhat extravagant terms. Seldom, however, did anything classifying as pure myth gain any currency among men who led, of necessity, a hard-working, practical existence. Such phantoms as the Seven Cities of Cibola and the Gran Quivira were notably lacking among the California forty-niners and their successors throughout the West.

The mountain of silver, if indeed it proves to be mythical, would be an exception. For there was a widely held belief that near the Colorado River, and vaguely south of the White Pine-Ely territory, there was a peak where ledges of silver ore were so massive and so rich that "mountain of silver" seemed a reasonable way to describe it.

After leaving Ely and driving 107 miles on U.S. 93, one reaches Pioche, a durable mining and trading town and the county seat of Lincoln County. On the way it is possible to make a short detour to the west and a cluster of old mine workings which include the highly profitable Bristol-Jackrabbit. Two mountain ranges, the Bristol and the Highland, raise their escarpments almost parallel to U.S. 93. On both sides of these mountains, but mainly on the western, or farther side, there have been mines dating back almost to Civil War times. Still farther west, across the wide Dry Lake Valley, are the Pahrock Mountains, promising from a gold-hunting standpoint but so dry and remote that any trip into their deep recesses should be carefully planned.

South of Pioche, Cathedral Gorge and Ryan State Park testify to the picturesque character of parts of this region. It is in the area south and east of this segment of U.S. 93, stretching an average distance of fifty miles to the Utah state line, that the massive silver deposit is reputed to lie.

The territory includes three principal mountain ranges, the Hiko, the Meadow Valley, and the Mormon. The Automobile Club of Southern California has done some detailed mapping in these empty spaces, and its Desert Area map traces the various dirt roads covered by its scout cars. The roads seem to lead nowhere in particular. One serves an airline beacon, another the single school in the entire area. In Nevada a public school will be maintained for as few as three pupils. Neither the auto club scouts nor the thoroughgoing Stanley Paher, whose ghost town research has been mentioned, have discovered old mining camps in this wasteland. It is and always has been difficult of access, and the problem of finding water, while not as severe as in the Arizona desert, is nevertheless a real one. There can be some true pioneering in this region.

It is possible that the mountain of silver report originated with the early Mormon colonizers. As the Latter-day Saints spread out from the Great Salt Lake, some parties were assigned to establish farms far

Sutter's sawmill on the American River, where flakes of gold in the tailrace set off the California gold rush.

In southwest Oregon miners found fabulously rich gravel, but would recover little or no gold if they panned as depicted here by an eastern artist.

Placer miners in the Mother Lode country. Ingenious "flutter wheel" poured water into sluice boxes for washing out gold.

Dry washing of gold-bearing sands by tossing them in the air, necessarily employed in the desert. The breeze winnowed out the lighter particles.

Elaborate flume systems, bringing water to placer workings, were developed by the California forty-niners and employed later in Colorado.

Artist's conception of how a bench deposit is worked. Geologic changes have left the auriferous area well above an existing stream.

to the southwest. In Nevada, what is now known as Moapa Valley was the focus of colonization. The largest town, St. Thomas, was abandoned by its six hundred inhabitants under unusual circumstances. In the middle 1860s the boundaries of the western states were quite uncertain and the Mormon settlers believed they were in their own state of Utah. When St. Thomas was adjudicated as lying in Nevada, Lincoln County authorities not only put all Mormon property on the current tax rolls, but made retroactive tax claims for previous years. It is said that a notice of tax sale was nailed to every house. The people of St. Thomas moved en masse back to Utah.

The Mormon colonizers were concerned with raising crops and cattle for survival. In their isolation there was no market for precious metals even if they had mined them, and at times the church was indifferent or actually hostile to expending manpower in this direction. Some of the Saints were quite capable prospectors, however, as discovery of the remote Potosi bonanza attests. Lacking a record of any other early penetration of the great unknown territory beyond the colonies, the discovery of something—of something impressive enough to be termed a "mountain of silver"—may reasonably be ascribed to some wandering Mormon. And the withdrawal of the Mormons from the region for many years, combined with church policy, may explain why the discovery never was followed up.

The searcher for this great deposit or other treasures has the option of only a single through road. This is one which leaves U.S. 93 near Caliente and rambles on southward for well over one hundred miles. It is a dirt track along winding Meadow Valley Wash, with the Meadow Valley Mountains rising on the west along the entire route. Old roads of unknown purpose branch off at various places.

One possible but not very convincing explanation of the silver mountain report may be mentioned. In the northwest part of the territory we are considering is the abandoned camp of Delamar. About $13 million in silver was recovered here during the last century from the great Delamar Mine and the nearby April Fool. This was at a great cost in human suffering. Both the type of ore mined and the process employed at the huge stamp mill were productive of silica dust in excessive amounts. Many miners sickened and died and Delamar was so widely condemned as a "maker of widows" that a wet milling process was instituted. Much of the rich ore came from a glory hole high on the mountain, which is still quite conspicuous. The silver ore at this spot occurred over such a large area that the original finder, sinking his pick here and there and always finding values, might well have used the expression "mountain of silver." The successful exploiter of the property, Captain John De Lamar of Montana, so far as can be learned, never associated his rich mine with the legendary mountain.

The long road along Meadow Valley Wash rejoins U.S. 93 near the Moapa River Indian Reservation and the town of Glendale. Not far south of here one enters the extensive Lake Mead Recreational Area. We are now at the end of our exploration of Nevada–except for one fairly large and very vacant district of low desert mountains, bearing the name of Virgin range.

If you plan ahead and get on the *east* side of Lake Mead and the Colorado River you will be in the Virgin Mountains. Here there was once an active little mining district in which the Key West Mine was the principal property. Twenty-two miles south of the Key West is Gold Butte, rising to 5,840 feet elevation. Gold Butte also is where all roads end. In this out-of-the-way corner of the southwest, favorable geology and good prospecting luck could combine to bring–who knows what?

If we bypass this side trip to the east of the lake and continue on U.S. 93, the route takes us to Las Vegas and the busier side of the lake, thence across Boulder Dam and into Mojave County, Arizona. Here we leave the so-called Silver State for one where mining has been chiefly for gold and copper.

7
Western Arizona

In turning from the semidesert state of Nevada to the mining possibilities of western Arizona, we pass into true desert country. Glimpses of such a landscape have been afforded by innumerable western motion pictures, in which the sand dunes, the giant cacti, and the craggy, barren mountains have formed the backgrounds. Some of the photography also has managed to convey the atmosphere of loneliness, aridity, and heat that pervades the scene.

From the prospecting viewpoint, the inhospitable structure of the desert is an asset. It is especially so for the beginner, the inexperienced amateur, for this is a terrain where most mineralization of any promise is clearly exposed on the surface of the earth. Sheer good luck in stumbling upon a visible deposit of this kind has been the origin of some great mines and some substantial fortunes. This characteristic of the desert country will not make the newcomer quite the equal of the grizzled old-timer or the trained geologist but it assuredly gives him better odds.

The territory to be considered under the western Arizona heading is enormous. You can choose a site for prospecting almost anywhere in an area of some 20,000 square miles, for nearly all this land is public domain and open to exploration. Mining history and geologic reports agree that scattered mineralization occurs throughout. Some of the showings must inevitably have been overlooked in earlier days. With the greater mobility and improved communications available to the modern prospector, a search for such undiscovered bonanzas would appear a good gamble.

For convenience we are defining western Arizona as virtually all of that state west of Phoenix, the principal population center. The remainder of the state has a somewhat different mining history and will be considered separately. The Colorado River forms our western boundary and the territory under consideration extends eastward for about one hundred miles. The extensive federal recreation areas associated with the Grand Canyon and Lake Mead may be taken as the northern boundary, and at the southern, we come to the Mexican line and the lonely, grave-strewn desert road known and feared as the Camino del Diablo.

The advantages of this country for the gold seeker are apparent when contrasted with conditions in states with a more generous rainfall. In the latter there is not only a heavier burden of soil lying over the bedrock, but most of this will be cultivated as farm or range land or covered with brush or forest. Except where occasional outcrops of rock push through and are visible, any underlying mineralization remains hidden. We have seen how discoveries in parts of California's Mother Lode country have been impeded by dense growths of manzanita, all but impenetrable unless one crawls on his hands and knees.

The Arizona desert on the contrary might be described by William Cullen Bryant's phrase, "rock-ribbed and ancient as the sun." There is some soil to be sure, supporting scattered growth of the towering sahuaro, the cholla, and other cacti, sparse bushes such as greasewood and cat's-claw, and a few hardy trees, including the mesquite, the ironwood, and the wispy smoke tree, which from a distance can be taken as sign of a friendly campfire. A very large proportion of the surface, however, consists of the basic, rocky structure of the earth. If some minute part of that structure reveals metals—or more precisely the streaked or mottled stains that indicate metals—these signs, in the Arizona desert, will commonly lie open to the eye of the prospector. It is a genuine advantage.

There are even rare spots where native metal itself may be seen in place—a silvery streak perhaps a foot long that would be lead, or the light yellow shine of iron sulphides, often crystallized into cubes. Such occurrences are quite exciting on first acquaintance, but in themselves are not to be taken too seriously. They testify that the spot has undeniable mineralization. Only additional exploration, digging, and sampling will tell whether it holds any real economic promise.

The gold seeker, whether he comes from the East or West, will likely arrive in western Arizona on Interstate 40, if headed for the northern part, or on Interstate 10 if he is more interested in the south. It also is possible to come via Las Vegas, crossing over Boulder Dam on U.S. 93 and driving about one hundred miles to Kingman, Arizona, an old-time mining center.

Once having arrived in this western desert, one need venture only a mile or two from pavement to find himself in a quite primitive land. There will be few signs of human occupancy, past or present. What you do see may date back many years, for this landscape was more populated a century ago than it is today. Old maps show such place names as Clip, Silent, Red Cloud, White Hills, Cyclopic, Secret Pass, Alamo Crossing, Golconda, and a dozen more. If their sites can be located at all in the encircling desert, they will be marked only by crumbling adobe walls and by tunnels, shafts, dumps of waste rock, and ponderous rusting machinery. These are the remnants of the mines that called such settlements into being and briefly sustained them.

These all but empty areas, at least those far removed from any town or highway, can be dangerous country for the careless or overconfident newcomer. In most sections of the United States, with their humid air, moist earth, and adequate rainfall, it is difficult to comprehend the perils of extreme thirst and dehydration that must be taken account of in desert prospecting. For one thing, any sustained exploration on foot is quite impracticable between the months of May and October. Official in-the-shade temperatures during the summer reach 100 degrees and higher—considerably higher—day after day. To tramp for hours under direct sunshine of such intensity, in a land where there is virtually no shade, is imprudent to put it mildly.

This may be illustrated by the case of a Midwesterner and the story he told after a six-mile hike under the July sun. Desert people alone are likely to find his account credible.

The man had come from Kansas City, Missouri, and though he had invested in a mine he was new to Arizona. Before leaving town this man had wisely placed a thermos jug of cool water in his automobile and had filled another jug with orange juice and cracked ice. After driving toward the old mine in which he was interested, he left his car where a desert road played out and walked the remaining three miles to the mine. When he got there, he somehow did not feel equal to making much of an inspection. Starting back, he already was vaguely suffering the effects of heat, thirst, and dehydration, and as he went on the symptoms became more recognizable and severe. He craved his water and cold orange juice as he had never craved such things before. In something of a panic, he began to speculate on how much they were worth to him. By the time his car was in sight he decided he would not sell those two jugs, right then and under these circumstances, for $50,000.

Some alarming tales of poisonous reptiles have come out of the southwestern desert but local people are not much concerned about danger from the fauna. The typical rattlesnake of the region is the sidewinder, which moves with a rapid, twisting motion but does not cover the ground very fast. The snake strikes low, and boots reaching up to the calf of the leg are good protection.

Of the two poisonous lizards known, the orange and black gila monster is the one native to western Arizona. If you see one, the warning is: hands off. Several years ago a scientist, Dr. Ernest R. Tinkham, was making a laboratory study of the gila monster and at length he was bitten. Although in great pain he was careful to remember and record his experiences. First, he had great difficulty in breaking the lizard's viselike grip on his hand. Then, in spite of emergency treatment by the freezing technique and prompt conveyance to a hospital, he had to undergo extensive treatment. There were ups and downs as his system struggled against the virulent toxin, and it was six days before physicians pronounced him out of danger. His hand and arm remained swollen and painful for a month.

Concerning the tarantula, the giant spider of forbidding aspect and various folk tales, there has been some controversy as to whether its bite is poisonous or as to whether it bites at all. Tarantulas are fairly common and the desert wanderer will see many of them. There appears to be no authenticated case of their harming anyone.

Two other denizens of this arid country may be mentioned. Prospectors in camp will ordinarily shake out their shoes and each piece of clothing in the morning to see whether a centipede or scorpion has crept into them. Either of these, the scorpion in particular, can inflict a painful little wound.

Except for the excessively hot summer, the climate of western Arizona is delightful, with agreeable temperatures day and night. To some, the unfailingly clear blue sky of this country may become monotonous. Yet at long intervals one will see a few clouds and at even longer intervals the parched land will receive a brief rainfall.

Any examination of the gold prospects of this region at once brings to mind the luckiest strike of them all, made in a geologically mysterious segment of the Weaver Mountains. The site became famous as Rich Hill and the gold deposit atop the hill was the fabulous Potato

Patch. Is there still a fortune to be made in that locality? Let us look first at the original story.

The year of the discovery was 1863 and the leader of the fortunate party of gold hunters was one Abraham H. Peeples. They were an adventurous group, determined to push beyond previously explored territory east of the Colorado River, apprehensive concerning water supply but sanguine as to finding gold. As guide they had hired the redoubtable Captain Pauline Weaver, a male despite his name, who was to become one of the legendary characters of the old Southwest.

We have seen earlier that a standard prospecting procedure is to follow along stream beds and watch for "float," pieces of rock showing promising mineralization. These are assumed to have weathered off from ore in place somewhere farther on and one tries to trace these indicators to their source.

The Peeples-Weaver party had followed this routine for some days and now were camped at the base of the unnamed and unexplored little peak that was to become famous. Next morning, either following the promise of some rich float or, as another version has it, rounding up their horses, four of the men climbed the peak. The summit area was flattened out in a singular, slightly concave expanse of considerable area and here and there on its surface was the unmistakable gleam of gold.

The big strike had been made. Nuggets lay on the surface and many more were very close to the surface. The auriferous layer of soil was only six to twenty-four inches deep but at all levels it yielded rich returns in coarse gold or in dust. Then came bedrock, where much gold was extracted from crevices with knives and spoons. The Potato Patch term was applied because digging up the soil for nuggets reminded some of the party of the back-home chore of digging up potatoes.

Some spots were fantastically rich. The leader, Peeples, told of taking out $7,000 one morning before breakfast. When the news spread and a representative of a mining publication came to report on the strike, he was invited to start digging and draw his own conclusions. In less than thirty minutes he had sifted or panned out gold to the value of $70. One of the nuggets weighed half an ounce, which would make it about the size of the old $10 gold piece.

The discoverers had the bonanza to themselves for a short while and every member of the party recovered what in those days must have been a tidy fortune. Then, as the news filtered out, the rush was on. The Potato Patch was worked and reworked and its known or discoverable treasure quickly exhausted. Latecomers spread out to the surrounding slopes and washes. Stream beds and benches were worked by dry-wash placering, some yielding good returns. The towns of Weaver, Octave, and Stanton sprang up. If one can credit the uncertain statistics of the day, the population of the district rose to 10,000. It was an untamed crowd, and the town of Stanton in particular contributed some of the bloodier episodes in the history of western gunslinging.

What if anything does the area offer today's gold hunter?

In the 1950s an experienced miner and soldier of fortune named William Esenwein took up residence at Rich Hill and spent considerable time exploring. He wrote a report with the somewhat enigmatic finding that the place "represents a compact bundle of possibilities for the aspiring prospector of today."

Specifically, Esenwein discovered a shelf high on the mountainside, about twenty by thirty feet in area, where gold-bearing gravel had accumulated to an estimated depth of ten feet. The place quite evidently had been overlooked by the early-day miners, being well concealed and difficult of access. Such gravel as he could pack down the mountain to where he had water for panning showed about $1.00 worth of color per pan. Gravel as rich as this constitutes a real discovery. Unfortunately the deposit could be reached only by an exhausting climb, and construction of even a burro trail appeared impossible. Though confident that he had found something of considerable value, Esenwein expressed little hope of capitalizing on his strike.

There may be more encouragement in a story from that pioneer lost-mine hunter, John D. Mitchell. The Yapavai Indians, he recounted, knew of the gold concentration on Rich Hill and for some inscrutable reason termed it the Little Antelope. Farther back in the range, said an old Indian, there was a Big Antelope.

The Weaver Mountains, which shelter the little peak and its three ghost camps, rise from the desert floor between the two small cities of Prescott and Wickenburg. These two points are on U.S. 89. About sixteen miles north of Wickenburg lies Congress, which is the takeoff point for the Rich Hill area. Ask how to get to the sites of the three old towns named—Weaver, Octave, and Stanton. They form a triangle at the base of the mountain. The round trip over unimproved desert roads will be something under twenty miles.

Having taken note of this early-day bonanza, let us turn to some more demonstrable present-day possibilities, documented to a considerable extent by government reports.

Exploration in the northern part of the territory covered in this chapter will center in the town of Kingman, which is reached by Interstate 40 from the east or west, and from the northwest by U.S. 93. Stretching out in all directions from Kingman is an expanse of all but empty country covering some 4,800 square miles. There is scattered but proved mineralization over the whole area and it has had some immensely profitable mines.

The old camps to look for in this region include Chloride, Cerbat, Mineral Park, White Hills, Cyclopic, Nestona, Signal City, Gold Road, Todd Basin, and Oatman. This list takes in the principal camps as far south as the Bill Williams River, a lonely desert stream flowing westward into the Colorado.

Few of these vanished mining centers appear on modern road maps. Some maps issued by the big oil companies have a special emblem for ghost towns and depict a considerable number. These selections on road maps are by no means complete. Such guides should be supplemented by the quadrangle maps of the U. S. Geological Survey, which are quite detailed. The contour lines on the government maps, normally showing each forty feet of elevation, are an additional help and many individual mines are shown by the standard symbol of crossed miners' picks. As in all travel in out-of-the-way places, local information, obtained by asking questions, is often more valuable than maps.

The signs of early mining activity that one encounters will range from shallow prospect holes with a small dump of excavated rock to the unmistakable mark of a really productive mine, the giant pile of tailings. Tailings are the residue of ore after it has been run through the mill and its gold or other metal content recovered. Their bulk is often augmented by barren rock that has necessarily been removed along with the ore. The size of the tailings dump is a rough index of the success of the operation. It is normally flat-topped with steeply sloping sides and shows up lighter in color than the hillside or other land surface behind it.

Where there is no foundation of a mill remaining or any other sign that a treatment plant ever was operated, the place may be worth investigation. This is often an indication that the ore in this mine, after being sorted out from the so-called gangue and country rock now visible on the dump, was valuable enough to be hauled away to a mill or smelter. Some of the vein matter in western Arizona mines was so rich that it could be packed on burros to the Colorado River, shipped downstream for transfer to an ocean-going vessel in the Gulf of California, and taken all the way to Swansea, Wales, for reduction. At such mines there could be profitable ore of milling grade remaining.

When an experienced prospector feels any interest in abandoned workings he will ordinarily prowl around the dump. Whether the operators shipped their ore or milled it on the site, some pieces of vein matter will have escaped. If these can be found and identified, they will show what lay beyond the yawning black mouth of the tunnel or down the dark mine shaft. Should the remnants carry good values, the question is: How much of such ore remains in place and how good is this remaining deposit? It is always possible that only the richer portion of the vein could be worked under early-day conditions and that with today's gold price and mechanized transportation the remaining deposit may have some real economic value.

Be wary about entering an old mine, and never do so unless you have a companion to await you outside. Foul air can render a man unconscious, then suffocate him. Rocks can jar loose from a tunnel's roof. Worst of all, one can step into winze, which is a shaft exactly equivalent to an abandoned well.

If one gets serious about taking over one of these old mines, a thorough examination in a skeptical rather than a buoyant frame of mind is recommended. With or without professional help, it must be established that there remains a substantial body of unmined ore, and of sufficient value, to promise that one will at least break even in the new operation. Samples for assay should be carefully taken to show the true average value of the entire vein. Even miners who know better are sometimes tempted to send in superior samples from a mine for the sake of a glowing assay report. Such sampling is of course valueless for any serious appraisal.

As the county seat of Mojave County, Kingman will have the records of mining locations filed through the long and hopeful years of desert prospecting. The mines that are patented, that is owned in fee simple like any other real estate, also will be on record. Unless things change very rapidly with the improving status of gold, there will be a list of patented claims that are tax delinquent, in short, mines that have been abandoned not only as to operation but as to ownership. These situations are well worth investigation. Some fine properties that later poured out fortunes under good management and adequate financing have been acquired at tax sales.

Among the big gold or silver producers that have flourished in the Kingman region are the Tennessee at Chloride, the Cerbat and Metallic Accident at Cerbat, the Gold Road at Gold Road, the McCracken near Greenwood City, the Indian Secret at White Hills, and the Tom Reed and United Eastern at Oatman. The last two are comparatively recent producers, having their peak operation in the two decades following World War I.

These better-known mines are mentioned as a guide to localities where heavy mineralization is a historical fact and hence where prospecting in the general area would seem reasonable. Such big mines will commonly be owned by corporations which have held on to them with an eye to the future. Even though inactive, some essential maintenance work may be carried on and a watchman kept on duty.

Before leaving the Kingman district and looking into the prospects south of the Bill Williams River, notice should be taken of the lost nugget field of the Coconino Plateau. Whether the deposit is as extensive and as opulent as the Potato Patch on Rich Hill remains to be seen, but it appears to be of the same type. By some unexplained geological freak, nuggets have been left exposed on the earth's surface and presumably are as ready to be gathered in today as they were when an ill-starred prospector came across them seventy years ago.

The site is roughly 120 miles northeast of Kingman. The Coconino Plateau proper occupies the northern part of an arid, unpopulated region of some 3,600 square miles bounded on the north by the Grand Canyon and on the south by Interstate 40. The closest town of any size is Flagstaff, Arizona, and the nugget field almost certainly lies west of U.S. 180, which is the highway linking Flagstaff with the principal resort center at the Grand Canyon. For the western limit of the search area we may take a sixty-seven-mile stretch of partly improved road running north from 40 and terminating at a canyon rim overlooking the hidden village of the Havasupai Indians. Old maps show some miscellaneous dirt roads in this desolate area with no discernible objective. On most modern highway maps, the region is a blank spot.

The bonanza is associated with an old mining man, well regarded in Flagstaff, known as Long Tom Watson. While clearing out an old cabin he had rented, Watson came across a letter. The writer's name has not come down to us and the letter had not been addressed or sent through the mail. It was clearly the attempt by a dying man to inform his brother of how he had discovered a marvelous gold deposit, of how he had filled a flour sack with nuggets and hidden it, of his fear of two strangers who had been following him, and of how he finally had had to shoot it out with the two men. He had been seriously wounded in the encounter, and it now appeared, at the time of the writing, that he was going to die. He had set down directions and on a separate sheet had attempted a crude map.

Watson was able to identify the cabin's former tenant and to learn that the man had indeed died of complications from a gunshot wound. So far, so good. The letter deserved careful study.

A disturbing statement in the narrative was that the flour sack and its gold had been hidden in a cave "behind a waterfall." In this dry country a waterfall was a great rarity. Watson reasoned, correctly as it turned out, that during the very brief spring rains the runoff had poured over some rocky ledge and the scene might well answer the waterfall description.

Handicapped by this seasonal factor as well as by the extreme difficulty of the country itself, Watson was engaged in his dogged search through first one spring season and then another. But he found the waterfall and he found the gold. They were in a distant and dangerous area even for an experienced desert man. Scooping up as much gold as he could carry in his pockets and perhaps tying some up in his bandanna, Watson now had to struggle back to civilization. He had suffered a painful injury to his knee and was otherwise near exhaustion. Luckily there was a single ranch far out in the desert solitude and Watson managed to make his way to rest and assistance at this haven. The ranch has been identified as that of one Martin Buggelin. It was a remote operation in very dry country, and the weary, injured man was fortunate to find help there.

Watson regained strength and was taken to Flagstaff. As late as the 1950s there were people in that town who had some personal knowledge of his exploit. They recalled that he was in the hospital for four months with his leg injury and that he paid for his treatment with gold nuggets. He sold other nuggets to equip himself for a second trip to the plateau. There was additional gold in the cave and he also felt, from the evidence of the letter, that the big prize, the field of nuggets, must be close-by.

But Watson's luck had run out. In two more heartbreaking forays, he was unable even to find the rocky ledge which, in that fateful spring, had been transformed into a waterfall. He was an old man, crippled, puzzled, and discouraged. Like the shadowy discoverer of the nugget field, Long Tom Watson was killed by a rifle shot, this one fired by his own hand.

The mining districts south of the Bill Williams River are reached by Interstate 10, which is the southernmost of the great travel arteries that extend from coast to coast. There also is an Interstate 8 serving the extreme southwest corner of Arizona. It branches off 10 at Casa Grande and extends to Yuma on the Colorado River and thence on to its terminus at San Diego.

Our present interest is in the territory along Interstate 10. This highway crosses the Colorado River near the sizable town of Blythe on the California side. On the Arizona bank is a tourist way station called Ehrenberg. The historic river port of Ehrenberg, which lay slightly north, is now obliterated except for the pioneer cemetery.

In this area any discussion of gold inevitably involves the opulent La Paz placers and their leaner but nevertheless auriferous surroundings. The activity centered on a flat stretch of alluvial ground on the east bank of the Colorado a few miles above Ehrenberg.

The boom at La Paz coincides roughly with the years of the Civil War. It was in 1861 that Captain Pauline Weaver, the same frontiersman identified with the Potato Patch strike, appeared in Yuma, exhibited gold nuggets, and brought first reports from the new diggings. A rush was soon on. The area of heavy enrichment was extensive and scores of men were able to locate claims from which takings of $100 a day were commonplace. Many yields were much higher, and on some days they were swollen by phenomenal nuggets in the range of ten to twenty ounces. Such a find by Juan Ferra weighed in at forty-seven ounces of solid gold.

The exploitation was hurried and intensive. By the third year it was apparent that the really profitable mining was over. La Paz lingered on as a river port until 1870, when the Colorado River changed its course and left the town two miles inland. In 1875 a visitor found it entirely abandoned, although the adobe buildings of a rather impressive main street remained standing. These also vanished when in 1912 the capricious river returned to the place and quickly dissolved the walls and foundations of sun-dried brick. There is virtually nothing left at the site today.

Has this old camp anything to offer the present-day miner? Or more precisely, was La Paz proper the only workable deposit in that area? Some encouragement may be drawn from the experiences of those who lingered on after the boom days, into the middle or late 1860s. It was found that all the little ravines to the east of La Paz contained gold and a few of them paid very well indeed. As the ground convenient to the river was worked out, the later miners pushed farther up these ravines and then on to more distant ones.

Here the task of recovering the gold became increasingly arduous. Either they had to use the laborious and imperfect method of dry washing or they must pack the gravel on burros or their own backs all the way to the river for the more effective system of panning or sluicing with water. It appears that most chose the latter procedure. What would now be called the law of diminishing returns was encountered. The more distant workings gradually were abandoned and little by little this hard-working crew left for other fields. One surmises not that the miners had reached the outer limits of gold values but rather had reached the limits of human endurance.

In these days of four-wheel-drive and caterpillar-type equipment, of self-contained motor homes and the like, this hinterland would seem to offer possibilities. Equipment of such types could carry enough water for panning out values for days on end. Operating over a tub or pail, water can be conserved and used over and over. Even a rocker, used in early California placering, would seem feasible, and it handles much more gravel than the miner's pan. Some gold almost certainly would be recovered. Another possibility is pocket hunting—that is, examining dry stream beds for rocky formations that form a natural catchment for alluvial gold.

There is solid evidence that gold occurs east of the old La Paz field. The findings of two extensive drilling operations have shown that the metal is disseminated in appreciable quantities over a very wide area.

One of these tests was made in 1910 by an O. L. Grimshaw and covered Goodman Wash, one of the more spacious gulches leading into La Paz's central producing area. Grimshaw's project was to pump water from the Colorado River and work Goodman Wash with heavy equipment. The tests encouraged him to go ahead, but the operation was barely organized when he met death in an automobile accident. Another, more extensive survey probed the very large gravel beds running north from the highway midway between Ehrenberg and Quartzsite. This was made in the late 1920s and resulted in an engineer's report that every one of the numerous drill holes showed gold values. Whatever hopeful recovery plan may have been formulated after these findings, it came to nothing. Like many another enterprise, it found itself paralyzed by the great depression of the 1930s.

The original La Paz diggings are within what is now the Colorado River Indian Reservation. Any mining must be done under contract with the tribal council. Besides any residue of rich placers, the reservation contains at least two lode mines which government geologists in the past considered quite promising, the Mammoth and the Dan Welch. Outside the reservation, north, south, and east, the territory is virtually all in the public domain.

East of the reservation there is a great quartz dyke which may well deserve further exploration. It originally was known as the Conquest Lode and in the late 1800s an extremely rich section was worked, and worked out, by Manuel Ravenna of San Bernardino, California. Now referred to locally as the Goodman Vein, it can be traced for a great distance over the desert surface, cropping out prominently in canyons and on hillsides. Except at the original workings it consists almost entirely of white quartz—"bull quartz" is the miners' term—and shows only meager mineralization. It has been by no means thoroughly explored as to feeder veins and possible local enrichment and there are responsible mining men who believe that hidden somewhere along the giant formation there exist the makings of another great mine.

Still farther east there are several proved gold districts to be considered, but for the moment we skip to the vicinity of Wickenburg, the rich Vulture Mine, and the puzzle of its lost vein.

Henry Wickenburg was an Austrian who arrived in Arizona in the early 1860s. We first hear of him in 1863, when he appears as a disappointed latecomer to the La Paz placers. With a single companion, he pushed eastward, prospecting as he went. Wickenburg one day sat down to rest on a protruding quartz outcrop and discovered his mine simply by looking down at his side. His rocky seat was studded with gold.

Having no resources to equip a major mining operation, he adopted the unusual expedient of selling ore as it still lay in the ground. Any miner could arrange to drill and blast away on the rich vein as long as he paid the owner $15 for each ton of ore removed. It was an agreeable system for Wickenburg, and the miners themselves prospered. The ore was ground in arrastras at the nearby Hassayampa River and yielded rich cleanups.

This went on for three years. Then Wickenburg was induced to sell a four fifths share of the Vulture to New York interests for $85,000. He received $20,000 in cash and that was the last money ever paid him. But the Austrian's tastes were simple and his ambition apparently limited. Ignored by the new owners, he built a cabin on the banks of the Hassayampa River and cultivated a garden in which he took great pride. He survived the tragic Hassayampa flood of 1890 and lived to an advanced age in the Arizona Pioneers Home at Prescott.

For its new owners, the Vulture promised to pour out millions. Buildings were erected at the mine, and a mill and other facilities established at the town that sprang up on the Hassayampa and grew into the Wickenburg of today. Ore values held up and for a time the mine was immensely profitable. Then the gold-studded vein suddenly came to an end against a vast intrusion of rock that was utterly barren.

A great deal of money and technical knowledge have been expended in searches for what might be called the missing half of the Vulture Mine. The dazzling vein, shifted by some cataclysmic forces of the past, has continued to elude the best expertise that can be employed. Perhaps it doesn't matter. The ways of the gold-bearing magmas that enrich a quartz lode are mysterious and even if the missing section is someday discovered there is no assurance that it will contain any gold whatever.

We have now examined some rich strikes and their associated present-day possibilities on the west and east sides of what, for convenient identification, has been labeled western Arizona. The region in between deserves considerable attention.

It may be said of this middle area that there are few places in the western mining states where the prospector has a better chance of coming upon a pocket, perhaps a very considerable pocket, of high-grade gold ore, but expert opinion is that the chances of finding a strong fissure vein that will make a great mine are somewhat slender. However, one can forget the prospect of great mines for a moment when some resident of the sleepy towns along U.S. 60 shows you a piece of white or rosy quartz heavily encrusted with virgin gold—"picture rock" as it is called—and you realize that half a gunnysack full of such ore could be cashed in for at least $10,000.

The historic mining and freighting road through this region ran from the Colorado River eastward through the towns of Quartzsite, Hope, Brenda, Vicksburg, Salome, Wenden, and Aguila and on to Wickenburg. There it turned southeast toward Phoenix and the main arteries to eastern and northern points. This route has become part of the federal highway system.

One wonders, for example, what has become of the high-grade

streak at the top of Mount Harquahala that was mined by a genial hermit who for thirty-two years lived near the very summit of that peak.

William B. Ellison was no embittered recluse. Except to note that his gold mine was supporting him, this otherwise sociable man from Illinois volunteered no reason for establishing and remaining for half his life in his lonely aerie. To bring his gold down to the town of Aguila meant a winding descent of several thousand feet on a narrow foot trail. On the climb back he was burdened with groceries and other supplies. Yet the townspeople say he made the trip regularly twice a month, and on occasion carried cement, iron pipe, tools, parts of machinery, and even lumber back up the trail.

In spite of its lofty location, Ellison's homesite had a good spring. He raised his own vegetables and kept a dog, a burro, and at one time a horse. His mine was two miles from his cabin along a rugged crest of the long, straggling mountain. The vein matter must have been fairly rich. One-man mining in hard rock is slow business, and the output necessarily was packed two miles back to his spring, where it was milled in a crude arrastra. Ellison once confided to a visitor that he made his strike while hunting deer and that at the beginning it ran 100 ounces of gold to a ton. That would be $2,000 a ton at the gold price of Ellison's time. And how much is a ton? With dense, mineralized rock such as gold-bearing quartz, a solid two-foot cube would come to approximately that weight.

In his late seventies, having broken a wrist and overstrained himself climbing the trail, Ellison decided it was time to return to civilization. With the accumulated takings from his mine, he spent his remaining days in a comfortable cottage at Aguila.

More important discoveries than Ellison's were made on or near the great mountain and called into being the town of Harquahala. The mineralization for miles around this ghost camp might be described as tantalizing. Some rich deposits have been found and lost, and other exciting surface showings have failed to make good on their early promise. Even mines of demonstrable value have suffered from litigation, absentee ownership, and erratic management and their true worth remains undetermined.

To reach Harquahala one leaves U.S. 60 at Salome, "where she danced." If only for the name, Salome calls for a brief description. In what he himself would call the "tin lizzie" days, a jocular pioneer named Dick Wick Hall established a gasoline station at the spot, sold "laughing gas," and adorned his station, the garage, and his Blue

Rock Inne with humorous slogans and depictions of Salome herself in action. These were days of unsurfaced, washboard roads and of considerable tire and mechanical trouble for the cross-country motorist. Hall's oasis about midway between Phoenix and the Colorado River must have been a welcome sight to the traveler, and the free-swinging buffoonery of the place seems to have been popular. As it turned out, Dick Wick Hall had talent of more than local appeal. His mimeographed *Salome Sun,* featuring tall tales of the desert country, gained a national following and for a time he contributed a full page of humor to a leading national magazine.

The ghost camp of Harquahala is about nine miles from Salome on a dirt road trending south. Another route, more easterly, passes the ghost of old Harrisburg, where there always has been a good water hole and where a party of early emigrants, never identified, were massacred by Indians.

A conspicuous feature of the Harquahala region is a long valley, sometimes narrow, sometimes quite wide, known as Centennial Wash. If one looks up any Harquahala properties he will find some on record not under "Harquahala" but under "Centennial District." The mines are chiefly up a little side canyon ending at Martin Peak. Stairstepping up the side of the barren mountain, the workings are quite conspicuous.

The most productive deposit of the lot, christened the Harquahala Bonanza, was also the first discovered. Three plodding prospectors trudging on foot, with supplies packed on burros, chanced on it in 1888. The ore at grass roots was very rich. As news of the strike got out there was the inevitable stampede and likely looking ground was staked out for a radius of more than ten miles. The three discoverers realized $70,000 in about three months from easily worked surface ore and then sold their claim. After several further sales and some litigation, the Bonanza and the Golden Eagle, another profitable mine, were taken over by an English syndicate. There was some highly successful mining under this management, but excavation by tunnel, crosscut, and stope was so extensive in the Bonanza that the ground clear up to the surface crashed down into the workings. The immense hole thus created can be inspected today and offers something of a subterranean view of how veins and stringers lace through the mountain.

Other successful mines nearby include the Bonanza Extension, Grand View, Summit Lode, and Narrow Gauge. The mineralization is so impressive at Harquahala that intermittent activity has continued through the years. Optimistic leasers have been the principal workers. Those who own any stake in the better properties have hung on to them doggedly, for they feel there still are fortunes to be made.

A little to the east of here is the Indian mine hunted by the late Captain Alfred Solano. This gentleman, whose family gave its name to Solano County, California, had a wide acquaintance among the Indians and Spanish-speaking people of the region. In 1932 one of his Indian friends told him of a rich outcrop of gold ore in a shallow canyon scarcely five miles from the small town of Wenden. It was described as being just off a desert trail leading south from there. Verbal directions, even among experienced desert men, can be misconstrued, and in successive explorations of the supposed canyon Captain Solano failed to discover anything of value.

Directions that were more effective were devised by the father of Palmer C. Ashley of Los Angeles when the younger man, jobless in the depression year of 1934, offered to help at the family gold mine, which was in this general region. The elder Ashley tied tin cans to brush and cacti along the sketchy trail to his diggings and these brought the son to the remote destination in Morgan City Wash, and to his own lost bonanza experience.

Morgan City Wash is a dry watercourse some twenty-three miles southeast of Wickenburg, the highway turnoff point being Wittmann. In the absence of any "Morgan City," one suspects a corruption of some early Spanish name, the suffix *cita* being a diminutive that could be confused with "city." The Ashley mine was a borderline operation, with the gold content at $20 an ounce just about paying the charges for shipping and milling. At today's higher value of gold, it might provide a modest living. But there was nothing borderline about young Ashley's discovery.

With his father in Phoenix for supplies, the son took a day off to check out a tip from an old prospector. The man had sketched out directions to a promising showing of galena, a typical ore of lead which may also carry considerable silver. Ashley duly found the spot, hacked off a few samples, started home, and somewhere along the way sat down for a rest and a welcome drink from his canteen. Beside him was an outcrop of white quartz with some rather bright red and green coloration. The brilliance struck his eye and he pried off a loose piece and put it into a bag with the galena.

When the elder Ashley returned the matter of the lead prospect and its possibilities was raised and the samples taken by son Palmer were emptied on the table. The older man picked up the colorful quartz and gasped. The under side was a net of fine wire gold.

That was their only sight of the rich ore. The Ashleys were to have the same experience of the desert's deceptiveness as many others before and after them. One hill, one ridge, one gully, one stand of mesquite looks much like another. Day after day they prowled the country to no avail. With times improving and more jobs available, father and son returned to city life and regular employment. Another lost bonanza, in or near Morgan City Wash, awaits a lucky man.

In these descriptions we have been south of U.S. 60. There is less of record but perhaps more of promise in the opposite direction. From the Salome-Wenden area a fair desert road leads north over the Harcuvar Mountains via Cunningham Pass and goes on into some of the most desolate and neglected territory in the state. With due precautions in the way of water, dependable transportation, and careful attention to direction, one can get into country that is a blank spot on the map and something of a blank spot as to known mineral resources. In short, this trans-Harcuvar region comes as close to virgin prospecting ground as one is likely to find.

One such area is the uncharted twenty-mile-wide Butler Valley to the west of the Cunningham Pass road. Another even larger area stretches northeasterly toward Ives Peak and the McCloud Mountains beyond. Smaller in extent but closer to civilization is McMillan Valley, just east of Cunningham Pass and on the near side of the Harcuvar range.

A conspicuous landmark visible from much of this country is Clara Peak, rising far to the northwest near the Bill Williams River. At the foot of Clara Peak, on its west side, is the ghost town of Swansea, named after the great mining and milling center in faraway Wales. Swansea was the location of a great copper mine with an elaborate smelter financed by French and Belgian capital.

Prospecting in such vacant country as this, one has to know his nearest source of gasoline and other supplies. For the Swansea-Cunningham Pass area this would be the town of Bouse. This settlement is on the Santa Fe Railroad and as a supply and shipping point for mines has had an importance quite out of proportion to its size. By highway, Bouse is some twenty miles north of Interstate 10, being reached by State Route 72.

There are a good many abandoned mines and prospect holes around the town, and since the land is almost 100 per cent in the public domain any discovery is a case of finders keepers. During the 1930s, for example, a young opportunist stumbled on a weathered surface deposit of almost pure iron. Investigation proved that this particular type was in demand, temporarily at least, for some manufacturing process in Los Angeles. The result was a simple and inexpensive mining enterprise. Chunks of iron were picked up off the ground and loaded in a truck. The truck was backed up a ramp into a railroad gondola car and the iron dumped out. When a full forty-ton load was accumulated, the next passing freight train would take the car to Los Angeles and a substantial payoff. That easy money is over, but the native deposit from which the surface fragments were derived might well justify the more difficult business of mining it at depth.

Just north of Bouse, near the town's tiny cemetery, is a hidden vein of quartz so laced with gold that specimens have been sold to collectors and museums. The discoverer was Jim Reed, who was in his seventies when, back in 1953, he told a little about his remarkable find.

Reed said that he was always careful to see that he was not followed to his diggings and on leaving them he covered the excavation with brush and dirt. There is reason to believe that the rich ledge is on or near a landmark known as Black Butte. This sprawling, barren hill also has produced semiprecious stones of several kinds. The cares of a standard mining operation did not appeal to the elderly Reed and his wife. The two remained satisfied with the small business of selling pieces of their high-grade ore as specimens. How much of such rock remains in the ground is unknown.

The Plomosa Mountains, lying between Bouse and the next town to the south, Quartzsite, have scattered but promising mineralization and we will come presently to their mysterious little black crag, which may hide a fortune in gold.

Quartzsite, south and a little west of the Plomosas, though of small and variable population, has long been an important place in the western desert. Since early days its bountiful Tyson's Well has been a boon to travelers, and the spot has been a home or a resting place for countless prospectors who, in the course of a century, have fanned out into the almost waterless desert surrounding.

Here at Quartzsite one can see a stone monument to Hi Jolly, properly Hadji Ali, the little Levantine who was brought to America to look after the camels at the Army's desert outposts. The camels constituted an interesting military transport experiment undertaken just before the Civil War. It was at a time when the Army was maintaining a number of outposts in the sparsely settled Far West. The beasts were imported from the Middle East and were in service several years hauling military supplies. At length the Camel Corps was discontinued and the animals were released into the desert to fend for themselves. Hi Jolly stayed on at Quartzsite and supported himself by placer mining until his death.

There is at least minor mineralization scattered almost everywhere around Quartzsite, but most gold seekers have been attracted to the country south of the town. The eighty-three miles of road leading from Quartzsite to Yuma, recently absorbed into the federal highway system as U.S. 95, run mostly through an utterly forsaken landscape. The northern part is known as La Posa Plain. Tracks take off from 95 toward the mountains on either side, but there is only one of these laterals that might be called an established road. This one branches off westward some twenty miles south of Quartzsite, goes through Weaver Pass in the Dome Rock Mountains, and ends at Cibola, an early settlement on the Colorado River where there usually has been a little cattle raising.

The principal mountain ranges west of the Yuma road are the Dome Rock, the Trigo, and the Chocolate, the latter not to be confused with California's Chocolate Mountains. To the east lie the Kofa and Castle Dome ranges, and if one presses eastward an additional thirty or forty miles from pavement he will come to the Little Horn, Tank, and Palomar mountain ranges and the Cemetery Hills. Closer to Yuma, still to the east of U.S. 95, one encounters the Muggins Mountains.

All of these mountains have produced gold and some contain old workings that have yielded millions. As to what fortunes may yet be discovered, today's prospector may glean some encouragement from the incident of the overlooked North Star Mine. If this great lode could lie unnoticed as long as it did, may there not be another such sleeper?

The scene was high in the Kofa Mountains, where the King of Arizona, one of the state's proudest bonanzas, was discovered in 1896. The mine was in active production within a year, and management, crew, freighters, and visitors went through their routines without any suspicion that only two miles away there was equally rich ore.

This went on for nearly ten years. The nearby deposit might still await discovery had not an Indian brought it to the attention of the King of Arizona operators. They went to the site and there, fully visible, was a strong outcrop of rich ore. The result was the immensely profitable North Star Mine. Like the King of Arizona, the North Star now has lapsed into inactive, caretaker status and maps show its supporting village, Polaris, as a ghost town.

If one gets around east of the Castle Dome range he will be within striking distance of the Little Horn Mountains. In the southwestern part of the Little Horns, and probably in their lower reaches, is a surface showing that a Tonto Apache described as "the richest gold ore in the world." A piece that was still in the possession of the Alvarado family of Yuma in 1918 would tend to support the claim. Witnesses said there was more gold than rock to the specimen.

The location is perhaps within a mile and a half of a road that was fairly well traveled in the 1870s and that can readily be identified today. It carried considerable army traffic. This was a decade in which various Apache tribes, more or less subdued by now, were being shifted around to various reservation sites, escorted by army detachments. One route followed, which ran north from the Gila River and past the Little Horns, was an unusually dry one and at a strategic point along one of the arid ravines the soldiers dug a cistern. It was fifteen feet deep, and by an ingenious combination of timbering, a stone collar, and a metal lid punched full of holes, the excavation would catch the runoff from the infrequent spring rains and largely protect the water from evaporation. The cistern was still functioning in 1957, when a photograph of it taken by Harold O. Weight was printed in *Desert* magazine.

This is one landmark in any search for the Apache gold. Another is Alamo Spring, which until recently at least was maintained by wildlife officials as a water hole for the desert fauna. The cistern is due east of the spring.

The extremely rich ledge was an Indian secret. It became known outside the tribe when the Apache, whose name has come down to us only as Pancho, decided to reveal its location to an Arizona rancher, Jose Alvarado, Sr. This was to be in return for many past kindnesses shown the Indian during a period when any beleaguered Apache could use a friend.

Alvarado was a member of an early Spanish family prominent in California affairs. At this time he was in advanced years and had a cattle ranch near the village of Paloma, Arizona. Pancho's offer was not to be ignored. With two ranch hands along to look after the pack animals, do the camp cooking, and handle chores in general, the feeble old gentleman set out under his Indian friend's guidance.

For reasons which need not concern us, hostility quickly developed between the Apache and Alvarado's two helpers. Pancho grew to hate the men and possibly he feared them. In any case, he changed his tactics. Resolved not to let the two ranch hands share his secret, he stole away from camp one night, went to his ledge, and broke off a convincing sample of the rich ore. Then he returned and cautiously awakened his old benefactor.

The party was camped at the old cistern at this time, and Pancho appears to have led Alvarado away only a short distance. This brought them to the mouth of a dry wash extending well back into the mountains. Here they stopped and as Alvarado remembered it the following directions were given: "A mile and a half up this wash to a little side wash–that is the place." Next morning the expedition headed home. Doubtless the directions were amplified during the return journey and more identifying detail attempted.

The strain of the initial trip on the aging cattleman had been severe and he never attempted a second one. The story and directions were conveyed to his son, Jose Alvarado, Jr., along with the visible evidence of the rich ore. The younger man was in the dairy business and appears to have been tied down by its concerns. The Apache gold was sometimes the subject of conversation among the Alvarados and their friends and one of the latter has related the story in some detail. Yet none seemed disposed to undertake a search. It is on record, to be sure, that two prospectors mined a small pocket of high-grade gold ore on the hillside just above the soldiers' old cistern. Neither the location nor the magnitude of this little strike conforms with Pancho's description. It would seem that the bonanza itself never has been found or even searched for with any diligence.

Gold ore can take many forms, but the yellow metal is not commonly associated with rock that is almost coal black. Yet somewhere near the sand dunes in the section of La Posa Plain north of Quartzsite is a little black crag, jutting up conspicuously from a mountainside, which is all or in part a gold formation. A local humorist has given the following directions for finding it:

"Get drunk in Ehrenberg and then get on your horse with just enough sense left to guide him to the Maraquita Mine [west-northwest

The Chile mill, similar to the arrastra, substituted a huge stone wheel for the drag stones in grinding up gold ore.

The hand-operated windlass was in universal use for sinking a discovery shaft. As the mine developed, such hoisting was mechanized.

Chinese using rockers to extract placer gold. The industrious Orientals successfully worked gravels that proved too lean for white men.

Shelter from Nevada's sharp winds is all that is offered by this sheep-pen hostelry. In new mining camps, sleeping accommodations were scarce.

Iowa Hill, California. The single main street and frame, false-fronted buildings were typical of all early-day mining camps.

A dump of this magnitude testifies to a highly successful mine. It consists of mill tailings plus barren rock necessarily removed with the ore.

Final operation at a successful mine. Gold recovered by the complicated treatment of the ore is molded into ingots for shipment to the mint.

Satiric magazine illustration shows eastern investors weighing an offer from a prospector. Wily miners sometimes sold worthless "coyote holes" for a good sum.

of Quartzsite]. Then get a sighting on Planet Peak [on the Bill Williams River] and head in that direction. When your horse begins to stumble around in the sand dunes, fall off. When you come to, pick up the black rocks that are scattered around. Scratch them off and you will see gold. The rocks have been weathered down from that black formation above you. You have found a good mine."

This was the routine followed by the hapless discoverer. He was the owner of an operating copper property at Planet Peak and his spree almost cost him his life. On coming to in the sand dunes he found himself stranded without water and with the intense heat of a summer's day ahead. He might well have perished had not his horse wandered on and been seen and recognized. Friends backtracked and rescued the man. The ore samples, which with a miner's instinct he had clung to, assayed at $750 a ton.

Neither the befuddled finder nor his friends ever were able to trace these rocks to their source. The scene of the rescue was not necessarily the scene of the gold discovery. Local opinion placed the black crag in the general latitude of Bouse and on some spur or detached erosion remnant of the Plomosa Mountains.

Faring south from Quartzsite along U.S. 95, one comes in a couple of miles to a formation known as Picture Rock or Point of Rocks. There once was a dependable seepage of good water here and the original Quartzsite settlement was at this point. In 1870 there occurred a cloudburst and flash flood. A well-established general store and several other structures were swept away. According to a Mrs. Jose Martinez, who was a friend of the store's owner, a safe there contained $50,000 in gold. The amount seems large but is not entirely incredible, for in those times payment over the counter was largely in nuggets or gold dust. Buried somewhere in the heavy burden of flood debris, the safe never has been recovered.

Traveling on south, one will see the Kofa Mountains to the east and the Dome Rock range to the west. They appear quite close in the clear desert air, but if one takes off on a side road, never more than a dirt trail winding through the giant sahuaros, he may travel as much as twenty miles to reach their base.

A somewhat inconspicuous feature of this landscape but one of possible importance is the line of great bajadas (pronounced bah-HAH-das) that lie against the base of the mountains. The bajadas are slopes of loose soil and gravel washed down from above and built up to their present volume over an unknown span of time. At canyon mouths they may take the form of an alluvial fan. The erosion of the mountainside included the erosion of auriferous veins and veinlets. As a result, in the opinion of some mining engineers, the bajadas contain in total an immense quantity of gold, offering in fact a brand-new source of the yellow metal overlooked by the early miners. That which has been recovered in sampling or in small individual workings has been aptly described as "short-haul gold." Unlike the worn, rounded nuggets that have been carried great distances by streams, these examples are rough and angular, which signifies that they have come from no great distance.

The mining engineer Addison W. Clark has written extensively of how the bajadas could be worked by bucket-line dredges and cites one such operation in Nevada. He quotes Dr. Olaf F. Jenkins of the California Division of Mines as believing "such desert placers . . . offer a practically virgin field for exploration, holding a potential wealth not yet known."

The bajadas scarcely offer an opportunity for the small miner unless he can get together a considerable sum of venture capital. Heavy equipment, water supply, and expert technical assistance would be among the requirements, all preceded of course by extensive and expensive sampling to make sure that the values are there in the first place.

The northern of the Kofa Mountains have a poor reputation as to mineralization. They are a jumbled, waterless mass that caused early prospectors much trouble and disappointment. Nevertheless a report persists that they shelter an outcrop of rusty quartz containing much visible gold. The account, difficult to trace to any identifiable discoverer, describes the vein as more than three feet wide and not a great distance southeast of Quartzsite. It is said to stand out prominently from the weathered rock around it, and an abortive attempt was once made to spot it from an airplane. In this connection it may be noted that the late Erle Stanley Gardner, the mystery writer, made a number of attempts to discover desert mineral deposits from a helicopter. His trials produced some interesting reading but, so far as he revealed, no bullion. Any serious search for this lode in the Kofas would necessarily be on foot and should be planned with great care.

Paralleling U.S. 95 on the west, just opposite the Kofas, are the Dome Rock Mountains. They offer considerably better inducements. There are several dependable tanks (*tinajas* to the Spanish-speaking population)—natural rock wells that hold water the year around. Some of the canyons in the Dome Rocks, visible from the highway, can be penetrated a short distance by motor car, while a four-wheel-drive vehicle can get one well back into the range at selected places.

Gold in small amounts has been brought out of these canyons year after year for at least a century. An Old Man Butler among others made a comfortable living there during a long lifetime. The human population always has been insignificant for so vast an expanse of territory. It would seem that such a scanty band could hardly have made any comprehensive surface examination of this rough and difficult country, that there must be many mineralized spots still to be searched out for the first time. Besides the possibility of completely new discoveries, the area displays a number of old shafts and tunnels. Most of them are shallow and show little promise, but those that were borderline operations in the old days may now deserve a second look.

Several reliable men who have penetrated deep into the Dome Rock range have reported a great dyke of a pinkish tinge but in places light gray, which is peppered with minute vugs of limonite and carries gold values throughout. The dyke matter is of low grade but the quantity is immense. Even if expert examination confirmed these reports, the deposit could be developed only by a large capital outlay. Milling would have to be on a grand scale and doubtless would entail piping water from the distant Colorado River.

To reach a district where highly profitable mines are a matter of record, we move on down U.S. 95 to the latitude of the Trigo Mountains on the Colorado River and the neighboring Chocolate range. Both of these lie considerably west of the highway and are partly obscured by a minor chain called the Middle Mountains.

It is in this country that the very rich Red Cloud silver mine was developed in the 1880s, along with the comparable Silver Clip and Black Rock. Somewhere along here, their stones now scattered, were the Celestine arrastras.

This report has referred several times to the arrastra as a primitive device for milling gold ore. Here at the scene of Celestine's large scale operations may be a good place to describe such a mill.

The basic structure was a circular floor of stone, commonly ten feet in diameter. Some were smaller and in El Dorado County, California, many measured sixteen feet across. In the center a heavy post was set solidly in the ground. From an axle atop this post, a strong beam extended out well beyond the circumference of the stone floor. A horse or mule was hitched to the outer end of the beam and he walked round and round the circle.

The ore was ground up by a large dragstone attached to the beam and thus was pulled round and round by the animal. It soon wore flat on the under side, increasing the area of its grinding action. Gold ore previously broken into chunks the size of a man's fist and smaller was dumped on the floor of the arrastra, typically about five hundred pounds at a time. A low wall around the outside, usually of rock, sometimes of heaped earth, kept the contents from spilling over.

When it appeared that the material was very close to being completely ground up, quicksilver was added. Miners call the liquid metal quicksilver or "quick" rather than by its scientific name of mercury. A little later water was added, sixteen buckets for a five-hundred-pound load, and the muddy, soupy result was agitated for a time longer by revolutions of the dragstone. The water then was either bailed out or run off through an opening. Another charge of ore then was shoveled in.

This process was repeated for days, sometimes for weeks, until the operator decided on a cleanup. The residue consisted of amalgam, which is a combination of gold and mercury, and the considerable amount of finely ground rock which had not been disposed of by bailing or drainage. Some of the best concentrations of the gold amalgam were in the crevices between the floor rocks. This precious material was carefully pried out and in some operations was recovered by removing the rocks themselves.

All that remained to be done now was pan out this residue, save the amalgam, and recover the gold by retorting. In the latter process, the mercury vaporizes under heat, leaving the gold behind.

The arrastra process was a slow one but surprisingly effective in saving the gold. A knowledgeable and careful operator would recover up to 95 per cent of the gold present, a record seldom equaled by the manufactured ore mills of the early days. The arrastra can be mechanized by installing a set of beveled gears in the center. There was one in Blythe, California, in the 1930s, powered by a large electric motor, which whirled the dragstones around about twenty times as fast as the patient horse or mule could pull them.

The Celestine operations were a remote, early-day operation that received little notice. We are told there was a group of arrastras but we do not know how many. One would like to know also where the ore bodies were situated that justified the somewhat ambitious project. The source or sources of this ore have been forgotten.

The region south of Quartzsite also is where not one but two very rich ore deposits were found—and lost—by the legendary desert rat John Nummel.

Nummel, a native of Germany, came to the desert country sometime before the turn of the century, found it to his liking, and remained there until he died in 1948. Much of the time he was employed at the Red Cloud Mine, whose great tailings dump and some rusted equipment are among the few reminders of man's activity in this isolated region. The location is about four miles northeast of Norton's Landing, an early port on the Colorado River. The closest modern population center is Yuma, Arizona.

Norton also worked at times at La Fortuna Mine in the Gila Mountains southeast of Yuma. He would make the forty-mile trip between the two points on foot, carrying a gallon canteen of water and refilling it at a known water hole along the route. Part of this trek coincided with an ancient Indian trail but Nummel ranged at varying distances from the path and prospected as he went.

His first good strike was made quite close to Yuma Wash, which is not near the city of Yuma but far to the north near the Red Cloud. The outcrop was a strong vein of yellowish-white quartz with native gold in little pinheads or clusters all over it. A single paloverde shaded the exposed ore body.

Nummel realized the value of his discovery but felt he could not go back immediately to the Red Cloud. For one thing, he had just had one of his many violent quarrels with the mine foreman. He explained later that in these disputes he never had come back at once to make his peace. To do so now, it appears, would have been so out of character that fellow workers would do some accurate guesswork as to his reason. An alternate job was waiting for him at La Fortuna. He concealed the rich ledge with rocks, pushed on to the water hole, and then went on to La Fortuna. When he had worked there for a while and had saved up a little stake he was ready to cash in on his find. He threw up his job again and headed north.

Once again the desert played its old, baffling trick on a lucky prospector. Nummel carefully retraced his way to the Red Cloud but nowhere on the trip could he locate his rich vein. Nor, in further attempts, was he ever able to return to it. A certain paloverde among hundreds of paloverdes, the concealing rocks arranged to look like a natural conformation of rocks–Nummel had defeated himself. Friends who knew him in later life say he tried again and again through the years, always hopeful, always disappointed.

Nummel's second discovery, this time silver ore of great value, came some thirty years after his first. At this time the Red Cloud had shut down and the old man was employed as watchman-caretaker. He was continuing to prospect and quite an accumulation of mineralized rock lay scattered around his cabin. One of these samples struck the eye of a visitor from Yuma, perhaps a more widely experienced mining man. Sent away for assay, it proved to be silver ore of extremely high grade. The interested friend and Nummel himself now wanted to know where it came from.

Here one is reminded of the rule that every prospector believes in but that so few of them practice. Each promising sample of rock, picked up from the ground or chipped off a vein, should be plainly identified as to its place of origin. One recommendation is for the prospector to carry a pocketful of small paper bags. The where and when of each find can then be written on the bag, together with other helpful notes–especially in the way of landmarks and directions. If the sample is worth saving at all it is worth labeling in some way.

The old man had relied on his memory, a memory that had to sort out nobody knows how many different specimens gathered at various times and at scattered locations. Searches were made, of course. But John Nummel's silver lode still lies there, open to the sky, awaiting not a luckier but a more painstaking discoverer.

8
The Spanish Mines

The mines of western Arizona were linked closely with navigation on the Colorado River and we have been examining them as something of a separate province. The eastern two thirds of the state that lie before us poured forth immensely greater wealth than the western part, but the values were chiefly in copper. The United Verde Mine at Jerome is the most notable example, and is credited with producing copper to the value of $300 million, certainly one of the richest operations in the annals of mining.

The gold and silver recovery in our present field cannot compete with figures of such magnitude, but there have been good precious metals mines quite rich enough to yield a fortune. There also have been spots where great mineral wealth was glimpsed briefly and never seen again, living on in legend, and perhaps in fact, to exert their magnetism on new generations of gold hunters.

Our trip down the western side of Arizona left us at Yuma, which is on the Colorado River and near the Mexican border. The principal route east from Yuma, very nearly the only one, is Interstate 8. This federal highway runs through sparsely settled country to Casa Grande. The town takes its name from a massive prehistoric ruin, a *casa grande* (big house), which stands nearby. Here Interstate 8 merges with Interstate 10. The next major stop is Tucson, a substantial city in its own right and the center of an expanding seasonal population in the winter, which, with parts of spring and fall, is also the time for prospecting in this region.

At Tucson one is seventy-three miles north of the international boundary and the twin border towns of Nogales, Arizona, and Nogales, Sonora. He also is on the Santa Cruz River, the scene of the state's first white settlements, the first missions, and the first mines. This is proved mineral country, with a history of important strikes of gold and silver. The river may be followed today by taking Interstate 19, past the missions San Xavier del Bac and Tumacacori, and past various side roads that beckon to old mines and old camps in differing stages of preservation.

Although the highway route outlined is the recommended way to reach this pioneer gold country, this report will follow a different course. In order to take in various far-flung mines, prospects, and present-day opportunities, it will be desirable to trace a route from Yuma along the once dreaded Camino del Diablo.

This trail across a wide and all but waterless desert is mentioned time and again in early western chronicles, but with little in the way of explicit narrative. Its sojourners were not the type to leave records—ill-advised Argonauts of the California rush, scalp hunters, filibusters, fugitives from Apache raids, and men who were in this unpopulated country for reasons into which, in the early West, it was unwise to inquire. It was a dangerous road but we have no statistics as to how dangerous it was in terms of human life. History seems to affirm, however, that more travelers perished on the Camino del Diablo than ever lost their lives crossing Death Valley or the snows of the Sierras.

A feature consistently mentioned in connection with the Camino is Tinajas Altas (high tanks), a series of seven rock cisterns created by nature, far up the side of a desert peak. The scanty rains of this climate would be held in these clefts for a considerable time, but only one was deep and narrow enough to resist evaporation the year around. This was the seventh, or very highest. Ben Humphreys, a prospector of the depression days of the 1930s, was told by Indians that many wayfarers, counting on water at Tinajas Altas, would find the shallow lower tanks dry and would stagger on to die of thirst. A climb to the lofty seventh tank, had they known of it, would have saved them.

Tinajas Altas is associated in Indian tradition with an abandoned church among the nearby sand dunes and, more relevant here, with a rich placer deposit. The story has some corroboration in old documents at the University of Arizona. The church may be the one known as the mission of Los Cuatro Evangelistos. If so it was a relatively late mission, established by the Franciscan friars who succeeded the Jesuits in the Arizona missions. The founding of a Los Cuatro Evangelistos is authenticated, but whether it was this church among the dunes is uncertain.

Years ago a Papago Indian rancher, Juan Orosco, sighted the belfry tower of the church protruding from a sand dune that had drifted over the remainder of the edifice and over whatever land, gold-bearing or otherwise, that lay around it. Orosco was fifteen years old at the time and in later life could give little in the way of directions except to say that he was not far, as desert distances go, from Tinajas Altas.

These water holes also figure in the Papago tradition regarding the mission and the mine. The placers had been worked by the Spanish fathers and the Indians in what was understood to be a partnership. The Indians were not as innocent as they once had been. By the time of the Franciscans' arrival the natives had learned the value of gold. The sand dunes, long stationary and regarded as a welcome shelter for the mission, the mine, and the Papago village, now began to move. They soon encroached on the gold workings and men had found no means, nor have they yet found any, of stopping the sand's inexorable march.

The tribal account is that in this situation the three missionaries in residence and two of their most trusted converts secretly decamped with the accumulated gold. Angry Papago miners followed them and in a two-day siege at Tinajas Altas all five of the fugitives were slain. No gold was found. The explanation from the Indian standpoint was this: The two red converts had been killed the first day of the attack. That night they were buried, for when the fight was over they were found lying in shallow graves. The crafty friars, Juan Orosco explained, had placed the treasure beneath the two bodies, knowing that Indian custom or superstitious fear would forbid their digging up the dead.

The Orosco ranch included a rare, tree-rimmed desert pond known and still mapped as Quitobaquito. The ranch passed out of the family's ownership when the Organ Pipe Cactus National Monument was created. In the course of years, have the sand dunes moved on, possibly uncovering the lonely church and the rich placers? If the departed Orosco family could be traced down, they might give some clues.

There are some four hundred or five hundred square miles of desolate country lying north of the Camino del Diablo that would seem well worth prospecting now that four-wheel-drive vehicles are available, capable of carrying ample water and supplies. Major gold discoveries have been made beyond the road's western terminus, and beyond

Sonoyta at its eastern end. The lack of discoveries in between is easily explained. Early travelers along the Camino were in transit and they were in a hurry, often desperately short of water and food. It is not surprising that, whatever the mineral possibilities, they passed by the desert ranges that jutted up on the horizon.

Two of the mountain ranges that lie quite close to the old road are the Copper and the Cabeza Prieta. Fred Wright, a veteran of the Boer War who came to Arizona and spent thirty-seven years mining and prospecting, had convincing information about a rich gold ledge at the south end of the Cabeza Prietas. He never was able to locate the prize, but during his long career along the border Wright consistently recovered gold by dry washing, lived well, and accumulated a comfortable retirement stake.

Also in this area are several detached little mountain ranges that should be mentioned as certainly containing mineralization and therefore inviting inspection. They are the Sierra Pinta, the Mohawk Mountains, the Eagle Tank Mountains, and the Diablo Hills. The approach would be by a road starting at O'Neil Pass on the boundary of Yuma and Pima counties and running north about forty miles to connect with Interstate 8. At this writing it is passable only by jeeps. The Eagle Tanks are best approached from the tiny settlement of Aztec on Interstate 8. Gold and silver have come from all these ranges but the records are hopelessly scattered and fragmentary.

Southeast of O'Neil Pass lie the Agua Dulce Mountains, which extend across the international boundary, and this brings us to the matter of Mexican territory and the importance of keeping on the American side.

Arizona's southern border was set by a treaty ratified in 1854 and usually referred to as the Gadsen Purchase. It is a straight line running southeasterly from the Colorado River almost to Nogales, then changing direction slightly and running due east. The Camino del Diablo roughly parallels the border in American territory. There is a similar road on the Mexican side. The latter is now paved and is designated No. 2 in our neighbor's highway system.

If a valuable discovery is made in doubtful territory, the suzerainty should of course be determined without delay. A discovery on the Mexican side is not necessarily fruitless. Americans can locate mining claims in Mexico, or "denounce" them as the term translates, but the law is different from ours and should be thoroughly understood.

One of the lost mines that would provide an instant fortune for its rediscoverer, the fabulous Planchas de Plata, is very close to the border and no one is sure in which country it is to be found. This bonanza has been mentioned in connection with the dubious Hardin silver strike in Nevada because in both cases the native metal was said to lie in a soft, gray, ashlike formation. The silver occurred in pure form and chiefly in irregular flat sheets. Mining consisted merely of gathering these pieces from the surface or digging or prying them out of the soft gray matrix. The formation at the Planchas de Plata extended in apparently dyke-like outcrops over several hillsides. The silver slabs, *planchas,* were taken from what an early written account by one Manuel Retes described as the "main deposit." This expression rather hints at additional, lesser occurrences, but Retes disappointingly fails to describe these latter or to state to what extent the far-flung gray intrusions were investigated.

The fame of the Planchas Mine was such as to receive notice in a number of early-day reports and personal chronicles, and did not escape the wide-ranging eye of the historian Bancroft. There are records of the "royal fifth" being paid to the Spanish crown, but the fortunate owners of the mine, if they were named in any of these records, have not been identified by later research.

The questions of why this rich location was abandoned, and of how so widely known a mine could become lost, have the answer common to many such puzzles: Indian raids. There is some evidence that silver was being mined and shipped between the years 1736 and 1741, and we know that the Planchas was definitely a lost mine in 1817. It was in that year that an ambitious expedition of some two hundred men started out to find it. Even this considerable force was turned back by Indians. There followed a period on which we have little information. Turbulent times under Spanish rule, then under the disorganized rule of a newly independent Mexico, left this part of the country quite unsafe and for a while a miner was scarcely to be found in wide areas of southern Arizona or in Sonora. The few who stayed on with a show of bravado commonly wound up at the bottom of a mine shaft, which the attacking natives would then cave in and conceal. The Apaches and the Yaquis were nearly always on the warpath and even the normally peaceful Pimas, often with some justification, would at times go on a murderous rampage.

The distinctive gray ash indications at the Planchas de Plata were not as distinctive to the early searchers as the written data make out. It appears that this identifying feature was not generally known until the Retes report was quoted in Hinton's *Handbook of Arizona,* published in 1878.

Some accounts, dates uncertain, attempt to give a clue to the Planchas location. Retes' bearing of 31½ degrees north, in the unlikely event that the latitude was correctly determined, would put the mine in Arizona if it lies east of Nogales, in Sonora if it is very far to the west. Directions in leagues also are involved and face the difficulty that the league was a variable unit. As the veteran desert sojourner Philip Bailey points out, the old chroniclers probably refer to a "traveling league," which can be translated into someone's rough estimate of the distances along his route. In the absence of clearer indications, the league can best be considered as equivalent to three English miles.

There also was a Bolas de Plata Mine, which produced an immense fortune for its owners, the Zambrono family. Here the native silver was dug up in ball form. This property was taken over by the king, and appears either to have been worked out or left idle once in the hands of colonial administrators. The site is unknown today. A discovery near the Agua Caliente River in Sonora in the 1870s yielded chunks of silver and for a time was thought to be either the Planchas or the Bolas rediscovered. Examined more carefully, the deposit did not fit the description of either great bonanza and moreover proved quite limited.

That chronicler of lost mines, a practical miner himself, John D. Mitchell, states that the pieces of native silver occurred in kaolin. There are bands of such material in the southern Arizona mountains, though one questions whether it classifies as kaolin in the usual sense, that is, that is the pure white clay that is made into porcelain. Is this what was meant by the "gray, ashlike material"? Indians have told Mitchell that after any rain their ancestors would scout along these kaolin streaks for chunks of silver that washed out of them.

One conclusion that appears warranted from all this is that nature has deposited sheets, balls, and other shapes of solid silver along the Mexican border. Some modern prospector with mechanized travel equipment might well come upon one of the historic deposits, or perhaps a similar treasure that was never reached by the plodding foot traveler of the past.

A rich gold vein that was lost owing to Indian attack and definitely is on the Arizona side of the boundary has become known as the Lost Jabonero Mine. The discovery was made by two Mexican miners from deep in the state of Sonora, one from Alamos, the other from the state capital, Hermosillo. The date by some accounts was 1830. It is known to have been before the California gold rush of 1849. The two men appeared at the border town of Sonoyta with such small amounts of ore as they could carry on a long and exhausting tramp from the western reaches of the Camino del Diablo. Witnesses said that massy gold made up a good third of the ore. The amount that could be carried on foot was small, and the pair welcomed a proposal from the local *jabonero*–Spanish for soapmaker–to supply tools, provisions, and burros for a return to the rich ledge.

Now a partnership of three men headed west, and near Quitobaquito passed a rancheria of the Areneno Indians. Their ample provisions and livestock did not pass unnoticed. As the mining expedition camped that night at Agua Salada, south of the border, there was a murderous attack and the Arenenos made off with their loot. The original miners were killed. The jabonero, though badly battered about the head and with his vision permanently impaired, recovered and made his way home.

Years later, now an old man and living in Los Angeles, the survivor could still recall the directions that had been shared with him. Among these people, memory, one may surmise, was possibly more accurate and long-lived than in a society of letters, notes, and memoranda. At any rate, the aged soapmaker conveyed his secret to a Mexican friend and dictated eleven rather precise steps to follow from Quitobaquito to the golden outcrop. The friend set these directions down in writing.

The Lost Jabonero appears to be in the Cabeza Prieta Mountains and at not a great distance from the Camino. Harold O. Weight, whose trips and reports have contributed a good deal to current knowledge of remote desert areas, has seen and analyzed this waybill. The key to the location consists of three peaks standing alone. The difficulty is that they appear this way–that is, three in number and apart from any other mountains–only when seen from a certain fork in the road. Now many, perhaps most, roads on the desert repeatedly branch out and reconverge, each of these separations possibly qualifying as a fork. Then there are other, more definite forks, as side roads branch off to mysterious destinations of long ago. Such trails through brush and cactus that served some old Indian camp or long-forgotten mining operation may be well preserved. Weight studied the clues with great care. The puzzle, he concluded, never would be solved by rational analysis. Once again, the prospector must call on the goddess Fortuna.

The Lost Jabonero is the final lost bonanza to be examined on the grim Diablo trail. We come next to a region with a concentration of known, abandoned gold and silver workings and also a concentration of legendary lost mines. This area consists of the previously mentioned Santa Cruz River Valley and its paralleling mountain ranges. In terms of modern population centers it is the district between Tucson and Nogales. Here the Jesuit fathers built the much-admired mission churches of Tumacacori and San Xavier del Bac, as well as a number of outlying *visitas*. Here they converted Pima and Papago Indians to the faith, irrigated arid desert to provide food and fodder, and, in an inhospitable desert, developed self-sustaining colonies. Here also, unless all tradition and a great deal of documentary evidence is to be rejected, they found and worked rich mines.

The extensive mineralization of this area is attested to by the output of gold and silver since American occupation. The Cerro Colorado, or Heintzelman Mine of checkered and tragic history was the most lavish producer. The Albatross, the Austerlitz, and others are on official record as good deposits, operated with success. Equally rich, but unknown as to exact location, were the lodes worked by the churchmen and their native helpers. All appear to have been silver mines. They included the Esmeralda, the Purisima Concepcion, the Escalante, the Carreta, the Guadalupe, and several others to which no precise names attach.

An additional allure for the many prospectors who have searched for the old diggings is the belief that in addition to rich ore in place the shafts and tunnels contain hidden treasure. This would consist, it is said, of gold or silver bars produced in crude smelting operations, of high-grade ore already mined and broken up, and in some cases gold and silver ornaments from the missions themselves.

Three principal crises in mission history are cited as reasons for the padres hastily hiding their treasures. In 1723 and again in 1751 there were Indian uprisings. The Jesuits fled for their lives and some few who stood steadfast before their altars were slaughtered. Later, in a period of peace with the Indians, the Jesuit order was expelled from New Spain, as has been noted earlier. Besides storing away their wealth under these circumstances, it is widely believed that the padres also skillfully concealed their mines. This is the standard theory of later gold seekers, possibly resulting from frustration in failing to find the old workings. Equally plausible is a belief that Indians covered them up. Aware that it was the precious metals that were drawing white men into their homeland, the native tribesmen employed this tactic in various mining districts.

Before tracing the early miners through the Santa Cruz Valley, note may be taken of an early mission farther west and the mystery of the gold mine operated there. The mission was San Marcelo, just within the Arizona border and near what is now the town of Sonoyta. The church was razed and three of its priests slain in the Indian uprising of 1751. The victims' names have been preserved as martyrs: Enrique Ruen, Tomas Rollo, and Francisco Xavier Saeta.

When peace was restored, the new settlement at San Marcelo was short-lived and the church was not rebuilt. If the old mine was reopened the event went unrecorded, as far as anyone has been able to discover. The tradition among later-day Papagos was that the mine was very rich and was situated in an arroya, or gulch, in Table Mountain. As the name suggests, this would be a flat-topped peak, possibly one visible from an early-day campsite known as Gunsight Well. The latter is a short distance south of the Ajo-Tucson highway. There is a story of a deer hunter who cut across country in this region and, going in the general direction of Ajo, came across extensive mine workings together with bronze machinery of an early Spanish type. His discovery was in a shallow, sandy wash and was marked by several mesquite trees. The hunter had been in the vicinity of Table Mountain, so the old mine's location presents no conflict with the sketchy Indian tradition. Or again, the tradition may not be so sketchy. The destruction of San Marcelo Mission is the type of story that the red men will hand down from generation to generation in minute detail. Today's Papagos may know more about the mine at San Marcelo than they care to tell.

The Jesuit mine that could be most positively identified if rediscovered is doubtless the Purisima Concepcion, for that name is carved in rock directly over the portal. The only other identifying carving that searchers of old records find mentioned is a flat rock having a cross on the visible, upper side. When turned over, the carved letters CCD-TD would be found. This marker was to signify that within a radius of 100 *varas,* or something close to 100 English yards, a mine would be found. The locations of the stone, the Concepcion Mine and Our Lady of Guadalupe Mine all were said to be almost due south of Tumacacori Mission. In the early 1960s, two mine and treasure hunters, Jerry Jenkins and Allan L. Pearce, found a massive boulder in this area with the much-weathered inscription: M200. What this may indicate is not known. No mine workings were visible in the vicinity.

The data of supposed Jesuit origin that crop up in modern searches are generally said to have been copied from church or government records contemporary with the mining itself. Certain it is that some followers of the golden gleam have had the time and money, and presumably enough knowledge of the Spanish language, to study old archives in Mexico City and even in Madrid. After such an investment it is understandable that the findings have been pretty well kept secret. Such searchers need local help, however, and various old-timers, especially among the Spanish-speaking population, have picked up and passed on such information as has been quoted here.

It might have been better for some of the investigators if they had been more open and candid with local people. A number made a simple mistake in following directions that were based on Tumacacori Mission. This impressive old structure, now a national monument, has seen various map-carrying individuals arrive at the mission, sight to the south, the east, and the southeast, make notes, and then take off into the desert. They all were on the wrong track. The references to mine locations in the old documents referred to an older Tumacacori Mission situated about twenty-five miles northwest of the existing one. The original church was sometimes called Tumacacori de las Cerritas (or Sierritas) in reference to a nearby mountain range.

One of the best of the now hidden mines is considered to be the Escalante. It is associated with Mission San Xavier del Bac, the much-admired "White Dove of the Missions," which is still standing. Tradition places the Escalante in the Santa Catalina Mountains, some twenty miles from the mission and northeast of present-day Tucson. Most of this range is protected from spreading development by being within the Coronado National Forest. A natural landmark is the Ventana, a windowlike hole in a rock formation. This was said to be visible from the mouth of the Escalante tunnel, which was "one league" to the northwest.

A *derrotero* to the Escalante, which was being optimistically followed by a well-financed party in the 1960s, was reputed to be copied from a church document dated 1648. It gave the distances in kilometers. Some member of the search group was suddenly inspired to look up a bit of history. He learned that France, where the metric system originated, did not adopt the kilometer and kindred measurements until 1799.

The Escalante has been fictionalized as *The Mine With the Iron Door*. The intriguing touch of the iron door has a faint echo in fact. A cowboy wandering in old mission territory came upon an old tunnel that was indeed protected by a door. This one was made of heavy planks and was secured to the timbering by an iron hasp and a clumsy ancient padlock. No record is found of the cowboy or any of his hearers following up the discovery.

The Carreta Mine is another that has been listed among the lost Jesuit diggings. The name is a later-day appellation, arising from the fact that an old *carreta,* the clumsy wooden cart with solid wood wheels of mission days, is said to have been left near the otherwise nameless mine. The place is a canyon of the Ascosa Mountains. The Carreta workings may also contain a treasure trove. It is related that certain priests had a prescience of the Pima uprising of 1751, and that the churches both at Tumacacori and at Altar, a point well down in Sonora, had their valuables hauled to this mine and hidden away. The lesser Isabella Mine, whose location and surface ruins are known to a few old families, and a San Pedro Mine are believed to lie on the route over which the church treasure was hauled. Some students of the old records believe that once a major Jesuit mine is found and authenticated, a straight line run from San Xavier del Bac through the discovery point will be a guide to other valuable deposits. The locations would be more or less along a straight line, but this is considered due to mere chance, not to any known geological factor.

In the widespread territory associated with the Spanish missions and the Camino del Diablo, the mining center best known over the course of years would doubtless be Ajo. It is situated on Arizona State Route 85, roughly forty-five miles south of Interstate 8. Ajo is no ghost town. It is the site of the immense New Cornelia open-pit copper mine, a property of Phelps-Dodge Corporation.

Ajo lies between the Growler Mountains to the west and the Pozo Redondos to the east. Within relatively easy reach to the north are the Crater and Sauceda ranges. Apart from its copper, Ajo has long had an attraction for the wandering individual miner. There is silver, there are a few old Spanish mines, and there are many gulches in the outlying hills that yielded enough gold by dry washing and pocket hunting to pay day wages.

Ajo may well be the place of origin of the legend of Montezuma's treasure. The Ajo Mountains are some distance south of the town, below where State Route 85 divides. A fantastic crag in this range is known as Montezuma's Head. Here, according to Papago lore, there is a great treasure of gold hidden in a walled-up cavern. As noted earlier, there are two Montezumas, the Papago god and father image, and the Aztec emperor overcome by Cortez. There is report of some of the Aztec gold being transported north from Tenochtitlán for hiding, but the report is a very dim one. The Indian version has locally recovered placer gold being secreted at the behest of Montezuma or, to be more literal, at the behest of the tribal medicine man.

The fact that rich placers were worked by the Papagos in historical times is well substantiated. As noted, several arroyas in the Ajo area yielded excellent returns. The gold would be traded in at Caborca, a town just below the border and site of a handsome mission church that is still in use. A greedy alcalde of Caborca organized a force of Mexicans and took over the placers from the unresisting Papagos. The usurpers, however, reckoned without the marauding Apaches. The slaughter and flight of the Sonorans still is commemorated in tribal rituals.

Before leaving the land and lore of the Spanish mine country, it will be well to look at some mineral developments after acquisition of this area by the United States. A road connecting Arivaca with U.S. 89 at the ghost town of Kinsley takes one into the area of greatest American activity and near one of its greatest mines, the Cerro Colorado. The story of this operation is partly glamorous, partly tragic, and partly mystifying. Not the least part of the mystery is why no decisive finding has been made as to whether the precious ore body has been exhausted.

The Cerro Colorado contained a strong vein of *pelanque,* or silver-copper glance, which originally was excavated to a depth of about six feet by one or more Mexicans engaged in hauling freight between Tucson and Sonora. A piece of the ore that had been traded in at the general store of Soloman Warner at Tucson came to the attention of Major Sam Heintzelman of the Tucson army post. As the story has been handed down, Heintzelman promised a teamster named Ouidican a half interest in the mine and $500 in cash if he would show him the source of the ore. Arriving at the diggings, the major dismissed the unhappy guide with a fifty-cent piece. This was in 1856.

With a company organized, the site legally acquired under a mining claim, and $7,000-a-ton ore available at the grass roots, the Cerro Colorado quickly was in bonanza production. We hear little more of the callous major but he is assumed to have enjoyed both the income and fame of the rich property, for maps, reports, and the scenic lithographs that were popular at that period usually designate it not as the Cerro Colorado but as the Heintzelman Mine.

The luck turned. In 1860 the shaft caved in, burying fifteen Indian and Mexican miners working at the seventy-foot level. Superstitious workers refused to clear the shaft and most of the remaining mining force deserted. Even before the accident there had been an unexplained feeling among the men that some baleful threat hung over the operation. At this time, the comparatively shallow workings had yielded $3 million.

The mine was now placed in charge of two capable mining men, Charles D. and John Poston, a new work force recruited, and mining resumed. By this time, 1861, army troops had been withdrawn from Arizona to fight in the Civil War, and Cochise, the Apache chieftain, was on the warpath. The mine at times was under siege and some walls of the old fortifications may still be seen. Its downfall, however, was not accomplished by the Apaches. A foreman called Juanito and other miners were known to be pilfering the highest grade ore whenever opportunity offered. When Juanito was observed riding away with full saddlebags, an exasperated John Poston leveled his rifle and shot the man dead.

This threw the camp into confusion and by morning the miners had seized all the valuables they could carry and had headed toward their native Sonora. Here the story of the all but deserted mine quickly circulated and quickly inspired action. A band in which known outlaws joined some of the old workmen immediately rode to the Cerro Colorado and took over. John Poston, two German technical men, and eight other employees, all of whom had remained at the place, were murdered. The excavated ore and the bullion awaiting shipment were carried away and the surface structures largely demolished.

A report persists that in his long-continued peculations the foreman Juanito had accumulated high-grade or actual bullion to the value of $70,000, and that his hoard was buried somewhere near the mine. For a time treasure hunters appeared in the vicinity and searched without success for some indication of the hiding place. During the present century title to the mine has passed to one or another of several local people. The record gives no answer as to whether the supposed curse has been overcome or as to whether a body of the rich pelanque remains in the mine.

The Cerro Colorado is some distance south of the mountain range bearing the same name, and is identified by a small isolated hill called the Cerro Chiquito. The mining district of which it is a part spreads up Las Guijas Creek and through the Las Guijas Mountains. Remains of a stamp mill and of several arrastras exist along the creek and there are ancient Indian petroglyphs in the region. Among the formerly active mines are the Mary, Black Prince, and Silver Colorado. Both sides of the Las Guijas Mountains exhibit prospect holes and several old mines, the most important of the latter being the Pescara Vieja. It is all arid, typical desert country, but a number of natural rock tanks are known to the natives. The town of Arivaca is at no great distance from this mineralized area.

None of the known diggings in the Las Guijas district has been identified as the old John Clark Mine, and it would be well worth discovering. Clark was a St. Louis man who worked this silver vein during the best days of the Cerro Colorado. He sent a sizable shipment of high-grade east with one of the Heintzelman wagon trains and was said to have realized $80,000. Both Apaches and outlaw raiders from Sonora were making the district untenable for American miners after 1861, and Clark got out, to die later in the East. A Mrs. Mary Black, who had lived in the area and presumably was one of the fugitives, told John D. Mitchell years later that she personally had seen a pile of very rich ore at Clark's mine and understood it to amount to about forty tons. Clark shoveled the accumulation back into the mine shaft, removed the timber collar that shored up the mouth of the shaft, and allowed the sides to collapse. The buried vein and the mined ore as well were said to run 2,000 ounces of silver to the ton.

The Apaches were not subdued and contained within the San Carlos Reservation until 1886. It was too long an interval for the aging Clark. Neither the owner, his heirs, nor anyone with a waybill is known to have reclaimed the mine. The fragmentary information on it is that the spot was quite isolated and the lode itself was connected with one of the great fault fissures that run through the district. A partly caved shaft, filled further by the erosion of a hundred years, is not an attractive gamble. But if it happened to be the right one . . . ?

Our course has now taken us near Arizona's principal city, Phoenix, and near Phoenix is the eerie jumble of cliffs and crags known as the Superstition Mountains. Sometimes the name is written "Superstition Mountain," in the singular, and there is a certain compactness about the huge, rugged uplift that may justify one in regarding it as a single mountain. Somewhere in the interior lies the Lost Dutchman Mine.

A fair amount of literature has developed around this hidden bonanza and it is possibly the most celebrated treasure trove of the Western world. The story has its disputatious aspects, and the minutiae of clues, landmarks, personal identities, and so on have been argued in a way worthy of the Bacon-Shakespeare controversy. The rich gold deposit, though its existence can scarcely be challenged, cannot be seriously recommended to the modern prospector. The elusive ledge has been hunted for many years without success, sometimes by well-financed expeditions operating over a long period of time from a base camp, by nearby ranchers, and by other Arizonans who have been able to roam the Superstitions year after year. Unfortunately attempts also have been made by brash individuals who have met their deaths from thirst or by foul play.

This report must exclude any lengthy account of such searches and of the theories and clues connected with them. The principal facts about the mine may be related as follows:

The deposit was known for many years to the wealthy, baronial Peralta family of Chihuahua in northern Mexico. An occasional expedition, including peons from the Peralta estates and numerous pack animals, would wend northward, take out all the ore that could be transported, and return to Chihuahua. There is no inkling of how the Peraltas came into possession of the mine or of how many gold-laden caravans unloaded their wealth at the hacienda. We know only when the trips ceased. In 1847 a returning pack train was ambushed by Apaches. The Peralta scion in charge, Miguel, was slain along with all, or nearly all, of his company. Two young Mexicans, who several years later were reputed to be working the mine, have been considered survivors of the massacre. Otherwise, it is asked, how could they know about it?

Metal pieces from old Spanish harness, a sword or scabbard or two, and several pieces of extremely rich gold ore have been found at what evidently was the scene of the Apache attack. These discoveries have been of little help in locating the mine because the Mexican entourage already had emerged from the mountain canyons. The relics of the fight were found on one of the gentle slopes of detritus that lead up to the sheer, almost vertical walls of rock that the Superstitions push into the sky.

The Dutchman himself now enters the picture. It should be remarked that up through the early years of this century Americans commonly applied the term "Dutchman" to persons of German birth. There really were two men of German origin involved in the mine, Jacob Waltz and Jacob Weiser. Their two families had emigrated together in 1848, an unsettled, revolutionary year in much of Europe, and had made their homes in St. Louis, where the boys completed their schooling together.

In the 1870s we find the two friends in Arizona, now grown men and trying their fortune at both farming and mining. Waltz, among other things, took up a government homestead. With regard to their possession of the fabulous mine, let us dispose of the more sensational story first. This widely printed account has the two Germans prospecting, and sighting the mine from a high vantage point. Two Mexicans were busy at the place. Confident that this was an old Spanish mine and *ipso facto* a valuable one, the newcomers drew a bead on the miners and shot them dead. The story goes on that, once the great value of the ledge was apparent, Waltz killed his friend and partner.

The man who accumulated the most detailed information of all investigators in this field, Sims Ely of Phoenix, dismisses this tale in its entirety. His account is kinder to both men. Weiser, the partner, he relates, was badly wounded in the shoulder by an Apache arrow, and at a time when Waltz had gone out of the mountains for supplies. There appears to be ample evidence that Weiser made his way, in great pain, to the home of Dr. John D. Walker at or near Florence, Arizona. There he received treatment, but pneumonia had set in and the physician could not save him.

Sims's inquiries, which he patiently carried out over a period of years and embodied in a book, *The Lost Dutchman Mine,* also disposes of the murdered Mexicans story. Waltz and Weiser did not chance upon the mine. They had a buckskin map, obtained in the state of Chihuahua, which led them to the spot. How they came to have the map is unknown. There is a story, rather lacking in supporting evidence, that the two Germans had come to the help of one of the Peralta family at a local fiesta, when knives were flashing over a game of monte. The Peraltas may have given them the map as a reward. At any rate, says Sims, such a waybill was in their possession. It was surrendered to Dr. Walker by the dying Jacob Weiser, and at times was examined by men of some standing in the territorial capital.

Waltz brought out small lots of high-grade ore. It was nearly one half gold, according to people who saw it. He concealed it under the floor of the adobe house where he lived alone on the outskirts of Phoenix. In his later years he did not go back to the mine. It may have been age or ill health or an understandable fear of the Apaches. A case also has been made out that he blamed himself bitterly for Weiser's death, inasmuch as his partner had been left alone in the Superstitions while he went out for supplies. Waltz evidently had ample resources without further mining, for shortly before his death he had sent ore valued at about $1,400 to a San Francisco smelter. The returns from this were used in part to prevent foreclosure on a little bakery and soda fountain operated by a friend of his old age, Mrs. Helena Thomas, and her foster son, Reiney Petrasch, then a lad of about fifteen.

When he felt he was nearing his end, Jacob Waltz told these two where in his house to find the remainder of his hoard, and also tried to give them adequate directions for finding the mine. These directions were all oral. As far as is known, Waltz never drew a map.

The first search for the mine by Helena and Reiney was quite hopeless. Later they called in more competent help, members of Reiney's family, but the hunt was without success. These trips were only the start of a series of Lost Dutchman expeditions that have continued up to the present time. Their clues, tactics, and misadventures would take a long time in telling. The bare description of the bonanza, as given by Waltz himself, may be set down as follows:

There is something "tricky" about the location. With the best of directions, one could not walk right up to it, would not even guess that he was near unless he knew the "trick." The expression is Waltz's own. It has been interpreted as referring to some passage through the giant rock formations, all but invisible unless one knows the "trick" of finding it.

The diggings consist of two parts. One is a simple pit, gradually deepened as more and more ore was removed. It is believed to be conical instead of being a typical mine shaft. At a lower level there is a tunnel, intended to tap the rich ore at depth according to standard mining practice. Whether the tunnel was driven far enough to reach pay rock is not known. Waltz also said that each time he left the mine he concealed both openings as best he could.

It is interesting and perhaps significant that the intensive scanning of the Superstitions' rocky surface in search of the Lost Dutchman has failed to disclose any other gold deposit of value. Was this lost mine the single concentration that nature allotted the entire range? We can hardly say that there never has been any other profitable mining in these mountains, for such operations in the distant, unsettled past may simply have left no record. One document of rather hazy origin states that the Peralta family had eighteen mines, though these were not necessarily in the Superstitions. However, the many well-financed, often well-publicized expeditions that have gone into the range in recent years have found little or nothing of value. They have necessarily operated in the local spotlight and even a modest new mining venture could hardly have gone unnoticed.

If we go slightly outside the Superstitions, to the so-called Mormon Slope leading up to their western escarpment, there are solid records of gold deposits. The lower parts of the slope were being worked for placer even during Jacob Waltz's time. The discovery of high-grade lode outcrops appears to have been made about 1893, when prospector J. R. Morse sold a group of five claims for $20,000. The camp that grew up took the name of Goldfield, and it is not entirely a ghost town even today. Some of the ore was very rich, the gold being pounded out in a mortar and sent direct to the mint. Production has been placed unofficially at $4 million. The Mammoth and Bluebird mines were among the best.

Drilling of wells at Goldfield produced a welcome supply of water for milling and domestic use, but the high water table also doomed the mines. As the shafts went deeper they were flooded. Intermittent efforts have been made to pump the water out of the better properties. Although modern pumping equipment can handle most such situations, at Goldfield the revivals never worked out from the profit-and-loss standpoint.

The town of Florence on combined U.S. 80 and 89 has figured prominently in the Lost Dutchman riddle, as well as in other mining lore. It is surrounded by multicolored mountains with zones of mineralization that may be worth a renewed reconnaissance. Visible to the northwest across the Gila River is Poston Butte, where there are numerous abandoned mine workings. Many closely spaced ledges were discovered here during the nineteenth century, and the activity all but made Florence the capital of Arizona.

Poston Butte is named for Charles D. Poston, sometimes called the "Father of Arizona" and brother of the John Poston who was killed at the Cerro Colorado Mine. After holding important federal and territorial offices and making successful mine investments, Charles Poston traveled in India. There he was converted to the Parsee religion. The Parsee observances include worship of a sacred fire, considered an emblem of divinity. On his return to Arizona, Poston piously transformed the little butte beside the Gila into a shrine and ignited a fire on the summit which he hoped would be kept aflame forever. He was buried there under a pyramid of rocks, and older residents of Florence refer to the landmark by its one-time name, Parsee Hill.

9
Eastward in Arizona

The region to the northeast of Phoenix and the Superstitions is sufficiently mineralized to have attracted prospectors sporadically over the course of a century. There are hundreds of square miles of unspoiled, almost uninhabited country along the Verde River and Tonto Creek and Cherry Creek. We see broken mountain ranges, all with some history of gold mining, and there are wide depressions bearing such names as Bloody Basin and Tonto Basin. The southern part of the great expanse is known to miners as the Four Peaks district, although it is so designated on only a few older maps. A major landmark is Roosevelt Lake at the southeastern corner of the Four Peaks district.

The area can be traversed by State Route 87, which cuts straight through it in a north-south course. Of lateral roads leading off into prospecting territory, there is none that rates even a dotted line on a road map. A few rutted tracks have been left from early-day activity, but should be followed with caution.

At the tiny settlement of Rye on State Route 87, a Yaqui Indian known as Valentino exhibited a gold-studded piece of quartz that he had picked up at an old shaft in the lower part of the Four Peaks group. He told of a miner's pick, almost rusted away, lying at the site. From his description, the tool would seem to be of an old type used by the Spaniards. Valentino worked as a cowboy for the Bar-T ranch at Rye, and had made his discovery while pursuing a young bull that had wandered far outside the usual grazing ground.

The Yaqui stolidly refused several offers of partnership for working the old mine, and would give out little information about it. A few firebrands at the ranch then plotted to seize the Indian and force him to lead them to the spot. Warned by a friendly Mexican, Valentino fled from the ranch that night carrying his secret away from Rye and the Bar-T forever.

A good many years went by, then we hear of Valentino again. In 1930 a Mr. and Mrs. Henry Hardt of Phoenix were planning a trip to look for the old shaft and the possible fortune that lay in its depths. They had heard of it from Valentino himself. The Hardts had befriended the Indian in his old age. He had told them of the mine and at one time had promised to show it to them. After a terrifying dream of tribal vengeance, Valentino had taken back his offer. He had let a few hints as to its location slip out, but directions were admittedly meager. If the Hardts ever found and worked the mine, the fact escaped both the press and state mining officials.

Not at all meager are the directions to an equally rich mine amid the Four Peaks, but here there is another difficulty—deliberate concealment. A "chimney" of rusty auriferous quartz was tapped by the local Indians from time to time, resulting in a small open pit, and the Indians always carefully filled the pit in. The surface, it is said, was made to look natural by the common expedient of building a campfire on the site. However, rains soaking into such a fill often will cause it to subside, so that a shallow, unnatural depression might well identify this deposit to some present-day gold hunter.

No name attaches to this mine, but perhaps the name Iretaba should be applied. Iretaba was an Apache who, as a boy, removed and sold some of the ore and was punished by his tribe. Also aware of the location was Colby Thomas, a successful Arizona mining engineer who passed the information along to a friend, Ed Abbott of Phoenix. Abbott made an unsuccessful search for the mine about 1950. A waybill to this deposit would read as follows:

From Mesa, Arizona, go north on State Route 87, the Four Peaks access road to which reference has been made. Before reaching the small settlement of Sunflower, watch for a dirt road leading east. At times there have been signs at this turnoff directing one to Cottonwood ranch and to an old-time watering place known as Hughes Well. This well, of dubious water quality, is eleven miles from the pavement. There may still be some debris from a house that once stood there. This is the takeoff point for a search on foot. There are two washes converging at Hughes Well. Take the southern one, leading off toward the towering Four Peaks. Some five miles up the wash some evidence will be found of "an old digging," as it was described to Abbott, and above it on the north slope of the wash and quite high is a "bench," or level area. At this stage the gold hunter will find that time and natural growth have confused the clues. In Iretaba's day one stood on the bench and sighted a conspicuous paloverde tree. Twenty feet southwest of that tree was the rich, albeit carefully concealed, ore chimney. When Ed Abbott made his search this viewing point commanded quite a number of paloverde trees, no single one more conspicuous than the rest.

The mining man Colby Thomas fully believed in the bonanza. The Indian revealed the secret only when he was eighty-six years old, blind, and badly in need of money. His story was that years ago two white men had chanced on the tribal secret and that Iretaba's father was a member of the Apache war party that terminated their operations in the usual lethal way. Such ore as lay about was scattered down the wash and the mine opening concealed as already described.

Engineer Thomas, although eighty years old when he heard the story, went off at once on a one-man search. He picked up a piece of dazzling, gold-studded quartz in the wash above Hughes Well, which he took as confirmation of Iretaba's account. Three days' exploration in the area failed to locate the source. Feeling unequal to the rigors of further exploration, he turned the matter over to his younger friend Abbott. As we have seen, Abbott also returned to Phoenix empty-handed.

In this area the Verde River is the principal stream. Many promising reports come from along its lonely course and from the Mazatzal Mountains, which parallel it on the east. Unfortunately neither of the roads that also parallel the Verde run close to the river itself. From either State 87 on the east or Interstate 17, which is across the Verde to the west, the investigator has to do some arduous hiking and climbing to get to this interesting stream. It is an overnight tramp there and back, at a minimum.

At a lower, more accessible point on the Verde the Army once maintained a Fort McDowell. It is the locale for a story of two young soldiers who had a rich gold mine and thought that their army rifles would be sufficient safeguard against the Apaches. At this time, the year 1870, there was virtually open warfare between the Indians and all-white men in the region, military or civilian. The obstreperous tribe here was the Tonto Apaches. Even so, there were many so-called tame Indians who came and went freely at most of Arizona's towns and army posts. Sometimes their white acquaintances received a shock when such an individual was recognized in another capacity—as part of a raiding party.

In their tame role one or more Apaches had several times obtained tobacco and other supplies at Fort McDowell and paid with resplendent gold ore, consisting very largely of the precious metal itself. Excitement, but with little hope of sharing the wealth, was always stirred up among the soldiers by these transactions. Now two young troopers, their enlistments about to expire, believed they had pieced together adequate clues to the Tontos' secret mine.

The young men made their way into the Mazatzal Mountains via an army trail to a Camp Reno, which was perched high on the range and overlooked the great depression known as Tonto Basin. At this time Camp Reno was being phased out. The departing troops met the young adventurers on the trail, and the commanding officer tried desperately to talk the youths out of their venture. Tonto Apaches were roaming all over the mountains, led by a daring and tricky chieftain named Delshay.

The grim warning by the Camp Reno officer and his men went unheeded. The two parties went their respective ways, and, as years passed, anxious speculation as to the boys' welfare died away. By 1886 the troublesome Indian tribes had been overcome. Geronimo, Cochise, Mangas Coloradas, Delshay, and other leaders were dead, imprisoned, or immobilized on reservations along with their followers.

When spring rains had created some forage on the desert's soil it was now possible to run cattle and sheep. In 1891 two sheepherders were at their leisurely occupation when they came upon five skeletons. "On a high slope on the north side of Mount Ord" was the way they described the location. On two of the skeletons the dry desert air had preserved shreds of clothing, and there were perhaps buttons and buckles found. At any rate, the two herdsmen stated that the remains represented two soldiers and three Indians. A number of discharged rifle shells were scattered about. There also was a piece of gold-impregnated white quartz.

The phelgmatic sheepherders told of their findings in an exasperatingly vague way, could cite no landmark except Mount Ord, and seemed innocent of the fact that they had had a fortune in their grasp.

Old Fort McDowell offers another account of a rich Indian mine that is more specific as to the participants. It is known as the Dr. Thorne Lost Mine. The physician realized $6,000 from his brief acquaintance with the mine, and over the course of years he would take time from his practice of medicine at Lemitar, New Mexico, to recount his experience.

He was Abraham D. Thorne, M.D., a native of Philadelphia, who came to Fort McDowell as a civilian army physician in the 1860s. His father and two brothers also came West, and for years these three Thornes operated a prosperous store in San Jose, California, specializing in imports from the Orient. Dr. Thorne is said to have been a close friend of Kit Carson at the time the celebrated scout had been made an army colonel and entrusted with important military and administrative duties in turbulent Arizona.

While stationed at Fort McDowell, Thorne gratuitously treated various Apache visitors and found them especially grateful for his success in curing some eye trouble that was prevalent among the children. This may have earned him the tribe's gratitude. There is another story, however, that he was taken prisoner by the Apaches and held captive for years in remote mountain hideouts. At length, according to this version, there was a stubborn skin affliction that spread among the women of the tribe and Thorne managed to arrest it with a concoction brewed from certain roots and twigs.

For whichever reason, the tribesmen were grateful and wished to reward the white medicine man. They paid off with gold. If Thorne, in later years, straightened out these conflicting stories the information was not handed down by his family or friends. In the unsettled years with which we are dealing either account is plausible.

At any rate, Dr. Thorne was rewarded by the Apaches. One day at sundown the doctor was placed on a horse and blindfolded. A small party of mounted braves then led him on a wandering course to the tribe's secret mine. They apparently started near Fort McDowell, for Thorne said later that he was aware that the party was crossing and recrossing a nearby river. This of course was to confuse him. After leaving the stream they rode through the night for what he would judge to be twenty miles. It was after dawn when the destination was reached. His blindfold removed, the doctor found himself in a canyon that he had no way to identify, and at his feet was a pile of amazingly rich gold ore. He was assisted in a businesslike way in filling a sack, this sack was slung in front of his saddle, the blindfold was replaced, and the cavalcade started back.

As they were leaving the canyon the mask slipped a trifle. Thorne glimpsed a peak several miles to the south, topped by a very sharp spire. He also saw on the trail what appeared to be the ruin of a very old stone house. His alert guides quickly tightened the blindfold and these were his final clues. Disposed of in San Francisco, the tribe's gift enriched Thorne by nearly $6,000.

For years Thorne respected his implied bargain with the Apaches and made no effort to go back to their secret mine. Then, at the insistence of a brother, a fairly ambitious search was organized. Financed by the brother and a well-known mining man named Bob Groom, a party of eight well-armed men followed the physician's scanty directions. At a certain canyon mouth there was indeed a crumbling stone structure of great age. From this point, too, could be seen a distant rocky pinnacle rising against the sky, which conformed pretty well with Thorne's impression. Through the fine autumn weather the party methodically explored the canyon and its surroundings. There was no gold, there were no favorable indications of gold.

For the rest of his days Dr. Thorne was content with his rural medical practice and with his adobe cottage and little fruit ranch at Lemitar. Out of that one search came a development that later appeared significant. Bob Groom was a man well acquainted with this part of Arizona. After weighing Thorne's story and gathering all the other clues possible, Groom was convinced that the sharp, verticle peak glimpsed from under the doctor's slipping blindfold was Weaver's Needle. This did not mean much at the time. In later days Weaver's Needle became a landmark. It is in the midst of the Superstition Mountains, and if Groom was correct the Apache diggings could conceivably be identical with the Lost Dutchman.

Still another Apache mine is related to early army operations in Arizona. This one takes us, at the start, back to Yuma, where a young West Point graduate, Thomas McLean, was serving as quartermaster. There was active shipping on the Colorado River in those days. The young lieutenant handled much government freight. Whether as a victim of wily traders or as a willing participant, he became involved in some nineteenth-century form of racketeering. He escaped imprisonment but was cashiered out of the Army.

Possibly innocent, at least innocent in his own eyes, he turned away from his own race, married a girl of the Yuma Indian tribe, and became to all practical purposes a member of the tribe. In the river settlements he was known as "Yuma." For a living he called on his quartermaster experience and became a fairly successful trader. As a white man he could do astute bargaining for supplies at Tucson and other points. As an Indian he could safely haul or pack goods into the territories not only of the Yumas but of the Pimas and Papagos, and eventually of the more warlike Arivaipa Apaches. This last tribe paid for its purchases in gold.

Playing on the vanity of a young Apache chief named Eskimizin, who yielded to the lure of a fine gun, an ornate saddle, and other items, Yuma was successful in being shown the tribe's source of gold. It was a vein in a small craterlike depression, which in turn was high up on an exceptionally rugged peak. Weighing miscellaneous evidence, later searchers believed the spot to be near the present town of Winkelman on State Route 77. Victor Stoyanow, who searched for the mine in the 1960s, believed Crozier Peak to be a likely location.

With a few samples of the gold-encrusted quartz in his possession, Yuma took in a partner. This would appear to have been a General Walker, not otherwise identified, who accompanied Yuma to the mine on one occasion. In failing health, this officer took no further part in the matter. A more determined partner was a man named Crittenden. On a fast, secretive trip to the deposit, Yuma and Crittenden carried away thirty pounds of the vein matter. A piece was submitted for assay and it returned the amazing value of $51,000 a ton. Such a find could scarcely be kept secret in Tucson, and the names of six men have been recorded who saw and marveled at the rich ore. The Apaches also heard about it.

Yuma tried to cover up his mining activity by continuing his normal trading routine. With a loaded wagon, his Indian wife, and a group of friendly Papagos, he set out for some of the westerly villages to make his usual merchandising calls. The Apaches struck with their usual speed. Overtaking the caravan, a party of their horsemen ignored the helpless Papagos but dragged Yuma and his wife to the ground and clubbed them to death.

Crittenden, although his body never was found, is known to have met a similar fate. After Yuma's extinction, the partner went into hiding. By the year 1870 he apparently felt it safe to return and make another and larger haul from the great bonanza. So safe, in fact, that at Camp Grant, which was to be his point of departure, he talked freely about his plans. He left the camp well armed and riding a fine horse, and headed downstream along the San Pedro River. Ten days later an army patrol found his horse, tethered to a tree and near death of thirst. The man had managed to get only ten miles from the army post. Later Crittenden's revolver was discovered, all its chambers fired, evidence of a final and futile defense against those "Tigers of the Desert," the roving Apaches.

Several places have been mentioned in this report as having the reputation of dependable gold placer ground. This must be construed in a modest way. These places are regarded as dependable in the sense that a miner down on his luck can usually eke out a living by working their gravel through a long working day and with ample patience. In central Arizona, Black Canyon Creek has such a reputation. It is not to be confused with the Black River, farther to the east. Bumblebee, sixty miles north of Phoenix and just off Interstate 17, is the nearest settlement. A better approach is via an abandoned road taking off from Interstate 17 just north of the Agua Fria River and leading to the old Maggie lode mine. This will bring one closer to the auriferous gravel than a route through Bumblebee.

It will still be a long hike on foot to the narrow valley of Black Canyon Creek. Placer gold has been recovered all along this stream, from its confluence with the Agua Fria on the south to beyond where Arrastra Creek comes in from the east. One cannot depend on finding flowing water, or even water standing in pools, in these desert "creeks," despite the encouraging term. Sometimes there is water for panning, sometimes dry washing may be necessary.

A short way below Arrastra Creek the main stream meanders to form about two thirds of a circle. As might be expected, the *inner* edge of this curve is a likely spot for the deposition of gold, and parts of the little flood plain within the circle have been worked in the past. An old dam on the east side of the big curve helps identify the spot. In view of the known gold content of the gravel all along Black Canyon Creek, the modern prospector might do well to avoid the scene of any older workings and hope to discover some virgin deposit for himself.

There are several points of interest near, but not within, the placer area. The Carl Pleasant dam and lake on the Agua Fria is one, and nearby is Castle Hot Springs, a venerable "get away from it all" resort. Eighteen miles north of the dam is historically interesting Horse Thief Valley. The area in between consists mostly of the desolate Bradshaw Mountains, out of which have issued some tempting but fragmentary reports of gold discoveries.

The immediate area of Black Canyon Creek offers a vague account of a rich surface accumulation of gold. Unfortunately the information, dating from 1916, lacks the name of the rancher most closely concerned with the matter. We know that in 1916 two Mexicans, brothers and both clearly of advanced age, were befriended by a local rancher, possibly living at the old hamlet of Menafee. They were looking for some shallow diggings worked long ago by their father. The latter had been forced to leave the site when his water and food were exhausted, but the brothers said he had brought enough gold back to Mexico to keep his family in comfort. Now, in harder times for the family, his two sons were following a crude *derrotero* the old miner had left.

They were not successful. The rancher recalled that the location they were seeking was on a high, black mesa west of Black Canyon Creek. The discoverer, their father, had recognized the gold-bearing formation as an ancient stream bed. Such deposits, lifted by volcanic convulsions and now lying at a considerable elevation, are often a profitable source of gold, as we have seen earlier in this report. The original Mexican had made his haul from two excavations, each, he said, about the size of a wagon. As the story went around the sparsely settled Agua Fria country, the rancher, now an elderly man and too old to tramp the rugged canyons, recalled a deer hunt he had engaged in back in 1901. Atop just such a black mesa, he had seen and wondered at two old excavations of about the size described.

It would be disappointing to leave southern Arizona without reference to Tombstone. The little town has been well publicized in print and on film for its feuds, its outlaws, and its stalwart lawmen, real and fictional. Less well known is the matter of its mines and what happened to them.

Tombstone is situated about sixty miles east of the early mission settlements and about twenty-five miles above the Mexican border and its twin towns, one on each side of the line, both named Naco. Tombstone is scarcely a mining district to attract the individual prospector or small enterpriser today. Its mines were large operations and the best are now consolidated under corporate ownership. True, there may be possibilities in several little mountain ranges, more or less mineralized, which surround the town. These are the Dragoon, Huachuca, Burro, Mule, and Whetstone ranges. But it should be remarked that during Tombstone's great days, from 1880 to 1886, there were at least 20,000 men in the area. They were watching the Schieffelin brothers, the Williams and Oliver partnership, the assayer Richard Gird, and others becoming rich from lucky surface strikes. One must assume that all this country, including the ranges mentioned, got a pretty thorough going-over. As in other intensively prospected districts, the hope here would be that a good vein had been uncovered by nearly a century of erosion.

Tombstone was 100 per cent a silver camp. The great mines were the Tough Nut and Lucky Cuss (both discoveries of Ed Schieffelin), the Contention, and the Grand Central. Tourists, after viewing the Bird Cage Theater, the OK Corral, scene of the Earp-Clanton gun battle, and other reminders of the old days may peer into the Million Dollar Stope, a great gouge left in the earth by extraction of a compact body of extremely rich silver ore.

Although Tombstone lies on a desert plateau almost void of streams or springs, at the depth of only five hundred feet there is a remarkable water table. This water was the finish of the mines. The shafts flooded. One mine expended $300,000 for a giant Cornish pump of the type used on the Comstock Lode. It was found that the water seeped in from the entire saturated stratum, that the owners were trying to pump out not only their own mine but all the mines of the district. The effort was a failure. After several quiet years the better mines were consolidated and an even greater pumping system was set up. Fuel costs alone for this massive machinery ran $700 a day. It was found impossible to show a profit even with good ore, and when carelessness and general bad luck resulted in a new flooding and virtual ruin of the costly equipment, Tombstone's days of silver mining came to an end.

There are a few episodes in the Tombstone story that may be instructive.

One concerns Ed Schieffelin, recognized as the discoverer of the Tombstone deposits. He carried samples from his rich strike about with him for months, unsure of their value. He could not interest his own brothers in the specimens. It was only the practiced eye of the assayer Gird that saw possibilities in them. An assay showed the ore to run $2,000 a ton. That finding was the beginning of Schieffelin's success and the Tombstone boom.

Another point is that this first location, the Graveyard Mine, contained only a pocket of good ore. It might have been the end of things for the far from confident Ed and his hesitant brothers. But exploring near the Graveyard, Schieffelin discovered another outcrop. Here the ore ran $15,000 a ton and the deposit became the immensely profitable Lucky Cuss Mine.

There is an incident parallel to this that ends on the negative side. The very first digging in the Tombstone area was known as the Brunckow Mine. It was worked by a German, Frederick Brunckow, reputed to be a scientist in some field, who had emigrated because of political trouble. Brucknow was killed by Indians, probably in 1878. When such rich discoveries as the Lucky Cuss and Contention were made nearby, these original diggings were relocated, explored intensively, and worked by several different owners or leasers. No values sufficient to show a profit ever were found. Although Brucknow did not live to hear the bad news, he was working a formation that never could have developed into a profitable mine.

The great open-pit copper mine at Ajo has been mentioned and there are other great copper properties well distributed through Arizona. As we move north the dominance of the red metal in Arizona mining becomes more evident. In their first century of recorded production, Arizona mines poured out metals valued at more than $3 billion. Copper accounted for 85 per cent of this sum.

At first copper was regarded as an unwelcome intrusion in silver mines. The mining center of Globe takes its name from a domelike formation of silver ore. At depth the silver content all but disappeared but copper was present in large quantity. This deposit became the Old Dominion Mine, one of the earliest and most lucrative of the great copper properties. The workings are now flooded.

An even greater copper bonanza was discovered to the northwest, the vast deposits that ran through Mingus Mountain, on which stands the ghost city of Jerome. This was the United Verde, mentioned earlier as one of the most productive mines in history. So much of Mingus Mountain was hollowed out by United Verde's massive extraction of ore that the city of Jerome was endangered. Buildings twisted, tilted, and collapsed and some slid down the steeply sloping townsite. This destruction, together with the closing of the great mine, left the town all but deserted as to permanent residents, but its eerie rows of empty buildings have made it something of a tourist attraction.

There are other great copper mines in Arizona, all producing some gold and silver as subsidiary values, but they have little interest for the individual prospector. In the Jerome area, however, we are near the Coconino National Forest and a mystery that takes us back to gold. This concerns not only a lost mine but an entire lost village.

The area in question, in the southeastern part of Coconino County, contains many scenic features which, if not known to any large public, have been discovered by artists and photographers. There are striking red and white rock formations in Oak Creek Canyon, and the town of Sedona, nestled among them, has been a seasonal residence for several outstanding figures in contemporary art. The Tuzigoot ruins from a forgotten culture are nearby, and a little farther east are Montezuma's Castle and Montezuma's Well. About thirty miles north of Sedona looms Humphreys Peak, at 12,611 feet elevation the highest point in Arizona.

Interstate 17 runs through this region, but to get to most points of interest and into the heart of our gold-town puzzle one should swing on to the roughly parallel U.S. 89. Most of the country traversed will be in the Coconino National Forest. This is a higher, moister region, unlike the desert terrain that is typical of most of Arizona.

Somewhere in this large timber reserve, and certainly well apart from the colorful sedimentary rocks of Oak Creek Canyon, is an abandoned Spanish mining village of considerable size. It presents the paradox of having been photographed in quite recent times and yet to have eluded discovery.

It is speculated that the gold at this site was discovered during the Antonio de Espejo expedition of 1583, and that extensive mining operations were soon being organized. We know that the Opata Indians of Sonora at times were employed by Spanish church or lay enterprises to accompany them north, for many Opatas were experienced miners. Research by Gladwell Richardson indicates that a work force of perhaps two hundred of these tribesmen was organized under competent Spanish overseers to exploit Espejo's discovery. Such an ambitious project would account for the number of dwellings erected at the mine, so many that the abandoned settlement is sometimes referred to as the "Lost City of Coconino County."

There is some historic background to indicate that the gold veins were worked off and on from about 1720 until 1760. If correct, this activity would be more than a century after Espejo's visit, and we know nothing about what happened in the intervening time. The end doubtless came as it came to other Spanish mines, with an Indian onslaught. The hostile Hualpai tribe lived nearby and the place was not out of range of roving Apaches.

It was also an Indian attack, long after the Spanish operations, that made the existence of the place known to American miners and prospectors. Among the adventurers who spread through Arizona after the Mexican war was a party of four who prospected out of Tucson. They made the mistake of entering the Hualpais' domain. A sudden attack disposed of three of the intruders. The fourth man, named Cliff Haines, was apart from the others and managed to get to his horse and escape. He knew he would be pursued. Riding furiously through unfamiliar country, Haines found himself at the head of a long canyon that looked promising as an escape route, then found himself on the edge of a substantial settlement. It was clear at once that this was a ghost town. In his hasty inspection, he related later in Tucson, he saw several tunnels running into the canyonside, with the usual dumps of ore or waste rock in front of them. Buildings, some of stone, some of logs, stretched along both sides of a street. There were signs that, whenever the place may have been abandoned, the departure was sudden. Ancient wooden carts stood about, half overgrown with brush. Wild growth all over the street and nests of pack rats in the houses added to the desolation. Toward the lower part of the settlement Haines observed old machinery and some adobe structures that could have been primitive smelters, the *vasos* of early mining records. His impressions were necessarily gained hastily because he was fleeing for his life. He kept ahead of the pursuit and emerged from the Hualpai country with no wish to enter it again. This was in 1853, a long time before peace was imposed on the Indians.

In Sedona and Oak Creek Canyon you might bypass scenery and recreation for a time and seek out some old prospector. The old Spanish mining town remains a lively topic with such men. They can tell you of a John T. Squires, who knew Cliff Haines and may have obtained a map from him. Squires, they say, found and worked the old mine barely twenty years after Haines's adventure, but the inevitable Indian attack closed down the operation. Squires himself escaped on that occasion, only to meet his end through some unspecified violence in Taos, New Mexico.

Even less successful was another rediscoverer. This was W. O. Howard, a professional hunter. He came upon the lost village quite by accident, and at the time was merely curious about the deserted houses. He went on about his hunting. After Howard learned that the old dwellings marked the site of a probably valuable mine he spent several years in futile efforts to retrace his steps. The hunter was an experienced outdoor man. His failure emphasizes the difficulty of the location.

Later there were searches by men named Milton Farrell, Melvin Halliday, and Robert Cahill. An airplane pilot felt sure he had sighted the town and diggings from aloft, but on trying to reach the spot on foot he experienced only days of frustration. An old log bridge, said to be an authentic remnant of the Squires enterprise, was found by one search party. Such a bridge should be part of a trail, but beyond the rotted timbers lay only trackless forest. The isolated bridge was a puzzle in itself.

The photograph of the mysterious village was made by a visiting Easterner who was staying at Sedona. This man enjoyed long hikes through the woods and would pack supplies on his back for overnight camping. On one occasion he was out for several days and turned up in tiny Perkinsville, twenty-eight airline miles west of Sedona. The intervening country is rough and uninhabited. The hiker had little stamina left, but after resting he was eager to tell how he had come upon a deserted town far back in the forest. He had taken pictures of the place.

Back in Sedona he had his films developed and printed, and when he left for his home on the Atlantic seaboard he took the lot with him. His story, at the time, appears to have aroused little interest.

Now the photo shop proprietor, acting on some impulse he couldn't explain, had made several extra prints from the hiker's films. Still living in the Sedona area was an elderly miner who had been a member of John Squires's work force. The photo prints were shown to this old-timer. In one of them there appeared stone buildings against the background of a rocky canyon wall. The miner unhesitatingly declared that this was a picture of the old Spanish mining village.

These incidents, if correctly related, roughly indicate the field of search. As we know, the hiking camera carrier was staying at Sedona and his long trek from there carried him into Perkinsville. This is almost due west from Sedona. How far the man roamed north and south of such a direct line can only be guessed. A large-scale map, correlated with U. S. Geological Survey quadrangle maps, would give a good idea of the country between the two towns. It would seem to set reasonable limits, though still rather wide ones, for the search. If you visit Coconino County, Arizona's largest, with its vast extent of uninhabited forest, it will be agreed that any kind of limitation is helpful.

The northeastern section of Arizona, interesting to the archaeologist and containing much striking scenery, is not prospecting territory. The brilliantly colored sandstones which form the spectacular towers of Monument Valley and the cliffs of Canyon de Chelly are not formations that carry gold or silver. Moreover, Navajo and Hopi Indian reservations cover most of northeast Arizona. Their combined areas extend nearly 150 miles west from the New Mexico state line, and an equal distance southward from the border of Utah. U.S. 160 cuts diagonally across the Indian lands, but the gold hunter would be well advised to keep on traveling until this highway connects with U.S. 666 and enters the highly mineralized zones of southwestern Colorado.

In such an extensive region of sedimentary rocks there can of course occur igneous intrusions that would make ore. One example of this in the Navajo country is recalled by the names attached to two colossal red buttes that rise from the floor of Monument Valley. They are officially named Merrick and Mitchell buttes, and the names commemorate two early prospectors who, to their undoing, discovered the secret Navajo silver mine of Pish-la-ki. The most familiar pointers to the mine are the two Mitten Buttes. Both formations suggest mittens as to shape but it would take a hand eight hundred feet long to fill either of them. Across a wide wash from the Mittens is the butte bearing the name of Colorado miner Jack Merrick, and a little beyond is one called Mitchell Butte for his twenty-one-year-old partner.

Unlike Indians in many parts of the West, the Navajos of this region are not recorded as ever having any gold for barter. They did, however, appear at trading posts with jewelry fashioned out of native silver. Such ornaments of primitive design have had a wide sale for years throughout the Southwest and still are much in demand. When he left his Colorado diggings in 1880 for Navajo country, Jack Merrick must have known about Navajo silver and have had rather precise directions to Pish-la-ki. It never came to light how he had obtained this information. He returned to Durango, the trade center of southwestern Colorado, and displayed ore sparkling with visible silver and assaying $800 a ton.

To help work the mine, Merrick took young Mitchell into partnership. They packed up for their trip at the former's home in Rico, Colorado, and started on their way leaving their friends well aware of their plans. As weeks went by they were assumed to be mining somewhere out in the sandstone country. Then the mysterious Indian underground brought word back to Rico that the men had been killed.

At this time there was a peaceable, self-sufficient colony of sheep-raising Navajos living near the Mitten Buttes. Their leader was a venerable warrior named Hoskininni. Friends of Merrick now came to the reservation to investigate. Hoskininni readily confirmed the news heard in Rico and led the Coloradans to the remains of the slain men. The killings, the Indian declared, were the work of roving Utes from the north, not of his own people. Several bags of rich silver ore were found beside what vultures and coyotes had left of the bodies. Under the eyes of Hoskininni and a dozen impassive Navajos, the men from Rico decided to accept the chief's story—also to leave the ore and the mine alone.

There were sequels to the Merrick-Mitchell episode that confirmed the existence of the rich silver ledge. In 1939 the old chief's son, Hoskininni Begay, told Charles Kelly of Torrey, Utah, the story in essentially the same form as has just been related. He added that the location of Pish-la-ki had been known to only seven of the very old Indians of the colony and that the last survivor of the seven had tried, but unsuccessfully, to pass the information on as he lay dying. The present patriarch, Hoskininni Begay, who was eighty-one years old as he talked with Kelly, asserted that he had no idea how to reach the deposit and that the tribe now had no supply of native silver.

Yet in some way the secret was handed down. Randall Henderson, founder of *Desert* magazine, was told about it in 1950 when visiting the Harry Goulding trading post in Monument Valley.

Goulding was a lifelong resident of the reservation and a good friend of the Indians. He spoke the Navajo language fluently. Over the years one of his Indian customers had let it slip out little by little that he knew the way to Pish-la-ki. Goulding had no wish to seize the mine, something which in fact would have been a legal and practical impossibility. However, there came a time when Goulding tried to reopen the diggings from purely altruistic motives. It was a year of bad weather and hard times for the sheep-raising and corn-growing Navajo. Goulding did what he could to help, using the limited resources of his little trading post. The keeper of the silver secret, whose name was Yazzi, at last made a deal. The Indian, his family, and his neighbors were very near starvation. He agreed to let the trader supply him with several pack burros and rations for a three-week stay, so he could go dig out enough silver to support his family.

In six days Yazzi was back, on foot, and with no pack burros and no silver. It was the old story of the Indian taboo. In this case, the gods had shown their displeasure with a drenching rainstorm. Actually, the change of weather was a boon to the Navajos, their sheep, and their crops. But Yazzi interpreted it otherwise. He turned his back on Pish-la-ki and fled on foot. The Good Samaritan Goulding resignedly rode off and spent several weary days rounding up the burros. If he saw any evidence of a silver mine he never afterward mentioned it.

Some distance south of the Navajos' land and in country geologically more promising is the San Carlos Indian Reservation. An L. K. Thompson, ranching in the Salt River Valley near Roosevelt Dam just before the turn of the century, knew of a massive deposit of silver ore there. At that period the long and bloody Indian conflicts were fresh in memory, and government officials were intent on seeing that such tribes as had settled down on reservations were protected against intrusions by enterprising whites. Such an official was W. J. Ellis, acting Indian agent at San Carlos.

When a party of Mexicans arrived at his headquarters to look for a lost mine, they carried excellent credentials from the governor of Sonora. Ellis felt bound to allow them to search. He extracted a pledge, however, that should the deposit be in the clearly defined reservation boundaries they would make no claim to it and would peacefully depart.

The rancher Thompson, though an American, was related by marriage to the Pedro Encinas family of Sonora, one of whose forebears had long ago made the discovery. The deposit had been described as a series of veins on a mountainside, some of which were almost half silver. The original Encinas had been an old man when he made his strike. He then had rapidly deteriorated in health but before his death had left his family a clear set of directions. As a relative, Thompson had been enlisted in the project by the arriving Sonorans.

The directions proved quite accurate. Thompson said later that his group was able to go directly to the scene and that there was no question but that this was the Encinas discovery. There was no digging done. It would appear that some Indian agency official personnel were with the party to enforce the agreement, and that the silver veins were definitely inside the reservation boundaries. The Mexicans departed, Thompson went back to his ranch.

Now that Indian tribal councils can enter into contracts for various types of concessions on their lands, a present-day discoverer might do better than the disappointed explorers of the 1890s. The clue to finding the old waybill, if it still exists, is a slender one. Someone named Encinas, possibly living in the state of Sonora, with an ancestor named Pedro—has this person a mysterious old paper among the family heirlooms?

10
New Mexico and Texas

From the earliest historical times, New Mexico has produced gold and it also has produced some historic fictions concerning gold. It was in New Mexico that Coronado expected to find the golden Seven Cities of Cibola, and it was primarily New Mexico that was the lure for Marcos de Niza, Antonio de Espejo, and assorted later Spanish explorers.

Gold is indeed found in most of New Mexico's thirty-two counties, only the flat, arid Llano Estacado, or Staked Plain, in the southeastern section being unproductive and unpromising. The state's total yield of gold and silver is not notable. In an unusual proportion of lode mines the vein has been lost by faulting, the ore terminating against an intrusion of barren rock which miners refer to as a "horse." Placer deposits of substantial value usually will be worked regardless of adverse conditions, but in New Mexico even the placers were unduly handicapped. Water too frequently lay at an impossible distance and laborious dry washing was the only expedient. With New Mexico specifically in mind, Thomas A. Edison applied his genius to the design of dry-wash machines. With some skepticism, a published report may be noted which said that the Wizard of Menlo Park lost $2 million on the venture.

Silver has yielded about the same dollar value as gold in New Mexico. In its deposits of the white metal, New Mexico cannot be compared to such silver states as Nevada and Idaho. The possibility of finding and profiting from an old silver mine, however, may well be equal to that in more favored states. The reason is that New Mexico felt the full fury of raids by Geronimo, Victorio, Mangas Coloradas, and other Indian war chiefs. Many miners were slaughtered at their workings. In New Mexico, it could reasonably be argued, many a good mine was abandoned for this reason alone. Some such mines may be identified by patient search of early records; a few are definitely lost.

New Mexico's most celebrated lost mine, the scene of two successive massacres by Apaches of the Mogollon Plateau, was not one of silver. It might be described as a double gold mine. There was—and is—a creek in a remote canyon so rich that its discoverers panned out $60,000 in dust and nuggets in three weeks. And in the walls of that same canyon there were—and are—at least two veins of quartz inlaid with straggling designs of native gold, incredibly rich ore.

This site is known as the Adams Diggings and it constitutes one of the classic lost-mine mysteries of the West. A town in Catron County has optimistically taken the official name of Adams Diggings, but all New Mexico runs to flamboyant place names, some in English, some in the more expressive and musical Spanish. The only geographical point that definitely connects with the Adams adventure is seventy miles north of this town. It is Fort Wingate. This old frontier post is near Gallup, an important supply point in the sparsely settled west-central part of the state. Gallup is on transcontinental Interstate 40, and also on U.S. 666, a major north-south artery.

The man Adams seems to have left no first name in the record. He was the leader of the tragically successful expedition to the hidden canyon, and was one of three men who happened to be apart from their fellow miners when the Indians attacked. Of these survivors one was a prudent German who, at the first opportunity, took his share of the gold and returned to his native land. Another survivor, named Davidson, died soon after the episode. Adams lived on. It is from his telling and retelling his story at Milligan's Plaza and other New Mexico towns that the circumstantial and consistent account has come down to us. Corroboration of a kind comes from a Captain C. A. Shaw, former skipper of a merchant ship, who was a friend and backer of the troubled gold hunter over many years. Shaw and several less reliable persons brought out printed pamphlets telling the Adams Diggings story. Such informative mining pamphlets, purporting to give tips and directions to those heading West, were widely sold during the gold and silver rushes.

Adams was a teamster, not a miner. The stage of his life that is relevant here begins in 1864, when he found himself at a Pima village in Arizona with twelve horses that he was taking to Los Angeles. Also at the village was a party of California placer miners. They were excited.

A youth of mixed Indian and Spanish parentage had been telling them of gold in a distant canyon to the east. The tale would seem akin to those that led Coronado and others off on their quests, but a tale of gold will always get listeners. There was a difference in this case. The informant, dubbed Gotch Ear because of some malformation of that member, was telling the truth.

Had not Adams happened along with his twelve horses at this time the search could not have been made. It involved a long journey and Gotch Ear was frank to say so. The miners had a good store of provisions but few horses. For the use of Adams' providentially available mounts, they willingly put the owner in charge of the expedition.

The starting point is believed to have been about twenty miles north of the Gila River, at what is now Gila Bend, a town on Interstate 8. Early accounts have the trip taking only eight days. Later searchers, putting together various episodes and clues, believe the party must have been on the trail much longer. In any case, it is agreed that as they neared the hidden canyon they were in *malpais*. Literally "bad country," this is a term widely applied to bare, lava-covered expanses of the Southwest of no discernible use to man or beast.

Beyond the malpais, mountains could be seen. Gotch Ear told his companions that there would be only two more days of travel. They were to find the gold in a canyon. The entrance, their guide said, was well concealed and would never be discovered by chance. All his statements were soon confirmed. High in a steep mountainside, virtually a cliff, and apparently unbroken, there was a narrow cleft. Once through this, the travelers looked down upon a small green valley threaded by a winding creek.

Descent from their entrance was by an old Indian trail that followed the shape of a sprawling, irregular "Z." The name "Z Canyon" and "Zig Zag Canyon" were used later by Adams and other searchers. A slow, careful descent brought horses and men safely down to the banks of the creek. The sands of its stream bed were quickly being scooped up and panned for gold. Gotch Ear's story was quickly confirmed. This was indeed a bonanza.

Unlike certain other beneficiaries of Indian help, the Adams party kept their bargain with the guide. He was given two horses, a saddle, two $20 gold pieces, and several items of clothing that he fancied. The young half-breed then took his leave and was not heard of again.

It was about the first of September. For several months ahead the weather would be ideal and the elated miners were intent on making the most of it. Some of them were too intent. That is why the diggings are lost.

There were two narrow meadows along the creek that flowed through the canyon. They were separated by a small waterfall. The miners, panning out gravel and working on a log cabin in the lower meadow, became aware one day that Indians had arrived in the upper one. They proved to be Chiricahua Apaches and, if not friendly, they were not overtly hostile. A group of them quickly appeared at the workings and the warriors stood impassive as their leader pow-wowed in Spanish with the miners. At the end, it was understood that the white invaders could continue their gold washing, but under no circumstances were they to go upstream past the waterfall. It had been a critical situation. The truce was unexpected and thankfully received by the Adams party.

The splendid recovery from the stream gravels was pooled for later division and deposited under a stone in the partly finished cabin. Only one miner, the German, preferred to keep his gold to himself.

Now the original description by Gotch Ear back at the Pima village had dwelt less on placer gold than on rich veins which he said occurred in the walls of the canyon. He had seen a $20 gold piece in a miner's hand and had indicated, as the party understood him, that gold in masses of that same size were visible in the veins and could be dug out with a pick or even with a knife. Nothing of the sort had been discovered so far. Most of the party were busy recovering gold from the stream.

The Indians having silently disappeared, evidently through a second secret passage, an ill-starred miner ventured above the waterfall. He found a high-grade outcrop just as described by Gotch Ear and came back with a large piece of native gold. Adams, still in command, was furious. But the news was out, and from that time on a few men were often missing from the placer workings. Adams was pretty sure where they were.

By this time provisions were low and additional tools had been needed from the start. Two days before reaching the canyon, the group had passed some shallow ruts in the sand. Gotch Ear had called attention to them and remarked that this was a road leading to Fort Wingate. The early army posts regularly sold supplies to miners and passing travelers. Six men were now selected to take most of the horses and go to the fort and obtain a long-lasting stock of supplies. They of course were amply able to pay. Adams later said that the common pool of gold at this time amounted to about $60,000. The German miner asked to be added to the outgoing party. He made it clear that it would be a one-way trip for him. He took the Apaches seriously.

Eight days went by and Adams became uneasy. The morning of the ninth, he and the miner named Davidson climbed the zigzag trail to look out toward the distant road. The look was unnecessary. At the narrow, hidden entrance to the canyon, a spot well contrived for an ambush, the delay was explained. Five men and three horses lay dead, pierced by Indian arrows. The stunned pair had little time to consider the situation. There now came wild shouts from the canyon below. Creeping back a little way along the Z, keeping behind cover, Adams and Davidson looked down and saw the cabin in flames. Apaches swarmed around the fire, and it was clear that the remaining miners had been slaughtered.

This was the last glimpse that Adams ever had of the secret, gold-laden canyon. He and Davidson fled down the outer wall and struck out into the inhospitable malpais. They used its scant vegetation to hide during the day and traveled by night, ineffectively trying to set a course by the stars. It was a desperate flight, and although they eluded the Indians they almost perished of thirst and starvation. It was thirteen days after the massacre that they got close enough to Fort Wingate to be sighted by an army patrol.

The fate of the Adams party now stood as follows: Five men of the returning supply party lay dead at the entrance pass. All those in the canyon, believed to have numbered eleven, had been slain. A sixth member of the supply party, named Brewster, was unaccounted for, but as he never was heard of again it was concluded that he also was killed in the ambush but that his body lay out of view. The German had continued on from Fort Wingate, satisfied with such gold as he had. Years later he was found and interviewed in his native Heidelberg. Gotch Ear had long since departed. Adams and Davidson, both in deplorable physical condition, were cared for at Fort Wingate.

Davidson did not long survive the famishing, exhausting flight through the malpais. The tougher frame of Adams, the desert teamster, carried him through. Although subject to nightmares and occasional hallucinations, he returned to good health and lived to the age of ninety-three.

As was to be expected, Adams promoted and led several searches for the fabulous canyon. The early ones had to be paramilitary expeditions, for a battle with the Apaches appeared inevitable. Much later, the Southwest became fairly safe for the individual traveler. Adams' demise of a heart attack is said to have occurred when he was on a solitary tramp through the Fort Wingate country.

The very term Lost Adams Diggings tells us that none of the searches was successful. There is a report of a letter written to a sister shortly before he died by Davidson, in which he tries to give directions to the bonanza. It indicates that the diggings are in or near the area now included in the irregularly shaped Zuni Indian Reservation, south of Gallup. Geographically this is quite possible. The sister got a search under way but it seems to have been quite perfunctory, and the present whereabouts of Davidson's letter is unknown. The gold-laden creek could be a tributary of the Zuni River. The towns of Ramah and McGaffey are far up two separate branches of the Zuni and outside the reservation. For a search within the Indian lands, the tribal council would have to be approached.

Hints from another source, not connected with the Adams party, point to the upper reaches of the Black River. A Lieutenant W. H. Emory wrote in a report on an army survey that the Black "flows down from the mountains freighted with gold." The lower course of the river is in Arizona. Emory told of a party that went to its headwaters and found lavish placer deposits, but most of the party were killed by Indians. As far as it goes, this could describe the Adams group.

Adams himself placed most emphasis on finding the skeletons of the ambushed supply party. He plainly had no definite recollection of where to look. The catch in the matter, in Adams' time and also today, is that the scene of the ambush, which also is the entrance to the golden canyon, is well concealed. There is a narrow passage in the rocky mountainside, but it blends in with its surroundings and is not to be distinguished at any distance.

The malpais is not a tempting place to hunt. Except for a highly unlikely anomaly, it is devoid of mineral indications. On the omnipresent lava, rough, sharp surfaces make walking difficult and there are dangerous clefts much like the crevasses of a glacier. Yet, hardy desert men have gone in and out of the malpais safely enough, and until the Lost Adams Diggings are found other men will go too.

New Mexico is the nation's pioneer mining state. Spaniards worked a few gold mines in the late 1600s and unmistakable evidence of these early operations have been found northeast of Taos. Indian diggings go back into antiquity. The tribesmen were taking copper from the great Santa Rita deposit in southwest New Mexico when the Spaniards arrived, and turquoise, always prized by the Indians, was being mined near Santa Fe. All unnoticed by Americans in the East, Spanish and Indians alike were working the rich Cerrillos gold placers from 1828 onward. The two deposits lay between the present Albuquerque and Santa Fe in the Ortiz Mountains. In 1833 gold was found in place and the successful Ortiz Mine developed—more than a decade before Marshall's gold discovery in California.

A brief review of mining since New Mexico became United States soil may start in Taos and Colfax counties. They are in the far north, adjoining the Colorado state line. Taos itself was a small but important early-day freighting and stagecoach stop. It was a picturesque adobe village near a large Indian pueblo, and in recent times has had a colony of artists.

Beyond the lofty mountains that tower above the pueblo there were numerous discoveries in the 1860s and 1870s. Values were chiefly in gold. The Aztec Mine recorded $3 million in gross values. Another rich mine was the Mystic, but the record becomes blurred when we try to get at the details of its values and operation. The fact that successive owners of the Mystic were mysteriously decapitated may account for the sketchy bookkeeping.

The original partners in the Mystic Mine were named Stone and Ferguson. Both the prosperity of the operation and the curious series of events date from 1879. In that year the mine was operating profitably and both men could look forward to a handsome income. But late that year Stone went into the woods on an undisclosed mission and never was seen again. Prosperity did not benefit the remaining partner, Ferguson. He became addicted to drugs and died in a mental institution. Ferguson's half interest in the mine passed to his daughter Teresita, who from her mother's side of the family was said to be heiress to a vast Spanish land grant.

A few years later we find Teresita, together with an Englishman, Arthur Roche Manby, and a man named Wilkinson, operating the Mystic Mine with great success. Manby built a large home near the mine and furnished it in luxurious style. The man Wilkinson, of whose earlier life we know nothing, now disappeared. His body was eventually discovered, with the head chopped off. Manby survived his associate by seven years. Then he, too, was decapitated. The corpse was found in the comfortable library of his big house, the head found hidden behind a trunk in an adjoining room. We have noted that one of the original owners, Stone, disappeared and that his body never was found. Wherever it lies, one wonders whether it is with or without an attached head.

The English writer Harold T. Wilkins, whose investigative interests lay in treasure trove and unsolved, mysterious phenomena, tried to get to the bottom of the strange sequence of events in the Mystic's history. The FBI, the French Deuxieme Bureau, and the British Embassy in Washington were queried by Wilkins. His account raised the question, which has gone unanswered, as to why a ranking French detective was sent from Paris to distant New Mexico to investigate the case.

From the standpoint of this report, it can only be said that the Mystic was obviously a valuable mine. What has happened to the workings and to the ownership since the decisive termination of Manby's management is unknown. As for Teresita, she vanished from the Taos area, and one hopes she went to claim her hereditary land grant.

The principal camp in the mining district near Taos was Elizabethtown. It is now a ghost except for some summer recreational use. A companion camp, Eagle Nest, may be reached by U.S. 64, which also serves colorful Taos.

Much of the mining by early American arrivals in New Mexico was in the central part of the state. The old scenes may be approached by way of Socorro, a town on Interstate 25 which retains much of its early colonial atmosphere. The New Mexico School of Mines is situated here.

A little way to the west, near U.S. 60, are the old camps of Kelly and Magdalena. The Kelly and Graphic mines stand as examples of how a prospector may follow up an uncertain piece of information and come into a fortune.

An interesting ore sample had been picked up in this area in 1862 by one of a passing detachment of Union soldiers. Shown to young J. S. Hutchason, it inspired him to go exploring the region and he filed a claim on the Graphic outcrop. When he discovered an equally promising site nearby he generously turned it over to a friend, Andy Kelly, and the mine and camp took the latter name. Good silver-lead ore in these and other properties brought a sustained boom. The enterprising Hutchason operated the Graphic at a steady profit, took over the neglected Kelly mine, and finally sold his holdings for a good price. By chance, some miner became interested in a greenish rock that had been thrown on the mine dumps as worthless. It turned out to be a high-grade ore of zinc. This was the basis of a new boom for the Kelly-Magdalena area, featured by heavy investments by the Empire Zinc Company and other corporations. Here one recalls the Comstock Lode and the manner in which its rich silver ore was tossed aside until some passing Mexicans recognized the *plata* of their homeland.

Continuing west on U.S. 60 one comes to Pietown, on the crest of the Continental Divide. From this high point can be seen an immense tract to the northeast of Pietown that calls for exploration, an area of very close to 5,000 square miles. It is roughly rectangular, with the Continental Divide as the western boundary. The Ladron and Gallinas are the principal mountain ranges and a number of streams, not necessarily flowing at all seasons, traverse the area. Except in the vicinity of Albuquerque there is only one improved road entering the region. This is State Route 23, only fourteen miles in length and terminating at the ancient Indian village of Acoma, perched atop a sheer rocky mesa and often referred to as the "City in the Sky."

East of Socorro is the old mining camp of White Oaks, reached by U.S. 380. It is deserted today, the nearest inhabited place being Carrizozo. A stranger named Wilson, who had stopped for the day with some hospitable placer miners, discovered the first great gold deposit at White Oaks. This man made a climb up a nearby mountainside and, returning about dusk, showed his hosts some pieces of rock which he had knocked off an outcrop. The miners, Harry Baxter and Jack Winters, went into action. The weary stranger was induced to make the climb again, and by lantern light the necessary procedures were gone through to perfect a legal location of a claim. The property, which they named the Homestake, yielded more than half a million dollars in gold, and an extension, called the South Homestake, proved equally valuable.

At the time of discovery a nonchalant Wilson declined a one third interest in the mine. He had merely wanted to give his horse a rest, he said, and would be on his way. Baxter and Winters presented him with all the money they had, about $40, gave him an extra horse, a bottle of whiskey and a good rifle, and watched him vanish into the night. They surmised later that he was a fugitive, involved in the bloody Lincoln County cattle war which was going on at the time. The stranger "Wilson" had climbed the mountain to look over the landscape for pursuers. The discovery of a rich gold mine was incidental.

Did the inhabitants of White Oaks, numbering 4,000 while mining flourished, get discouraged too early? The tangled records of those times hint at a form of mass panic which may have left good mines deserted, left good ore in the ground.

A major objective of early mining camps and of White Oaks in particular was to have a railroad connection. Many a long branch line was built, either by a railroad corporation or independent capital. Such lines often were highly profitable as long as the mines served were producing and shipping a great deal of ore. The community also benefited because supplies could be cheaply shipped in.

For more than a decade mining and businessmen of White Oaks tried in vain to interest railroad officials and other investors in building such a line. At last, in 1899, a road incorporated as the El Paso & Northern extended tracks to within eight miles of the town. Its president was Charles B. Eddy, an astute financier and developer with wide interests in New Mexico. White Oaks waited expectantly for grading to start for the final link–then came the shattering announcement. There would be no White Oaks connection.

Considering Eddy's good reputation and especially considering his business acumen, the town fathers came to a disturbing conclusion: Eddy had some inside information and it was of a negative sort. In a mining camp this could mean only one thing–that the mines were playing out. Why build a railroad to haul ore if there soon would be no ore to haul?

Something of a panic ensued. Uneasy store owners began to sell or close their businesses and move away. The mines were soon affected. Like the merchants, operators decided to take their profits and close down. The atmosphere suggests that worthwhile, or at least borderline, workings may have gone inactive, and whether worthwhile or not, they have not been reopened. A large cemetery remains as proof of a once populous place. In other respects, White Oaks is a rapidly vanishing set of ruins.

In the general White Oaks area a number of small streams flow east and join the Pecos River. On one of these, or more likely on the Pecos itself, is an early Spanish bonanza known as La Mina Perdida. The Pecos is 735 miles long and twists through a vast area of country, some of it settled, most of it empty.

Tradition has many shipments of gold being made from the mine, some reaching Mexico City, some being buried en route because of impending attack by Indians. The best-informed searches appear to have been made from the village of Pecos, which was a point on the old Santa Fe Trail from Missouri to Spanish settlements in the

Southwest. This would put the mine in the far upper reaches of the Pecos, in about the latitude of Santa Fe.

In his *Coronado's Children,* the late J. Frank Dobie had the mine rediscovered by a young Jose Vaca, who lived in Pecos. It appears that there was some family interference with the boy which kept him from following up his good fortune. When he later tried to make his way back, armed with assurances from the village *viejos* that he had found the ancient prize, the youth was unable to find the mine or even to find his way to the fine trout stream with which he associated it.

As one traces the Pecos down its long course to its confluence with the Rio Grande east of Big Bend National Park, he encounters many slender reports of mineral deposits, as often as not confused with tales of buried treasure. The sources of all this hidden gold range from rich ore or bullion shipments through plundered army payrolls and the personal wealth of the Emperor Maximilian to treasure hoarded by the two Montezumas—one the Aztec ruler of historical record, the other a legendary god or ancestral chieftain of the Jicarilla Apaches. Along the Pecos there may well have been more treasure hunting than honest prospecting.

There was one discovery in the middle Pecos country that came up to expectations. This was the great series of underground vaults that became known as the Carlsbad Caverns and are now administered as a national park. The town of Carlsbad, fifteen miles distant, is reached by U.S. 62 and U.S. 285. Due west of here are the Guadalupe Mountains, also the location of a national park and of old Ben Sublette's highly improbable but fully authenticated gold mine. Since Ben worked out of Odessa, Texas, and probably mined in the Texas segment of the range, his story will be told in connection with that state.

To reach the next major mining district of New Mexico one must drive around the Guadalupes and the adjoining Sacramento range to the regionally important city of Las Cruces. Mining in the Las Cruces region centered in the camp of Organ, and the jagged Organ Mountains form an effective landmark. The best mine was the Torpedo, a silver-lead property. Hugh Stevenson, the owner, was tempted to join the 1849 gold rush to California, but decided to stay with the Torpedo and work it on a small scale. Over the years he took out $90,000. Interest in the Organ Mountains developed slowly. A mild rush to the district occurred in 1893 and other ore bodies were brought into production. None of the miners was lucky enough to discover the richest deposit of all in the Organ district, the hidden gold mine of Father La Rue.

This humble cleric of northern Mexico felt called upon to rescue his flock from the effects of a severe drought which threatened them with almost inevitable starvation. Their settlement possibly was near Samalayuca; certainly it was somewhere in the northern part of Chihuahua. A dying prospector to whom the padre administered the last rites had told him of a high-grade gold vein he had been working. It was two days' journey north of El Paso, the man said. His description of the flowing water and green banks of the river to be followed, the Rio Grande, doubtless impressed La Rue considerably more than the gold. The priest now led his followers to a new home. The distinctive spires of the Organs identified the general location and the mine was found.

Organ is twelve miles from the Rio Grande and it is evident that some immediate sustenance and care for the pilgrims' goats and emaciated cattle was the first concern. They therefore settled on the river. In due course, however, the men got busy at the mine and began to accumulate gold.

How word of the new colony and its wealth ever reached Mexico City is unknown. The official version given out later was that the officials there knew nothing of the gold mine, that church authorities were concerned about not hearing from Father La Rue, and that the government sent a detachment of soldiers on an errand of mercy. If such was true, it sent the wrong men.

Apparently the colony kept sentinels posted along any line of approach to warn of possible Indian attack. When a runner told of white soldiers approaching, the good father had no illusions as to what would happen. He ordered all gold hidden in the mine tunnel and the opening well concealed. The Spanish soldiers, whatever their ostensible mission, knew about the gold. The priest was killed, and ultimately all the rest of his flock, but in the whole bloody episode the military failed to locate the hidden mine. It is still searched for today.

New Mexico's most extensive mining was in the southwestern part of the state. On the thesis that an area of proved mineralization offers better opportunities than a dubious one, it may be recommended as the favored place for renewed prospecting today.

The great Santa Rita copper mine, which was in sustained production as early as 1804, stimulated prospecting for miles around. The additional discoveries proved to be chiefly of gold and silver. They gave rise to the camps of Hillsboro, Kingston, Percha City, Gold Dust,

Hanover, Chloride Flat, and Pinos Altos. Another camp, Silver City, has survived and grown into a modern trade center.

To reach this area, turn off New Mexico's major north-south artery, Interstate 25, at Elephant Butte reservoir on to State Route 90. You will be heading west toward the mountain barrier called the Black range in the north and the Mimbres range in the south. All these mountains are of interest to the prospector, for many profitable mines have been developed there, and history indicates that some excellent properties were abandoned. The reason is quite clear. They were in country especially vulnerable to Indian attack. As late as 1885 workers at isolated mines were surprised and killed and even the substantial town of Kingston was menaced. After the so-called pacification of the Apaches the mines were dealt another blow. The principal values in this district, aside from Santa Rita's copper, were in silver. With the disastrous drop in price of the white metal in 1893, many of the mines in the Silver City region, as in many other localities mentioned earlier in this report, had to close down. It may take a painstaking search of old records, plus equally patient inquiry among older residents left in the district, to determine which individual properties are worth serious examination.

The investigator of old mines in New Mexico has one source of information that may be unique with this state. In 1965 a *Directory of Mines of New Mexico* was brought out, compiled by Lucien A. File. This appears to be all-inclusive, both geographically and historically. There are ninety-six pages of listings with about forty-five mines being named on each page, a total of more than 4,000. Where the compiler has found any references in state or federal reports, these are noted after each listing. There also is a list, with addresses so far as known, of some 1,500 "mining companies and corporations."

There is one intriguing mine in the Silver City district which landed its promoters in jail but which might have been, and may still be, a potential bonanza. To the prosecution, in a trial for using the mails to defraud, the property in question was a natural cave containing no precious metals of value. To the defendants it was an old Spanish mine with huge reserves of gold ore.

The rediscoverers, if such they were, were led to the place by a friendly Indian. They were George Dubois and his son Lee. The year was 1904. Father and son went to Denver and organized a mining corporation which immediately attracted capital and associates. A dentist, Dr. R. C. Hunt, invested $10,000 and was made president.

Several substantial businessmen bought blocks of stock and became officers or directors. The group then made the mistake of attempting a heavy, high-pressure sales campaign in the eastern states, and the advertisements and promises of the Spanish Bullion Mining Company came to the attention of federal authorities.

In 1907 the responsible corporation officials went on trial. In the early part of the century there had been thousands of mining stock promotions, ranging from the highly speculative to the outright fraudulent. A reaction had set in and, as was noted in a chapter on Nevada, such practices were finally curbed by passage of the so-called blue sky laws.

The Spanish Bullion trial was one of the early consequences. It appears to be the occasion when the phrase "swindling of widows and orphans" was originated. The prosecution went into the case armed not only with sympathy for the widows and orphans, if any, who held stock in the mine, but with confessions of fraud by the two Duboises, father and son. The confessions were immediately repudiated as having been obtained by duress.

Then came conflicting testimony by engineers who had inspected the mine. The defense expert, named Lindeman, was a highly regarded engineer, as his father had been before him. He testified that there was $600,000 worth of gold ore in sight and that the ultimate yield of the mine could well be $20 million. He also stated that there was ample evidence that this was an old Spanish mine, still rich, and abandoned for reasons unknown. It was a distressing ruling for the defense when the prosecution was allowed to introduce exhibits of worthless rock from the mine while the defense could not present samples of what they insisted was commercial grade ore. The verdict was guilty. The federal judge delivered a stiff lecture on fraudulent stock promotions but imposed only two thirty-day jail sentences and a scattering of fines. After all, these were great days in the mining West, and in few places were the excitement of the game and enthusiasm for mines and miners stronger than in the Rocky Mountain metropolis of Denver.

From Silver City, U.S. 180 runs northwest to the Mogollon Mining District, near the Arizona state line. Besides the ghost camp of Mogollon itself there are old mines clustered around Claremont, Cooney, Alma, and Glenwood. It is a rugged country of deep canyons and few roads. There is one unusual route, however, that will take you around a great semicircle of more than a hundred miles running through unpopulated and unspoiled country. In the center of the half circle is the

Gila Cliff Dwellings National Monument, but this is reached by a separate route from Silver City. The unpaved but improved semicircular road crosses a number of tributaries of the Gila River and touches the upper spurs of the Black Mountains. Designated State Route 78 in one section and 61 in another, this entry into Mogollon's back country was not available to earlier prospectors. What such easy access may open up in the way of new mining is awaited with interest.

The Mogollon name attaches to an immense plateau and various other natural features in addition to the old town and a tribal division of the Apaches. It is pronounced in various ways. To retain the original sound, accent the last syllable, with the "long" sound of the O. The double 1, as in all Mexican Spanish, is pronounced as Y. Mogollon is the surname of Juan I. Flores Mogollon, who in 1712 was appointed captain-general of a Spanish territory in the Southwest at least as large as all the thirteen American colonies. The Britannica has overlooked Mogollon the individual, and one speculates as to what kind of man, what kind of career, impressed his name on so much western scenery.

Mining in the Mogollon district is associated with the Cooney brothers, James and Michael. The former, as an army sergeant mapping trails near old Fort Bayard, stumbled upon a ledge of good gold ore. He held his tongue and waited for his enlistment to expire. The mine was on what is now called Mineral Creek and his first shipments ran $200 a ton. The Apaches at this time were troublesome but not on the warpath. In a comparatively minor incident involving protection of a farm at Alma, a young Indian was killed and James Cooney, rightly or wrongly, was said to have fired the fatal shot. The youth turned out to be one Torribeo, son-in-law of the Apache chieftain Victorio. The ferocity and strategic ability of Victorio, now a part of history, were soon to be known. Miners at the Cooney operation were immediately slaughtered. Cooney himself was hunted down and killed. The rampage continued. On one occasion a force of a hundred soldiers was lured into a canyon by the wily Apache and wiped out to the last man.

The Cooney mine was taken over by brother Michael, who had been with the customs service in New Orleans. He had been in the Army and was addressed as captain. He made a fair success of this and other mine ventures but did not amass any great wealth. The fascination of the country and the game of hunting for gold took him on many exploratory rambles. In 1914, well advanced in years, he overtaxed himself and was unable to escape an early autumn snowfall in Sycamore Canyon.

One account pictures Captain Cooney as a lone hermitlike wanderer in his later life and as having taken up a dogged search for the Lost Adams Diggings. Actually he was not a recluse, served a term in the state legislature, maintained a home in Socorro, and took an affectionate leave of his wife and son on starting what proved to be his last trip. The occasion of this lone journey soon came to light. Some years earlier a miner whom he had grubstaked reported a rich gold discovery. This man, Turner, reported to his grubstaker but before he could take Cooney to the exact location he was murdered. When the captain's body was found after the spring thaw of 1915, it was in the general area that Turner had described. Cooney, unwisely as it turned out, had determined to find the spot.

When we cross from New Mexico into Texas we enter a state rich in petroleum and other economic assets but where production of gold and silver, to date at least, has been far from impressive. There was one mine, however, that was quite impressive. This was the astonishing Presidio, which, according to University of Texas reports, yielded more silver than all the other mines of Texas combined, a total of 20,282,000 ounces. At today's prices that would mean $100 million. The great deposit was in the Chinati Mountains, just across the border from Mexico. The nearest town of any size is Marla, county seat of Presidio County, and anyone hoping to match the great discovery might make this his headquarters. From Marla south to the old Presidio one is in barren, imperfectly explored country. There is a U.S. 67 running from Marla to the border and the Mexican village of Ojinaga, a distance of fifty-nine miles. State Route 118 takes a much longer course to get to the same point, but at Mile 104 it passes the entrance to Big Bend National Park, where some supplies and amenities are available.

West of U.S. 67 there is a great, straggling triangle lying along the Rio Grande River containing no towns but traversed by several primitive, neglected roads dating back to the Spaniards. Within its 120-mile extent we find the Quitman Mountains, the Eagles, the Van Horns, and the Chinatis, as well as the smaller Devil's Ridge and the extensive Sierra Vieja. Ruined adobes and some place names that correspond with those in early Spanish mining reports, and particularly the old roads, hint at some things of interest in this neglected region. Official mining reports do not date back to these early operations and thus give us little to go on. General history deals with scenes of greater activity far to the northeast. The triangle may be said to terminate where Interstate 10 reaches the Rio Grande at the town of Esperanza. From there westward one would be in slightly more settled country along the eighty-seven miles of Interstate 10 leading to El Paso, the principal city of this western extension of the Lone Star State.

El Paso would be the reasonable point of entry from New Mexico. Here, closer to the mineralization of the Rocky Mountain chain, we can take a more optimistic view regarding the gold and silver of Texas than is warranted by the over-all picture. For El Paso is the takeoff point for the Lost Padre Mine and for the strange deposit of gold that brought comfort and security to the long-suffering family of Old Ben Sublette.

The Lost Padre is in Texas but to locate it we cross the Rio Grande to the Mexican city of Juarez. In Juarez is the venerable church of Nuestra Señora de la Guadalupe. It has a tower from which a great bell has rolled out its solemn tones since 1659. If one ascends to the belfry at sunrise and looks north toward American territory and the Franklin Mountains, he may get a fleeting glimpse of the old mine; so runs a tradition that traces back to the padres themselves, in this case Jesuit missionaries.

When the Jesuit order was expelled from New Spain the priests are said to have sealed a great hoard of gold and silver in the mine, filled in the tunnel, and concealed the entrance. Some flaw, some failure, in this camouflage is the reason for the sunrise observation. At a certain time of year, for several days the sun's rays strike one of the Franklins' southernmost peaks in a way that reveals the old workings. The sun's direction ranges through 23 degrees in the course of the year and on what dates the observation should be made never has been determined.

We know a little more about Sublette's mine, considered still to be holding a fortune in gold. It appears to be in the Guadalupe Mountains, which rise about ninety miles east of El Paso and are reached by U.S. 62-180. On some maps the southern part of the range is titled the Delaware Mountains.

It may have been nature, it may have been early Spanish miners, but somehow a singular rocky pit has been created in the distant recesses of this mountain country. Other wanderers may have seen it, but Sublette was the only one who explored it. The floor could be reached only by descending with the help of a rope. At the bottom was a cavernous opening in the rock wall at one end. This could be an old mine tunnel, it could be a natural cave. Whatever its origin, it has been the source of very rich gold ore.

The deposit, or one of similar value, certainly was known to an old colonial family in the state of Chihuahua. Even after the Texas Pacific railroad and hundreds of American ranchers opened up the far west of Texas there were quiet and presumably profitable incursions from south of the border to certain old mines. Many believe that it was one of these old workings that was discovered by Old Ben, rather than any virgin ledge.

By all accounts, Sublette's family had endured years of hardship as the head of the family would alternately work for wages and then spend profitless weeks or months tramping through this or that desert mountain range looking for gold. After the railroad had pushed its tracks through arid west Texas we find Sublette living first at Monahan, county seat of Ward County, and then at Odessa, a larger place a little east along the line. Sublette would work off and on for the railroad.

From Odessa it is 140 miles to the Guadalupe Mountains. Exploring that range, one may be in either New Mexico or Texas. Old Ben's base was a long way off, but it was in the Guadalupes that Sublette had kept prospecting. In his time, and also today, a logical entrance to the range would be along Delaware Creek, starting from its confluence with the Pecos River near the large lake formed by Red Bluff Dam.

Once he had made his strike, Sublette would bring in about a thousand dollars in gold each time he absented himself from Odessa. He would sell the gold and deposit the funds in the local bank, replenishing the account as needed by making another trip to the mountains. His long-impoverished family was now living well.

Many attempts were made to trail Old Ben but he invariably managed to disappear. Even less successful were the attempts to buy into a partnership or to glean some hint of the location by the old routine of free drinks and adroit questions. The older he got, the more proud he became of his discovery and the more secretive. In a dying attempt to pass information along to his son, he did describe a pit, or steep-walled depression, that was very difficult to enter. This conformed with words dropped in some of his earlier, boastful accounts. It also conformed with the picture of an old Mexican mine as dimly preserved in local tradition. Whether identical or not, Sublette's mine evidently will not be found by standard prospecting methods. It will be found by someone who chances on a strange pit, with a dark, tunnel-like aperture at one end, and who has the curiosity, the courage, and the ingenuity, as Sublette had, to get down into the place and investigate.

The mining experiences that loom largest in the Texas story go back to the early eighteenth century. Except for the Presidio Mine and a few much lesser operations, the important working of precious metal deposits appears to have been done by the early Spaniards. The scenes of their activity were near the present town of Menard, county seat of Menard County. This is about 120 miles, as the crow flies, northwest of San Antonio. Menard is on the San Saba River, and it is along this stream and the neighboring Llano River that men have searched for rich colonial silver mines and for hoards of hidden silver in the form of massive ingots. So persuasive are the records of mining in this particular area that the Texas historian J. Frank Dobie groups the legends of Spanish workings together under the term "the Lost San Saba Mine."

The discovery of silver in Texas is credited to Don Bernardo de Miranda, commander and public official in the years when Mexico and the American Southwest were both under Spanish rule. In 1756, Miranda made a mining reconnaissance and issued a report telling of deposits of native silver so pure that Indians were gouging out the metal and fashioning ornaments from it. The fictitious tales that had lured Coronado and others on their futile expeditions had created a skeptical attitude in Mexico City toward any northern treasures. Miranda had samples to show. In modern mining terms, his specimens ran two hundred ounces of silver to the ton. They came from a short tunnel to which Miranda, following the custom of Spanish explorers, had attached a grandiose name, Cave of Saint Joseph of Alcazar. The digging has not been identified by later searchers but it is known from good evidence that it was somewhere on Honey Creek, a tributary of the Llano River.

After much waiting, Miranda got action from the all-powerful viceroy at Mexico City. A presidio with a garrison of a hundred men was established in the silver region. The site chosen was not at or near St. Joseph's cave but on the San Saba River, sixty miles to the northwest. The remains of this old fort survived into modern times. They are a mile upstream from Menard.

Mines were developed in this new area also. Unquestionably some silver ore was produced, treated in *vasos,* or primitive smelters, and bars of bullion sent to Mexico City. Perhaps a great deal was produced. The legend is of many mule trains making the trip, of elaborate precautions to guard them, and of occasional attacks by bandits or Indians. The closing of the mines was due chiefly to Indian attacks, not to the Texans' struggle for independence. The Apaches had been driven out of the San Saba country by the Comanches, a tribe so fierce and so feared that the term "wild Comanche" was a colloquialism in the Middle West well into the present century.

Gold production in Texas has been very meager. A survey of the state's economic geology by E. H. Sellards and C. L. Baker of the University of Texas notes that gold ore in place appears confined to Mason and Llano counties, and that it is found only in quartz veinlets and stringers. This does not give the prospector much encouragement, since such thin deposits are usually unprofitable to mine. It is interesting, however, that these two counties are adjoined on the west, north, and east by counties that are convincingly linked with the Spanish silver mines. The "Llano Uplift," as this region is termed, is evidently a silver, not a gold, country.

The paucity of gold in Texas is further indicated by the lack of any significant placer mining. We have seen that in other states it is placer gold that is first discovered and is first—and most easily—worked. Are there any overlooked auriferous gravels in this large state? Sellards and Baker say it is "possible," but warn that the deposits would be expensive to develop. In the locations they have in mind, the water that would carry the gold is an "underflow" through the sands and gravel.

"It is precisely these places where placer gold would be found," they write, "but the lone prospector, without considerable capital, cannot even prospect these places, let alone work them. It is necessary to dig a large number of holes down to bedrock with auger or other types of prospecting drills or, if shafts are sunk, to shut the water off with caissons. Such prospecting is expensive."

For anyone willing to cope with such difficulties, the locations favored by the university geologists may be listed. They are the Llano River and Sandy Creek (both tributaries of Texas' own Colorado River), Crabapple and Cole creeks, and the "main" Sandy Creek below Enchanted Rock.

Unless new discoveries are made or more encouraging geologic information is forthcoming, the prospector in Texas may well consider that his best chances lie with old Spanish or Indian mines. James Bowie, one of the state's traditional heroes, believed he knew the exact location of one of the mines of the Lipan Indians, necessarily a rich one, it was assumed, since the tribesmen were bringing chunks of pure silver into San Antonio for trade. A party of miners was organized by Bowie and his brother Rezin. The group was ambushed in Calf Creek Canyon, somewhere east of the original San Saba fort. Although the white men had the best of an all-day battle, their presence in the Indian country was untenable and they returned to San Antonio. James Bowie told a good deal about the expedition but never revealed whether, at the time of the Indian attack, they were at or near the mine. A short time later he was among 180 Texans killed in the siege of the Alamo. County Road 42, leading south from Brady, in McCulloch County, will take one fairly close to Calf Creek.

Various other discoveries appear in the Texas story. There is said to be a cave with silver-encrusted walls and ceiling near the present town of Lampasas. Silver Creek, branching off the San Saba, appears on such convincing waybills to this cave that several well-organized searches have been made there. A similar cave, near the Colorado River and above the confluence of the San Saba, was described by a former captive of the Comanches. From other Indian sources come stories of silver on the Llano and Frio rivers and on Los Moros Creek, all in the general San Saba country. Far up the Llano on its north fork was another of the suddenly deserted mines, and here, too, much silver is said to have been stored. An Indian mine was said by a Comanche named Yellow Wolf to be on the north fork of the San Gabriel River in what is now Williamson County.

A matter of practical importance to today's prospector is the fact that vast realms of once open land passed into private hands through early land grants. Under the Guadalupe Hidalgo Treaty the United States agreed to honor these old Spanish conveyances. Some of them were for tracts of immense size. Much undeveloped country that would be public domain in other western states was pre-empted by lavish and uninformed donations by the Spanish crown. The original grants are largely broken up now, but the resultant parcels are private property. Nearly all the later accounts of searches for old mines or buried treasure tell of preliminary negotiations with ranch owners. The prospector would wish to have a definite and promising lead in mind before undertaking exploration under these conditions.

11
The Pacific Northwest

The first gold in the Northwest was discovered, but not recognized as gold, by a wagon train party of midwestern farmers. The year was 1845, three years before Marshall's discovery of gold in California.

The party was a true pioneering expedition. Word of fine agricultural possibilities along Oregon's Willamette River had been brought back to civilization by trappers and missionaries. In land-conscious early America this was exciting news. From the farmlands of the Ohio, Mississippi, and Missouri valleys a large emigrant party was organized. It was made up of small farmers and their families, of landless farm hands, and of a few would-be farmers of adventurous bent. The lure was that of wide, well-watered homesteads to be had for the taking, and in this the emigrants were not disappointed. After many hardships along the Oregon Trail, a dim trace laid out by early explorers, they reached the fertile soil of western Oregon and settled down to a rugged but satisfactory life of farming and cattle raising. These solid achievements, while commendable, would hardly give the 1845 trek a place in history. The Midwesterners' journey is remembered today only because of an incident along the route—the matter of the Lost Blue Bucket Mine.

This was an opulent deposit of yellow gold glittering in a dry creek bed. It was seen and ignored by the agriculturists and never has been sighted again. Along with the Lost Port Orford Meteorite, the Blue Bucket is one of the Northwest's favorite mysteries, always worth a discussion, occasionally worth still another hopeful search.

These two stories can best be told when we come to their geographical locations. Taking up first the historical sequence of Oregon's mining development, we go to the southwest corner of the state, where two gold-bearing mountain ranges, the Klamath and the Siskiyou, cut across the Oregon-California state line.

As might be expected, the mining here was carried on by an overflow from the California mines, by men who had been either late or unlucky along the Mother Lode. The original discovery, however, was made not by miners but by sailors. A small ship, name and destination unknown, had been driven ashore near the present site of Crescent City, California. What led the stranded seamen to start on a northeasterly tramp into unknown country can only be surmised. We know the approximate route they took, and it can be followed today on U.S. 199 between Crescent City and the Oregon Caves National Monument. Like the fabled crew of the Argo, these sailors seem to have had gold on their minds, for they panned the sands of various creeks as they passed. At last they found placer gold in liberal quantity. It was on a stream, a tributary of the Rogue, that now bears the name of Illinois River. The camp they set up was called Sailors Diggings, but later took the name of Waldo, and it so appears on maps. News of the strike circulated quickly and miners from California soon arrived in force.

There was a good deal of gold to be had at this early stage. In this part of the Siskiyous virtually all the streams showed colors. The principal camp turned out to be Jacksonville, on the Applegate River. Only a short distance off Interstate 5, it is a desirable base for any present-day prospecting. A better-known town in the old mining region is Ashland, the scene of an annual Shakespearean festival that attracts attendance from many states. Amid the cultural activity the old Ashland mine stands largely forgotten. It was discovered in 1890 and was worked continuously for half a century, succumbing finally to the War Production Board closing order of 1942. Perhaps the Ashland should not stand entirely forgotten. State geologists have traced the vein for more than a mile beyond the old workings. In this massive formation, is the old Ashland location the only one of promise?

As miners left the streams to look for gold ore in place there were many rich strikes, mostly of the pocket variety. In their *Gold and Silver in Oregon,* Howard C. Brooks and Len Ramp, state geologists, tell of some of the lucky finds.

The Hicks pocket was a small one but it yielded $1,000 in gold for two hours' work. On Thompson Creek in Josephine County, Ray Briggs struck a pocket worth $32,000 and before cleaning out the limited ore body struck another one worth $18,000. His father, David

Briggs, was photographed holding a sheet of gold about the size of a dinner plate, just as it came out of the ground. Two miles west of the present reservoir on Jackson Creek the partners Johnson and Bowden dug into pockets paying $30,000 and $60,000. The so-called Revenue Pocket, on Kane Creek, set a near record, the Rhotan brothers encountering a concentration of gold worth nearly $100,000. The Gold Hill discovery, almost too large to be described as a "pocket," was excavated by a trench five feet wide and two hundred feet long. The ore that lay within those dimensions yielded $700,000.

Such surface discoveries were exhilarating, but few of them developed into real mines. Values did not extend to depth, and this statement, with a few notable exceptions, can be applied to gold occurrences throughout the state.

Besides the placer gold of the creeks and bars there were two other types encountered in southwestern Oregon. One consisted of the bench deposits, lying high above the streams. Many of these were torn down and washed by hydraulic mining but by no means all of them. The other type was placer gold found in the sands of the ocean shore. The town of Gold Beach derived its name from this asset. The gold, it was generally conceded, had no connection with the ocean as such, but had been washed down by the Rogue River and trapped in the sands. There was a much richer occurrence about fifty miles to the north, where the Coquille River entered the sea. Here the brothers Charles and Peter Grauleaux investigated extensive beds of black sand and found them heavily impregnated with flour gold. Recovery of the fine particles required skill and care, but so abundant was the gold that the brothers took out up to $1,500 a day. Thousands of miners were soon working along the beach, a preview of what would develop at the fabulous seaside placers at Nome, Alaska.

For a very rich placer mine which is by no means exhausted, one has only to find a cabin which was occupied in the early 1860s by two French miners. It was far out in wild country, somewhere south of Cottage Grove, a town on Interstate 5. In their first summer there, the two Frenchmen built a cabin and accumulated a large amount of gold in dust and nuggets. Their takings were freely displayed at the county seat of Eugene, where they exchanged gold for as much coin and currency as the merchants could spare. They spent the winter, presumably in some luxury, in San Francisco, then returned to their mining.

Their nemesis was an Indian woman whom they had engaged for household chores. What kind of ill treatment, real or imagined, the woman suffered is not related, but one night she fled in terror into the woods. She was found and cared for by a detail of soldiers under Captain F. B. Sprague. In due course she returned to her home in the Klamath Indian Reservation. Here she repeated her story to her fellow tribesmen and it called for vengeance. The Indian war party was led by a brother of the aggrieved squaw. Guided to the miners' remote cabin, the Indians disposed themselves facing the single door and roused the occupants with a war whoop. As the two men emerged with rifles they were shot down in their tracks.

A tribal taboo was placed on the site. Although by this date the Indians understood the value of gold, those on the reservation repeatedly stated that none of the red men ever returned to the site. To the soldiers who cared for the fugitive woman, she had described the cabin as being built against a large, vertical rock face that formed the rear wall. In front of the cabin, a short distance away, was a small meadow where the Frenchmen had grazed their horses.

Even on the basis of these slight clues there have been searches for the old cabin, for the prospective rewards appear very great. State Route 138 from Roseburg will take one along the southern edge of the favored area, while several dirt roads out of Cottage Grove thread through the northern and middle sections.

If you are on 138 and do any searching near Steamboat Mountain, be alert for another log cabin that is also a pointer to some waiting gold. This one is definitely on Steamboat Mountain and evidently was abandoned in panic, for personal belongings and even a little scattering of gold nuggets were left lying about. In 1870 Constantine Magruder of Central Point came upon the place when hunting deer. He appropriated the gold, then made a long and not very observant trip back to camp and his companion on the hunt.

Back home, friends convinced Magruder that he had found the original Lost Cabin Mine. This was a fabled bonanza of the 1850s, which for two successive summers had enriched a miner known as "Old Set 'Em Up" because of his liberality around the saloons of Yreka, California. Magruder never was to know whether this was so, nor was he ever able, like Old Set 'Em Up, to buy drinks for the house. In several weary and exasperating tramps through Steamboat's broken terrain, he failed to find his way back to the long-deserted dwelling.

For prospecting in southwest Oregon the most pleasant, but not necessarily the most productive, plan would be to choose as a base one of the attractive coastal towns on U.S. 101. This highway runs along the ocean from California clear to Astoria at the mouth of the Columbia River. The streams flowing down from the mineralized back country include the three forks of the Coquille, the Elk, the Chetko, and, as an added attraction, the Sixes.

Should you go up the Sixes River, it would be wise to enlarge your gold and silver objectives and include the mineral pallasite, which at times has sold for $10, $20, and up to $100 a pound. There are tons of pallasite in that country, all in one big mass, the Port Orford meteorite. Besides being rewarded for its intrinsic value, the discoverer will receive a reward and honors from the Smithsonian Institution.

A shooting star coursing across the sky is a meteor, but once the mass has struck the ground it is known as a meteorite. Port Orford's celestial visitor has been seen only in the latter, static capacity, and so far as is known has been seen by only one man. He was Dr. John Evans, a professional geologist in the employ of the Department of the Interior. Many "geological reconnaissances" were made by government field men in the years succeeding the California gold discovery, preliminaries to the valuable contour mapping that was to follow, and Evans was engaged in one of these.

The Evans trip started at Port Orford in 1856 and took him inland to the highest peaks of the Coast range. Collection of rock samples, of geological interest and not necessarily of mineral value, was part of his assignment. In a batch of such specimens sent East there was a piece of meteoric origin. Dr. Charles T. Jackson, an authority in the field, identified it as not only of meteoric origin but as representing a very rare type of extra-terrestrial visitor, being composed of pallasite. His finding was corroborated by a recognized expert in Vienna. Inquiries to Evans revealed that the fragment came from a very large piece of the same material.

In a limited scientific circle Evans' discovery was of great interest. Several organizations planned an expedition to the site but funds were hard to come by. Congress, concerned with the dissensions which would soon lead to civil war, refused to appropriate funds.

In Vienna an astronomer gave out a precise fix in latitude and longitude which he affirmed would lead to the meteorite. It appeared that a European party might cross the Atlantic on the strength of these promising directions. Hopes subsided when it was discovered, by conferring with American colleagues, that the announced bearing was merely the location of the town of Port Orford, not of the meteorite itself. Evans estimated that his discovery was made forty miles from

Port Orford, but whether he meant forty miles airline or by his own irregular trail was never determined.

The geologist himself confidently wrote his eastern associates that he could lead them to the place. He described the visible section of the meteorite as protruding four or five feet from the ground. The mass was about four feet thick viewed from one side, slightly narrower viewed from the other. It rose from a grassy spot on the west side of what he called a "bald mountain" and which also was the highest peak in the area.

Evans died unexpectedly in 1861. Copies of the day-by-day log of his exploratory journey may be inspected either in Washington, D.C., or in Portland, Oregon. The record has been studied by interested astronomers and by others with purely mercenary motives. One of the most sophisticated searches was made in 1932 by the widely recognized meteor authority H. H. Nininger. He concluded that owing to the scanty directions, the nature of the country, and the erosion and changing ground cover over the years, no systematic search would succeed. The Port Orford meteorite would be discovered only by sheer chance.

When one leaves the coastal mountains of Oregon and moves east, the country changes. The southeastern part of the state is sparsely populated and near desert. There is little proved mineralization in this barren, windy region but there have been a few startling discoveries that will be described later. Toward the north one gets into more promising territory. The productive area of record consists of a broad band extending along the Washington state line. The logical base for prospecting trips is Baker, on Interstate 80 North.

The placers of this northeastern region were opened up by a party of miners and nonminers from the Willamette Valley who were hopeful of finding the Lost Blue Bucket Mine. Earliest diggings appear to have been on the John Day River. Other streams found to be auriferous were the Powder River, the Wallowa, the Umatilla, the Grande Ronde, and particularly some of their tributaries. A few places were extremely rich, but the region as a whole did not contribute many spectacular success stories comparable to those in the western diggings. It nevertheless supported thousands of men through the 1860s. The normal shift to lode mining developed during this period and there were a number of profitable mines.

Some of the old camps that invite renewed inspection are Sumpter, Granite, Bourne, McEwen, Auburn, and Salisbury in the Baker district and Sparta, Cornucopia, Carson, Richland, and Halfway, all east of Baker and near the Idaho state line.

Much of this country is included in national forests, and except where an enclave is set aside as a wilderness area or an Indian reservation, it is open to prospecting, location, and mining. The rather empty country outside the forest boundaries is largely public domain, but in considering any particular mining enterprise one should ascertain whether the site has passed into private hands through homesteading or purchase.

Oregon is correctly pictured as a green, heavily forested state with ample rainfall. Yet a full quarter of its area, the southeastern section already mentioned, is something very close to desert, an extension of the arid surface of northwestern Nevada. A few rich gold deposits have been found and lost in this section. Towns and paved roads are few and even on recent maps many of its roads are represented only by dotted lines. U.S. 95 cuts across the far southeastern corner of the Oregon desert, while U.S. 395 runs through the isolated little mountain ranges and sand dunes of its western sector. The town of Burns, county seat of Harney County, looks out on a vast, lonely landscape that extends 120 miles south to the Nevada border and eighty miles east to Idaho. Burns is the only supply point of any size in the whole region. At scattered places, ground water has been tapped at fairly shallow depth and extensive cattle spreads developed, along with some sheep raising.

This little-known country shelters a gold deposit described by those who saw it as something that "could make a hundred men rich." The long and lonely Owyhee River is the avenue to this bonanza. The Owyhee is a twisting, little-explored stream that flows northward roughly parallel to the Idaho state line. A modern dam has created a long lake on the river and there is a state campground on its east shore that cannot be a great distance from where the strike was made.

Oregon newspapers have uncovered what they could about the lost mine and the Oregon *Journal* of Portland once published a map representing its best judgment as to the mine's location. The western writer Ruby El Hult has made a valiant attempt to straighten out the published clues in her *Lost Mines and Treasures of the Pacific Northwest.*

The discovery was an incident of the Bannock War. The Bannock Indian tribe had gone on the warpath against the miners and other settlers in southern Idaho, and the situation at Boise and its neighboring settlements was critical. In adjoining Oregon there was a cavalry troop stationed at Fort Harney, which old maps show to have been ten miles east of Burns. A relief column of some eighty mounted men set out from Harney to give urgently needed help. Their destination lay almost due east. Whether they rode cross country or followed the easier northerly course now traversed by U.S. 20 is something we do not know.

A critical question for the lost-mine hunter is how far such a cavalry force could travel in a single day. For it was at their first overnight encampment that the rich discovery was made. Three young soldiers wandering about in the lingering daylight came upon the outcrop, visible on a slope rising from an unidentified dry stream bed. They said the vein was about eighteen inches wide and that the exposed matter was about half rock and about half gold. As we shall see, this extravagant statement was confirmed by persons who later handled pieces of the ore.

The three young cavalrymen perforce moved on the next day with their company. At Dunaway, on the Snake River, men and horses had to be ferried across in installments, and here an identifiable figure enters the story. He was a man named Keeney, operator of the ferry. The three soldiers confided in him. There were indications later that Keeney, whose maturity and ownership of the ferry may have impressed the youths, had some agreement with them for working the mine. It is evident that they told him a great deal.

None of the three discoverers was heard of again. There were heavy casualties when soldiers and settlers clashed with the Indians, and Ms. El Hult, weighing the evidence, concludes that they were among the dead. The ferry operator, Keeney, may have known this to be a fact, for he soon went out alone to look for the dazzling ledge. His information guided him correctly up to a point. He had found where the troopers had camped on that first night out of Fort Harney. At least he felt certain that he had it identified. If correct, he should have found the outcrop within a reasonable distance, for it was said to be plainly visible. He did not find it.

The Oregon *Journal*'s map, based on information from several sources, places the discovery at thirty miles south of Vale, a town on U.S. 20 near the Idaho state line. It is perhaps significant that the ferryman Keeney settled in Vale for the latter part of his life. Local people who accepted the *Journal*'s approximate location mentioned that in such an area Dry Creek Mountain would be a good landmark.

Though no supporting evidence comes to hand, it was further stated that the soldiers' outcrop was in a gully extending down from that mountain. Hints dropped by their fellow townsman Keeney? Or did one of the soldiers come back?

More direct evidence that the rich ledge existed was furnished in 1912 by a young sheepherder named Victor A. Casmyer. His employers brought Casmyer into Vale from the desert country to the south, suffering from Rocky Mountain spotted fever. This is a stubborn disease that is spread by a vicious little tick, and the young man was in serious condition. The victim was cared for by a Good Samaritan named Billy Huffman, owner of a livery stable. Huffman put him in a room at the little Vale hotel and the patient's bedroll of blankets and personal articles was stowed away at the stable. Casmyer had found gold and he made several efforts to tell his benefactor about it. Huffman did not take this rambling very seriously. About all that he recalled of it was that the discovery was made as the herdsman took his flock on a slow circuit of water holes and the few scant creeks where the banks offered some forage.

Young Casmyer did not recover. After his death, his bedroll was opened by Huffman and a bag of amazing ore was revealed. It was, witnesses said, actually half gold. As the story circulated, both townspeople and various newcomers sallied into the vaguely defined sheep ranges to the south. Attempts were made to find what course Casmyer would normally follow in his duties. If his immediate overseer, Peter Rambau, was able to be of any help in this it does not appear in the record. A county judge, Sewell Stanton, was long intrigued by the Casmyer episode, and he both led and financed several search parties. There is general agreement on this: The ledge of the three soldiers and the sheepherder's ledge are one and the same.

The tiny town of Vale has also been connected with the Lost Blue Bucket Mine. Vale is on the Malheur River, whose North Fork was favored in early accounts as the probable location of that rich placer. Later investigation has pointed to other territory, and a search organized by Erle Stanley Gardner, whose interest in desert riddles already has been noted, explored the region near Wagontire and the Lost Forest, a hundred miles away from the Malheur. The Lost Forest is an unusual growth of ponderosa pines which flourishes in the midst of this almost desert terrain.

The Blue Bucket Mine is Oregon's counterpart to the Lost Dutchman Mine of Arizona and Pegleg Smith's gold-strewn butte in California. Many columns of print have been filled with arguments as to which emigrant party, which cross-country route, even the gruesome question of which graves, should be related to the celebrated bonanza. It will be necessary to select and condense from this material.

There were six wagon trains, separately organized and under separate commands, pulled up at Fort Boise in August 1845 ready for the final, difficult part of the journey to the lush Willamette Valley near the Pacific Coast. A young trapper named Stephen Meek professed to know a cutoff that would save about two hundred miles of travel along the Oregon Trail. He was engaged as guide by a large contingent of the emigrants and paid a fee of $5.00 per wagon. It was an unfortunate decision. The guide became hopelessly lost in the desert country, men and animals became desperate for water. A diary kept by wagon-master Samuel Parker contains the grim entries, "Buried four persons here," "buried three today," and so on.

At one stop, where the caravan halted and water remaining in pools along a dry creek afforded some relief, travelers picked up some heavy yellow "rocks." A fetching version of an otherwise dismal tale has a flaxen-haired child garnering these objects and filling her little tin play bucket with them. Actually the bucket involved appears to have been a substantial one made of wooden staves and painted blue because this individual group of travelers had painted all their equipment blue for identification. Whatever the receptacle, its contents attracted no attention at the time the dispirited emigrants broke camp and moved on.

It may seem incredible that nobody recognized the yellow objects as gold. One piece, it was solemnly affirmed by a woman emigrant, Mrs. Chapman, was so large that she later used it as a door stop. These were farm people, and the only native gold so far known in the United States was that taken from some minor deposits in Georgia. The true nature of the metal they had found became known to members of the wagon train party in 1849 or possibly 1850. They were now a settled farm colony in the Willamette Valley, and some members were shown nuggets from the California diggings.

The deposit in eastern Oregon was now vividly recalled. Several of the younger and more adventurous men determined on an immediate attempt to backtrack on their route. The names of Dan Herren and other members of the Herren family figure prominently in this over-confident expedition. They were soon disillusioned. The "Meek Cutoff" had not been a recognizable route. The course had been a wandering one, generally toward the west but shifting with the erratic directions from the guide. The passage of four years had blurred impressions of the various wagon-train stops. Even with adequate recollection of landmarks, this trip back into a desert where many of their companions had perished would have been a difficult and dangerous one. Moreover, by this time the Bannock and Paiute Indians were aroused by successive invasions of their homeland, and it would be years before whites would have any degree of safety there. The initial search failed and it set a pattern of futility for miscellaneous trials that were to follow.

Later hunters for the Blue Bucket relied on involved data derived from a few diaries, from letters, and from recollections handed down verbally to a younger generation. The mass of evidence as to landmarks, distances, personnel, and miscellaneous factors would challenge a computer. One suspicion arising from all this is that the gold was not on the Meek Cutoff at all, that it was found by members of one of the more cautious parties that stuck to the Oregon Trail.

If you care to stay with the Herrens and their contemporaries, who after all had made the overland journey and had some, if rather dim, memory of their course, then the North Fork of the Malheur River, or on one of the North Fork's tributaries, is where you should look. Of these tributaries, Bully Creek, plus the two streams that unite to form Bully Creek, Clover and Gopher by name, seem the most promising.

There is one thing on which Blue Bucket enthusiasts are unanimous: the gold is still there. Later placer discoveries along the old trails have been given a critical review and ruled out. From the town of Vale, U.S. 20 will take you to a little place called Harper. Then, by going eleven miles north on a dirt road, you will be where Clover, Gopher, and Bully creeks come together. There are still some unworked bars and benches along their courses. And among them, possibly the elusive deposit itself, the Lost Blue Bucket Mine.

Across the Columbia River, the state of Washington presents a picture of scattered gold distribution similar to Oregon's. Production has been smaller in Washington and the state's gold mining has been overshadowed by the much richer and more active districts across the international boundary in British Columbia.

The effects of this Canadian mining may afford some useful hints for the present-day prospector in Washington. Discoveries on the Fraser River inspired a major gold rush to southern British Columbia and there were succeeding booms at the rich placers of Cariboo and Kootenay. Miners in the state of Washington were conveniently close to these new opportunities. The region around the present town of Colville was largely deserted, as were many mines and exploratory diggings in the Okanogan district. Both these areas are unmistakably mineralized. The abandonment of so many workings during the Canadian stampede raises interesting questions as to what "sleepers" may exist in the way of genuinely promising mines. The districts certainly are not mined out, for up to 1965 Okanogan County had yielded only $2 million in gold, while the output in Stevens County, of which Colville is the county seat, was barely $1 million.

These and other gold districts will be examined in more detail, but first we may take note of a tremendous natural catastrophe that changed the face of much of Washington, an event long suspected by obscure scientists and only recently confirmed by photographs taken from a height of 570 miles by our circling weather satellite.

"The greatest flood known to man" was described in theory half a century ago by J. Harlen Bretz, a geologist on the faculty of the University of Chicago. The physical changes to be expected from such a deluge, Bretz said, could not be identified. They would be so vast as to be unrecognizable from the ground or even from an airplane. Bretz's conception remained plausible but unproved. The greater scope of the satellite cameras, however, has depicted such effects clearly, and apparently they leave no room for skepticism.

The time of the occurrence was 18,000 to 20,000 years ago when the great ice cap lying over North America was melting and receding. A stubborn barrier of ice existed near the present Idaho-Montana state line, blocking a natural drainage channel to the west. Behind the barrier and adjoining mountains an immense lake accumulated from the ice melt and the heavy annual precipitation, a lake to be measured not in the standard acre feet but in cubic miles of water. At length the barrier gave way. The flood that poured across the Idaho panhandle and into Washington is described by the United States Geological Survey as equalling the combined flow of all the rivers of the world, multiplied by ten.

The surface of eastern Washington was scarred to an extent quite unknown, to date at least, in any other part of the hemisphere. No appraisal is yet available concerning the effect of this flood on mineral deposits, but it is perhaps significant that the first discovery of gold in Washington was in the state's far northeast corner, outside the range of the flood, and that additional discoveries for the most part also lay well to the north.

The basic constitution of the country, quite independent of the flood, could of course be the sole factor involved. The Geological Survey in a recent report says: "The Washington geologic map shows ore-making rocks to be mostly in the north central region next to the Canadian border. At points they extend into the middle of the state." Chelan County, whose straggling extent places it in both northern and central Washington, leads all other counties in gold production, with $35 million recorded.

In no state other than Washington do we find the initial discovery of gold heralded by the blast of a cannon. The little artillery piece was old, rusty, and dangerous to load and fire, but one Angus MacDonald, chief factotum of a Hudson's Bay trading post near Colville, took the chance.

The trading post had been in operation since 1826 with the single-minded policy of buying and shipping furs. Neither trappers nor staff had shown any interest in minerals. Then, in 1855, a recently hired employee who had been in the California gold country stooped to drink from the Colville River and saw some promise in the black sand along the bank. Black sand does not necessarily contain gold but in a new country it is always worth examining. The Hudson's Bay man is said to have used his hat for a gold pan and to have recovered flakes of some size. It was a convincing discovery. When he reported to MacDonald, who as "chief clerk" was in charge of the post, the Scotsman ran up the Union Jack and let the cannon roar a challenge to California and its Mother Lode.

This was a time of general gold excitement resulting from the California discoveries and there was a fair-sized rush to the new field. Values were found in streams within a wide radius of Colville. Although the wide distribution of gold was impressive, nearly all the locations were rather lean as to content. Most of the miners made little more than day wages from long hours of arduous digging and sluicing. Such work went on for two years; then came word of extremely rich strikes on the Fraser River in Canada. Being already close to the international border, miners of the Colville district were in a position to get in early on the Fraser. The few who remained behind were also soon lured north by the sensational takings reported from the great Cariboo placer field.

Earlier parts of this report have noted the great amount of mining in the early 1860s in spite of the fact that the country was in the throes of the Civil War. During this period Washington, with its scanty evidence of mineral riches, was largely ignored. The extreme difficulty of transportation to the northern, most promising part of the state was a factor, as were recurrent Indian troubles. The significant mineral discoveries and serious mining date from the 1880s. The normal transition from declining placer deposits to hard-rock mining was slow.

One of the first enterprisers to make a search for a lode mine was well rewarded. In 1885 a practical miner named A. E. Benoist made a location on Old Dominion Mountain, a little east of Colville. It was well up on the peak's 5,774 feet of elevation, and Benoist had to brave the skepticism of his two prospecting partners to climb up to the place. High-grade ore, chiefly of silver and lead, with some gold, was taken out at grass roots. The Old Dominion, as the mine also was called, yielded $500,000 in the first year and continued in profitable operation for decades. There were other good mines in this silver-lead zone, notably the Daisy, Eagle, Young America, and Silver Crown.

Examining the Old Dominion-Colville district today, one may identify the productive areas by the old camps, some of them ghost towns, some with a few lingering inhabitants. They include Kettle Falls, Marcus, Bossburg, Daisy, Addy, and Orient. A little more remote and farther to the west is Republic, which owed its existence to the profitable Republic Mine.

The sheet of water seen from some of these camps is Franklin D. Roosevelt Lake. It is an extremely long, crooked reservoir created by Grand Coulee Dam, which backs up the water of the Columbia River. Much of the territory along its shore is known to be mineralized and much has not been fully explored. The question may be raised as to whether trips by boat might take one to previously unvisited spots, just as Lake Mead has extended navigable waters into several remote canyons in Arizona.

Another proved zone of mineralization lies along the Okanogan River and its tributaries, about sixty miles west of the Colville district. First discoveries were made along a smaller stream, the Similkameen, not by miners but by surveyors determining the Canadian-American boundary line, which recently had been agreed upon after the critical "Fifty-four Forty or Fight" dispute with Great Britain. Gold recoveries here and on the nearby Okanogan eventually gave rise to the town of Oroville, near the border, and, to the south, Tonasket, Omak, Brewster, and others.

Development was slowed by the fact that most of the finds were on Indian reservation land. In 1886, by one of the numerous treaty revisions worked out with the red men, the Okanogan district became public domain. There was now extensive development and substantial outside capital was attracted. Lode mines at Ruby and Conconully, far back in the mountains from the Okanogan, had long and successful histories.

There has been speculation that valuable discoveries, or rediscoveries, might be made on the slopes of Mount Chopaka. The peak is west of Lake Osoyoos, a long, scenic body of water that is intersected by the international boundary. Very promising veins of gold-bearing quartz were found on Chopaka in the early 1870s. The miners soon learned that they were on Indian land and in 1873 federal troops removed them. After the land release of 1886 it appears that few of the original claimants returned. Their settlement, Chopaka City, has disappeared but a dirt road will take one to Nighthawk, which is quite close to the site. The general area and the still flourishing town of Oroville may be reached by U.S. 97.

In driving north on this highway one might wish to pause at Blewett Pass, about halfway between the little cities of Ellenburg and Wenatchee. In the 1890s extremely high-grade gold ore was discovered near the pass, and mine dumps and scattered debris tell of considerable activity there in the past.

Blewett Pass is in the Wenatchee Mountains, which are part of the eastern slope of the extensive Cascade range. This chain, running north and south through Washington, was a formidable barrier to the development of the inland mineralized areas. It was late in the nineteenth century before it was realized that the Cascades possessed mining potential of their own. A number of mines were developed above Wenatchee, centering in the old camp of Peshastin. These were on the eastern side of the Cascades. There followed belated workings on the west. The important mines were on the Skyhomish River, which flows into Puget Sound at Everett. There was some placer gold. Ore found in place was largely silver-lead, and successive discoveries gave rise to the settlements of Sultan, Monroe, Startup, Gold Bar, Mineral City, Galena, and Index. Construction of the Great Northern Railway through the Skyhomish Valley brought payroll money into the camps, and on completion enabled the mines to ship ore and concentrates cheaply to a smelter at Everett. Some of the mines here were rich enough to survive the collapse of silver prices in 1893, and production continued into the early 1900s.

The most productive mine by far in the state of Washington has been the Holden in Chelan County. When it closed down in 1958 it had produced 212 million pounds of copper, 40 million pounds of zinc, 2 million ounces of silver, and 600,000 ounces of gold.

Mining in the coastal areas of Washington has been of little importance. With the Oregon beach placers in mind, prospectors were alert to similar occurrences in Washington. Some auriferous sands were found along the shore in Clallam County but the recovery was small. Clallam, however, which forms the northern part of the scenic Olympic Peninsula, conceals somewhere in its wilder regions a rich lost mine. It is a combination placer and lode deposit that the veteran mining man Fred Rynerson was convinced would yield him a fortune.

Rynerson lived in San Diego and alternately was interested in gold and gems. He had successfully mined tourmaline and other semiprecious stones and also traded in them. He set down his reminiscences of fifty-four years of prospecting for a book that was published posthumously as *Exploring and Mining for Gems and Gold in the West*. He evidently was in the Northwest on business when he was pressed by friends to lend his experience to a search for the Olympic riches.

The discovery had been made by an outdoorsman named Jim Dailey, who had been grubstaked to roam the interior of the peninsula in search of a rumored iron deposit. He also was instructed to be on the alert for any large beds of black sand. His backer was in some ill-defined ceramic business and doubtless had use for such sand.

The story of the deposit given Rynerson ran as follows: Dailey had followed instructions and no more. He had found a very large bed of black sand—"tons and tons of it," he said—and had brought back a sample. He had found no obvious occurrence of iron ore, but returned with several specimens broken from a massive outcrop of unusual-looking rock which he thought might have some iron content. The fact laid before Rynerson was that both sand and rock, when assayed, proved very rich in gold. Rynerson agreed to take charge of a search and Dailey agreed to go along.

At this time the Forest Service required all hikers—and they so classified the prospecting party—to be accompanied by a licensed guide if they entered the Olympic Mountains. Such a guide had been with Dailey throughout his trip. It was a serious setback to Rynerson and his companions when they found that this individual had moved away. They necessarily engaged another guide, but now they were entirely dependent on Jim Dailey, his memory, and his woodsmanship for retracing his original wanderings. Dailey tried hard and drove himself to the point of collapse, but the search was an utter failure.

Dailey wanted no more of this and Rynerson had to return to his gem business in San Diego. After some interval, Dailey settled in San Diego and he looked up his erstwhile companion of the trail. Rynerson's prospecting instincts were still strong, and in repeated conversations with Dailey he pieced together a fragmentary but promising set of directions. In 1924 he tried his luck again. A friend living in Seattle, Leet Elliott, joined him. Elliott was no stranger to Washington's mountains and proved a dependable ally.

This search started from Port Angeles on the north coast of the peninsula, which can be reached by U.S. 101. Today's gold hunter could doubtless locate an essential landmark, Coxe's Valley, at the head of Morris Creek. In 1924 a deserted cabin stood here, a large hole in its roof. One is now in high country, above the milder climate of sea level. Rynerson and Elliott were at Coxe's Valley in November and observed ice forming in the creek by four o'clock in the afternoon. Higher on the mountain, drifted snow lay ten feet deep.

Beyond Coxe's Valley the verbal waybill they were following called for a steep climb, 3,000 feet of elevation in three miles. This was done on foot, with supplies carried by backpack. The trail had long since become too difficult for a pack horse. After the climb, there was a descent to Wolf Creek and a chance to check up on clues.

Rynerson had been patient in eliciting all the detail he could from Jim Dailey, and had promptly written down every iota of evidence. The result of the Wolf Creek checkup was encouraging. From their camp they could see a "gooseneck-shaped" mountain in the distance, a landmark that Dailey had accented. Just what this puzzling term meant has never been explained, but it must have been well understood locally. When Rynerson had asked several old residents of Seattle about it, they had answered that yes, there was such a peak in the northern Olympics. Other features also were falling into place. There was an expanse of burned and fallen timber such as Dailey had described and from their lofty situation three distinct ridges sloped downward. The ridges also were among the specifications.

The two men chose to follow the middle ridge. Later it was admitted that in this respect the directions were not clear. The correct choice of a ridge did not appear in Rynerson's notes, nor could he dredge it up from his memory. They went down the middle ridge slowly, looking all the while but failing to see anything of mineralogical interest. At the base of the ridge was a flowing stream, one which they never were able to identify as they later pored over maps. It had been raining during most of their trip down the ridge. The downpour now became heavy. They were near exhaustion and had little food left. It was necessary to get back to civilization.

Through the rest of his life Rynerson remained convinced that the gold of the black sand and the gold ore that cropped out in massive form somewhere below Wolf Creek actually existed and that their value was not exaggerated. He had spent many hours talking with Dailey. He had seen the assay reports and samples of the vein matter. A sophistication concerning mines and mining stories, born of decades of experience in the southwestern desert, enabled him to weigh the evidence, guardedly but fairly. A third expedition, however, was more than he could undertake. His interests had turned chiefly to gemstones, and he was beyond the age when laborious mountain climbing for days in succession could be lightly undertaken. But with good wishes to any younger searchers, he passed on the clue of the "gooseneck mountain" and such other information as might help.

12
Idaho

The gold of Idaho was discovered by a romantic frontiersman who was searching for a gigantic diamond. The gem was said to be embedded high on a mountainside. Belief in the great crystal was widespread among the Nez Perce Indians and to this day the nature of the phenomenon—whatever it was they had seen—has not been explained. Even at night the great "eye," as the natives termed it, gathered enough light to give out some faint scintillation, to puzzle and fascinate the tribesmen gathered below in some as yet unidentified canyon.

In 1860, Walla Walla, in the eastern part of Washington, was the farthest outpost of the civilization that had grown up along the Pacific northwest coast. The lands making up Idaho and Montana were *terra incognita,* and unlike most such regions in the early West were not productive even of rumors. (A survey party had found gold on the Coeur d'Alene River in 1854 but the news had not traveled far.)

Among the adventurers gathered at Walla Walla was Elias D. Pierce, who, for some vague exploit in the past, bore the courtesy title of "captain." Various stories of "the Shining Eye" were put together by Pierce and to him they spelled diamond. He interested five other men in going in search of it. The party tried to ascend the Clearwater River, a long stream with many branches which joins the Snake at the present site of Lewiston. Indians, whether protecting the great eye or instinctively hostile, turned the searchers back.

It was natural for such men to prospect as they went along, and they returned to Walla Walla with about $80 worth of placer gold. It had been taken along the Clearwater. A minor discovery at first glance, it was not minor to the sophisticated old-time mining men. A new gold country had been found and the rush began. Captain Pierce was in the forefront, leading a placer mining party of thirty-five men.

The Shining Eye seems to have lost its interest from that point, but a curious story of Idaho diamonds is encountered in going over old newspapers. The scene is the little town of Post Falls, which is situated on Interstate 90 near the Washington state line. A Kootenai County newspaperman, Clement Wilkins, relates that in the early part of this century a Richard Owens came West from Georgia and acquired a small farm about five miles south of Post Falls. It lay on the south bank of the Spokane River. It was not very successful as a farm, but, in 1907, Owens startled the townspeople of Post Falls by exhibiting a handful of diamonds. They were genuine, for they were so appraised and sold. The only question in people's minds was whether they were native to Idaho or whether Owens had some financial scheme in mind and had salted his property with gems obtained elsewhere. The "Great Diamond Hoax" that had rocked San Francisco financial circles some years before was known even in Post Falls, and in seeking capital to mine his supposed gem field the farmer obtained only one fleeting investor. Disappointed, he left the area but retained ownership of his farm.

After an absence of several years, during which he tried to accumulate a stake by gold mining in California and elsewhere, Owens returned. The original "pocket," as he described it, had been exhausted when he dug out his first batch of diamonds. Now, like the fictional optimist in *God's Little Acre,* he began putting down holes about his property as inspiration dictated. Such industry would argue the truth of his story, the sincerity of his conviction. The activity went on until "Diamond Dick," as he was locally known, died in 1946. No discoveries beyond the original one of 1907 ever came to light.

Idaho is shaped somewhat like an upright triangle, broad at the south and tapering off in the north. There has been profitable mining in the south, but the great mines of Idaho, some of them still active, have been in the narrow, north segment of the state. The great Coeur d'Alene district in that region is well known throughout the mining world.

This once remote area is now served by Interstate 90, northernmost of our great east-west traffic arteries, which crosses the Continental Divide near Butte, Montana. Ahead lies Coeur d'Alene, a name which applies to a town, a lake, and a rich silver-mining district. In approaching this complex, Interstate 90 crosses, at a single point, the

Bitterroot Mountains, the Idaho-Montana state line, and the line separating Mountain and Pacific time zones.

Once arrived at Coeur d'Alene, and having appreciated the theatrical scenery of this mountain and lake land, the gold hunter may be less interested in the big operating mines than in a speculative search for several older mines, mines which are very rich and very much lost. The start would be a scenic fifty-six-mile drive along the east shore of Lake Coeur d'Alene on U.S. 95 Alternate. This leads to the town of St. Maries and a little farther east is a smaller place, St. Joe.

Either of these points could be a base for either lost mine searches or original prospecting in a large expanse of wild and only lightly explored country. Most of the area lies in Shoshone County and the St. Joe National Forest, and through it runs a strange, neglected river also called the St. Joe. Old-timers in Wallace, the county seat, or Kellogg, a major mining center, will tell you that the banks of the long–extremely long–St. Joe never have had a real going-over by either prospectors or geologists. A glance at the map showing the stream's lonely course along the unpopulated base of the Bitter Root Mountains makes this understandable. Maps published as late as the 1940s do not even depict the St. Joe in its entirety. It is doubtful that it ever has been explored clear to its source.

The most promising of the lost mines of the region could be on this stream, but one story favors the next river to the west, the St. Maries, and a tributary of the St. Maries, Canyon Creek. This is considered the probable site of a very early discovery, an extremely rich silver vein. In Idaho it has become known as the Old Spanish Mine, a name given with some basis to many an unidentified *antigua* in the Southwest. Its application in northern Idaho appears rather fanciful. The rich lode has been worked only by a short incline, but it appears that this excavation connects with a natural cavern, for we hear of a stream of running water alongside the lode which disappears into a dark void too dangerous to explore.

Northwestern journalists have traced interest in the mine as far back as the 1870s, and to a James Williams, who came to Idaho with an antiquated waybill to the diggings. Williams had spent some time in Santa Fe, New Mexico, where many church records were kept, and he had been acquainted with some of the Spanish-speaking priests there. Whatever their origin, Williams possessed a map and additional papers relating to the mine. He spent two summers in unsuccessful searches. The elderly man may not have been able to reach even the vicinity of the mine, for on the St. Maries River the terrain was a jumble of fallen trees and thick undergrowth. Canyon Creek was unnavigable even by canoe.

On Williams' death the quest descended to a nephew, also named James Williams. The younger man found the mine. He collected broken ore from the workings, took careful note of the surroundings, and on his return journey left guides in the way of rock cairns and blazed trees.

His trip was made in 1882, and through a tragic mischance he became a victim of the rough frontier justice of the time. On his way back to civilization, he spent a night in an isolated, empty cabin. The place happened to have been a hideout for horse thieves in the recent past. It also happened that a valuable horse had been stolen from a nearby ranch a few nights before Williams' arrival. A grim posse of ranchers surrounded the miner in his ill-chosen resting place and his lost-mine story had little appeal for them. He paid the penalty of a summarily convicted rustler.

From 1860 onward major placers were being discovered in Idaho. The earliest has been mentioned in relating the story of Captain Pierce. These deposits were along the Clearwater River and its three forks. In 1861 they were sending gold worth $100,000 to Portland, Oregon, each month for conversion into coin or currency. Lewiston, on the Washington border, became the permanent settlement for the booming region, but the real center was Orofino. This is now a ghost camp, and the present Orofino, county seat of Clearwater County, is forty miles from the original site.

A second placer area was connected with discovery of the Salmon River. From Indians the miners had heard of a great river system to the south and some of the red men's accounts hinted of gold. The chancy nature of prospecting is exemplified by the first exploration here. A party of twenty-three hopeful men spent most of a summer along the Salmon without finding gold of any significance. They had to get home before the onset of winter, and had concluded that nothing would be found in this country. There was some final, cheerless panning at their campsite on Miller Creek, said to have been done on a bet. Suddenly all were seeing liberal colors in their pans. Definitely a strike had at last been made.

Next season the same party returned and for a time the members were each recovering four to six ounces of gold a day, or roughly $100 per day per man. Other miners appeared and the influx soon swelled to a great rush. The town of Florence grew up. Even with thousands of men in the district, there were good takings for all, and the historian Bancroft wrote that a claim yielding only $12 a day "was held in small esteem." The richest gravels were in Baboon Gulch, named for a Hollander who for some obscure reason was known as "the Baboon." Cleanups of well over $1,000 a day were common.

An even more lucrative camp than Florence was Warren. The concentrated deposits there were discovered by James Warren, an eastern college graduate who had been in western Idaho for some time and bore a poor reputation. He is described by one contemporary writer as "a man of loose morals" and by another as "a shiftless individual and tinhorn gambler." Whatever his ethical deficiencies, Warren made his weary way up Big Creek, a stream "that never seems to come to an end," and found a series of narrow but very deep gravel bars carrying high values. Even the top layers were rich in gold and values increased rapidly at depth. Yields were spectacular at bedrock, but of more lasting importance was the fact that the auriferous deposits extended on and on. The Warren diggings therefore outlasted those of Florence by years.

Among the characters attracted by the Salmon River rushes was the poet Joaquin Miller. He had not yet assumed this pen name and was known as Charlie. Miners found his first name hard to handle–Cincinnatus. At the age of twenty-five, and under his real surname of Heine, Miller was appointed to a judgeship in Oregon. It was during his four years on the bench that he began writing verse.

Farther south a principal placer field was developed in Boise Basin, which lies thirty miles northeast of the state capital of Boise and is favored by a comparatively mild climate. A hard-working placer miner named Moses Splawn had been told of gold in that area by a friendly Indian. To reach the place, Splawn joined a large party that was searching for the Lost Blue Bucket Mine, inspired by some clue placing it not in Oregon but farther east. Splawn and seven others withdrew from the group and abandoned any interest in the legendary Blue Bucket. The Indian's story was of more interest. They soon united with a similarly minded group headed by George Grimes. It was this businesslike band that discovered and first worked the fabulous Boise Basin placers. The gold-bearing gravel was very extensive. In the early 1860s it supported a population of 20,000 and in the first four bonanza years produced $24 million. The historic spot set aside as the "Grimes Massacre Site" recalls the fact that George Grimes, the co-discoverer, was slain by Indians. The term "massacre" rather magnifies the episode. Indians did attack, but only Grimes and a single tribesman were killed in the fight.

Should you be traveling through this region on U.S. 93, stop at Challis, county seat of Custer County, for you will be near the lost mine of Isaac T. Swim. Unlike Charles Warren, Swim enjoyed an excellent reputation. In Challis he was known as "a man of his word." In living up to that reputation he lost both his mine and his life.

Swim's ledge, rich in gold, was revealed when a towering old fir tree was blown down in a storm. Most prospectors will give at least passing attention to the pits left by overturned trees, as well as to earth thrown up by burrowing animals. Both types of disturbance in the soil have occasionally revealed gold. In this case, part of the tree's great root system was lifted clear of the earth and the roots held pieces of gold-flecked quartz. At the bottom of the new cavity was the disintegrated surface of the vein, also showing free gold in generous amounts.

Isaac Swim was prospecting under a grubstake agreement with two friends at the county seat, apparently small businessmen. He left a location notice, gathered as much of the ore as he could carry, and, as so often happened in the Idaho mining scene, had to hasten home ahead of the winter snows.

Back at Challis, Swim and his associates had to wait out the season and in so small a place his discovery could not be kept secret. When he started back in the spring he was accompanied by at least a dozen uninvited townsmen. Relations between these men and the discoverer appear to have been amicable. Perhaps Swim had no choice. The party is known to have passed through what is now the Stanley Basin Recreation Area. A flooded stream brought them to a halt. This must have been either the Middle Fork or the Yankee Fork of the Salmon River.

The dangerous obstacle caused delay, impatience, and some dissension. It was finally agreed that Swim would ride on alone and make sure the sketchy mining location he had made for himself and his two backers was in good legal shape. This would involve pacing off the boundaries and placing six posts or stone monuments. He would then return and guide the others, so they could start their own explorations.

It was a decision born of impatience and it proved too hasty. Swim rode off downstream to look for a possible crossing. It was assumed in camp that he had got across but day after day went by and the man of his word was failing to keep his word. When the river level fell it was possible to make a search. Wedged under flood debris downstream was Swim's gray saddle horse. It was weeks later, far down on the Salmon itself, that a man's body was pulled from Dead Man's Hole. It could not be identified, but neither Swim nor Swim's pit of rich gold ore was ever seen again.

The historic placer fields of Idaho cannot be recommended to today's prospector. They were worked intensively, and when the gravel became too lean for the claimants to continue they were taken over by Chinese. As we have seen in California, the Orientals had a genius, or perhaps only the perseverance, to recover virtually the last speck of gold.

A rich but limited placer stream far to the north is of interest because it almost became a Shangri-la of well-financed communal living and freethinking. Its intended founder was A. J. Pritchard. In 1882 he was one of the few prospectors who cared to penetrate the trailless, heavily wooded Coeur d'Alene country. In due time he sampled a stream and took $42 in gold from a single pan. He knew that in this remote creek, later named for him, he had a bonanza.

With ample funds from his mining, Pritchard now spent some time in the Middle West lecturing and organizing. He held what appear to have been agnostic and economically radical views and recruited quite a following. He next returned to northern Idaho, laid out a townsite named Evolution, and located many mining claims in the names of his followers. Through direct letters and through notices in the *Truth Seeker,* a widely read magazine of the time, Pritchard summoned his adherents to join him. He envisioned a population of 15,000 for Evolution. A great many people responded, current accounts telling of migration from Illinois, Michigan, and Wisconsin and even from the Atlantic seaboard. Unfortunately the summons, particularly its reference to the rich placer grounds, was also read by miners throughout the Northwest's declining gold-washing camps. Arriving miners took exception, possibly with both moral and legal justification, to the tying up of valuable claims by locating them in absentia for the Midwesterners. There was some gunplay and much litigation.

Fate of the Pritchard colony is somewhat clouded. There was a freethinking group that flourished for a while, not at Evolution but at Eagle City. Local control lay not with the communal group but with the unattached majority of the residents, who were there for mining alone. How Pritchard emerged financially is not known. His name is preserved in Pritchard Creek and the ghost town of Pritchard, and historians credit him with opening up the Coeur d'Alene country, whose massive dykes of silver and lead have made it one of the most productive and most durable of all American mining districts.

13
Montana and the Black Hills

Montana shared in the intensive mining exploration of the 1860s, for it had some very rich placer streams. They were nearly all in the western part of the state. The first known digging into auriferous gravel was done with broken antlers of the elk. The primitive tools, lying beside workings about four feet deep, were found in a ravine near the present city of Butte by two prospectors of the early rush, G. O. Humphreys and William Allen. This is traditionally Montana's first mine. It is only a guess as to whether the excavations were made by Indians or trappers.

The first Montana miner known to us by name was an uncommunicative old man named John Silverthorne, who operated near the central part of the state, not a great distance from the present city of Great Falls. At this time there was a Fort Benton at the head of navigation on the Missouri River. It was a lonely place, and the soldiers saw few other white men—a rarc, passing missionary or a trapper or fur trader in need of supplies. Silverthorne called at the post in 1856, sorely in need of provisions.

In Montana, as elsewhere, the isolated army posts ran something of a store for the accommodation of both troops and civilians. When the old man offered gold dust for payment, it was a strange medium to the civilian storekeeper. The latter doubtfully accepted the miner's pouch in exchange for $12 worth of goods. When he sent the gold to St. Louis with his next order he was credited with $1,500.

Montana's productive placers were fairly near, but not necessarily derived from, the Bitterroot Mountains, which separate the state from Idaho. They lie largely in what are now Beaverhead, Madison, Silver Bow, and Deer Lodge counties. The capital city of Helena, a little farther north, was originally a placer miners' settlement in Last Chance Gulch, and Last Chance remains the name of a principal business street.

The discoveries of the early 1860s brought thousands of men into the state and some achieved spectacular success. The good ground, even much very poor ground, was worked intensively. Later, hydraulic operations and gold dredging, either of which can make a profit out of very lean deposits, tore up many a ravine and even some of the deserted camps. In her comprehensive book on Montana ghost towns, *Montana Pay Dirt,* Muriel Sibell Wolle describes the very large areas of land rendered useless by these types of pre-environmental mining. Possibilities of a new strike in these heavily worked Montana placer fields seem very slim indeed.

Above many of the gold-laden streams there were outcrops of good ore, and profitable and long-lived lode mines were developed. There also were some failures and some puzzles. Among the latter was the mysterious fiasco of Quigley.

It is a town all but forgotten now, but in 1895 its central thoroughfare, Broadway, was perhaps the liveliest spot in the state, and its Golden Sceptre Mine was hailed—at least in stock-selling correspondence—as one of the great bonanzas of the Western world. The location was in the general Missoula region, the nearest railroad point being Bonita, on the Northern Pacific railroad. Today the highways running closest would be State Route 200 and its connecting County Road 209. There is also an old, graded roadbed to Quigley for a railway that never was built.

At one time Quigley's prospects looked so good that investors put a million and a half dollars into the Golden Sceptre Mining Company. Newspapers said that President Grover Cleveland was among the stockholders, as well as whatever Colgate then headed the corporation of that name.

As always there was an entrepreneur, in this case a jovial and well-liked Montana man named George Babcock. There was also a state of the law that permitted the most audacious stock promotions and afforded little redress for outright fraud. What became of the Golden Sceptre investments of President Cleveland and others never has been explained. Babcock himself does not appear to have been dishonest. He was well regarded by his contemporaries, showed no signs of personal affluence after the Quigley collapse, and died a poor man. As the Golden Sceptre operations got under way, he built and settled his family in a fine house at Quigley. This would indicate that he expected his income to flow from successful mining, not from skimming off the proceeds from stock sales.

Why then did the entire population of Quigley hastily quit the town and the mines in 1896?

By this date the peril of Indian attack was long past. The disaster of fire, so common in western gold towns, was not visited on Quigley. There is no record of earthquake, of destructive flood or landslide, of an epidemic of disease. Yet something had sent the residents scurrying. When an A. T. Morgan of Butte rode horseback into the place in 1898 he found not only a deserted town but one which had been deserted if not in panic at least in a great hurry. Houses had been left with their furnishings almost intact, stores with their fixtures still there, even with some stocks that should have been well worth salvaging.

In 1920 there was a Joe Daigle and his family living at the old ghost town, not mining but raising potatoes. The plowing up of Broadway for a potato patch led Eiluf Rue of the Dillon *Examiner* to go into the history of the once lively street and its surroundings. His account was reprinted in *Shallow Diggings,* a collection of Montaniana compiled by Jean Davis. Rue stated that there was simply no gold in the Quigley mines. Yet he tells of a carload of ore shipped to Denver that resulted in the Golden Sceptre being declared "the bonanza of the century." Some agency, not necessarily the smelter at Denver, stated further that the shipment contained gold "in almost unprecedented quantities."

The mystery of Quigley's abandonment remains. Besides the Sceptre, the Jumbo and the Brewster were considered good properties and, considering the size of the town that grew up, there were no doubt other mines whose names have not been preserved. Were all these discoveries only surface blowouts? We have the costly preparations for large-scale mining which surely had some basis in professional engineering reports. Besides the houses and business facilities of the town itself, a huge stamp mill was constructed, a route graded for a railroad connection that was confidently expected. This would not be the first time, of course, that a treatment plant was built at great expense before adequate ore reserves were blocked out. It was a fairly common mistake. But the Quigley operation, if a mistake, was a far more inclusive one than represented by an ill-advised stamp mill. Charges paid the Northern Pacific for Quigley freight set down at Bonita, twelve miles away, were said to exceed $300,000.

The manufacturer Colgate was one of the heaviest investors. Did he go into the enterprise, lending his name as well as his money to it, without convincing engineering reports or other proof that here was a real gold district? The financial panic of 1897 could well explain a sudden withdrawal of working capital, except that Quigley's collapse came a full year earlier. The whole venture remains a puzzle. Also a puzzle is whether the cryptic geology of Quigley's mountainside is of interest only to the historian or whether it deserves a new and more expert examination.

A contrasting story, one of success, concerns the Drum Lummon Mine at Marysville. Thomas Cruse was a young Irish immigrant who had spent several years on grubstakes, prospecting on the eastern side of the Continental Divide northwest of Helena. In spite of his engaging personality he had about worn out his welcome with storekeepers and other possible backers. This was in the mid-1870s. Virtually all the rich placers had been discovered and staked by this time, and Cruse failed to find a new one. Near the end of his resources, he decided to make a try for ore in place. Ascending what later was known as Marysville Mountain, he found something promising and, with more perspicacity than he had been given credit for, started a tunnel to strike the vein at depth. He was now on his own. He would work determinedly on a mediocre placer claim that had been given him, then use the hard-won gold to buy the provisions, powder, and tools needed for another spell of work at his lode mine.

When he reached the ore body it proved rich in gold, profitable and self-sustaining from the start. The name Drum Lummon was that of his home parish in Ireland, the camp's name of Marysville was in honor of the first woman to live in the settlement. Young Cruse mined about $140,000 in gold for himself, then sold to an English company for $1.5 million. It proved one of London's more fortunate investments. Bullion was shipped in the amount of some $40,000 each month and over the years the distant stockholders received more than $15 million in dividends.

Thirty miles southwest of Marysville is one of the important early-day settlements of Montana, still a flourishing little city and almost surrounded by mineralized territory. This is Deer Lodge, now reached by Interstate 90, which in this region is combined with U.S. 10. Various inactive mines both to the east and west might be of interest under present conditions, and there are two extremely rich quartz veins, known and lost in earlier days and never rediscovered.

Situated in a wide, high, well-watered meadow, Deer Lodge was a trading post in the early 1850s, almost a decade before gold caused a stampede into the region. A bountiful spring, pouring out warm water and emitting clouds of steam, had made it a gathering place for Indians. When the various streams in the area began to yield fortunes in placer gold, Deer Lodge was the natural gathering place, and it soon became the supply center. The copper magnate William A. Clark displayed his business acumen even as a young man by opening a bank at Deer Lodge. During the few years that he managed it, before moving on to bigger things at Butte and Anaconda, more than a million dollars in gold was handled by the little institution.

Many small settlements grew up around Deer Lodge. The placer ground all through this country has been so thoroughly worked and reworked that today's prospector would be well advised to ignore the gravels and visit the few camps associated with lode mines. One which may have been neglected is Danielsville, a small, almost vanished aerie in the Flint Creek Mountains west of a highway point called Galen. Danielsville is reached by driving twelve miles from Galen to the ruins of Racetrack, then walking and climbing some six additional miles.

The town took its name from the three Daniels brothers. In 1901 they came upon a truly gigantic gold-bearing formation, measuring eighteen feet wide at the point of discovery. The average value across the vein has not been recorded, but the elaborate preparations for mining the Golden Leaf, as the claim was called, indicated full confidence in heavy and profitable production. A townsite was laid out, a costly wagon road down into the valley was begun, cabins for miners built, and a twenty-stamp mill placed on order. Then the whole enterprise, for unaccountable reasons, collapsed. Anyone wishing, for profit or otherwise, to probe into the matter may still find a large amount of ore in the Golden Leaf dump, never milled, never shipped.

Others may find it more inviting to search for a smaller but richer outcrop in the Danielsville area. The story has come down to us as follows:

An elderly Frenchman found promising float in the lower reaches of the Flint Creek Mountains around Danielsville. He was successful in tracing it, but the trail led upward and upward. Finally the outcrop, flecked with visible gold, was located. It was near the summit of the highest peak in that part of the range. His samples, carried into Deer Lodge, admitted no argument—he had made a real strike.

Because of his age and the difficulty of the high location, the discoverer was willing to enter a partnership. Several men from Anaconda made the arrangement with him, and insisted on inspecting the place at once, although it was early winter. It was a poor time to go into the high mountains, but the discoverer agreed. Snowfall slowed the party's progress and altered the appearance of the mountain landscape. The vein itself possibly lay under a drift by this time. The Frenchman was unable to show the outcrop, and the prospective investors, disappointed and now somewhat skeptical, turned back.

The weather moderated a little, and the elderly miner unwisely set out to vindicate himself. With the arrival of summer and the melting of the mountain snowpack, it seemed to the men in Anaconda that another attempt was worthwhile. On arriving in Deer Lodge, they learned that their guide had not been seen for several months. Their only course was to retrace as well as they could the ascent made during the snowstorm. When well into the mountains they found the body of the old Frenchman, clearly a victim of the Montana winter. The land surrounding the grim scene showed no mineralization. Anyway, it was admitted, the old man's description had placed the mine much higher up than this, near the top of the range.

Another gold lode that may be hunted in this area has taken on the name of the Lost Deadman Mine. The deceased, also the mine's discoverer, was a heavy-drinking, solitary old woodsman named Thomas O. Spring. The name sometimes appeared as Springer. His ore was a disintegrated, reddish-brown quartz. There was one occasion in the town of Deer Lodge when one of Spring's few friends, Sam Scott, watched the miner put a double handful of ore into a small mortar, pound it up, then pan out a full ounce of gold. At another time Spring brought in two ordinary tomato cans filled with the ore and recovered two ounces of gold from the contents.

As long as he was in funds the old miner drank heavily, but not heavily enough to give away his secret. He once let Scott, who was the owner of a small hotel, accompany him part of the way to his mine because Scott was going that direction anyway on a hunting trip. The hotel man was pretty well convinced that the rich ledge was on or near Boggs Creek. This is a small tributary to Cottonwood Creek, a stream familiar to early placer miners.

Spring's mine became a lost mine in 1872. In the course of one of his prolonged sprees, he yielded to the blandishments of two newly arrived mining men and agreed to go grouse hunting with them. What the newcomers' intentions were, once they got the old man into the mountains, would seem obvious. Reasonably sober, Spring accompanied his new friends far into the hills. If there was indeed a plot, nothing sinister had developed as they made camp near the close of the first day. Spring took his shotgun and wandered off to try his hunting luck. He failed to return when dinner was ready. After a little longer wait, his companions set out on a frantic search. They soon found the old man, seated against a tree, his gun lying across his lap. He was dead of natural causes.

If ever sighted, the Deadman Mine will consist of a small pit with hand tools lying at the bottom, possibly with native gold gleaming in the exposed red-brown quartz. That is the way Spring said he always left the place. The hole was not covered or camouflaged, a legal claim was never filed on the mine. Like the Dutchman of Superstition Mountain fame, Spring felt certain that "nobody but me could ever find it."

Directly across the ridge from Deer Lodge is the old mining camp of Philipsburg. It was named for Philip Deidesheimer, a German engineer who devised the "square set" system of timbering for the great mines of the Comstock Lode and enjoyed an international reputation. There were many good mines, gold, silver, lead, copper, and manganese in the Philipsburg district, and at least a dozen subsidiary camps existed for a time. Much of this mining was large scale and well financed. There may be something left there for today's miner, but it will take patience and money to evaluate any of these old properties. Many are in corporate ownership.

One thing that Philipsburg could not boast, with all its diverse mineralization, was a deposit of sapphires. For this unusual product of Montana mining we turn to the region south of Great Falls, the Little Belt Mountains, and the vanished camp of Yogo.

The old mining centers of Monarch and Niehart in the Little Belts can be reached today by U.S. 89, but one must cross the range by trail to reach Yogo Creek and its parent Judith River. Gold washings in these two streams were disappointing. A Chicago physician, Dr. J. A. Bovet, invested $38,000 in a ditch to bring water to some overrated bench deposits near Yogo. This resulted in a total loss, but since his associates brought him into their later gem mining it is possible that he recovered his money. Yield from the sapphire workings has occasionally been described in extravagant terms, but it certainly was not less than $3 million.

It appears that Jake Hoover, a rancher who settled near the Judith in 1870 was the discoverer of the stones. Hoover's personal history is a checkered one. In the middle of the sapphire excitement he fled to Alaska to escape his wife, a violent harridan who had twice tried to murder him. There was a pursuit. The woman followed Hoover to the Alaska port of Valdez, tried to catch up with him in the wild back country, then gave up and returned to Montana and a bigamous marriage to another rancher in the district. When she tried to kill this man, he fired first. Husband number one was safe at last.

It was while panning for gold at Yogo that Hoover observed the clear, light blue pebbles. He would toss or push them over the side of his pan, the usual way of getting rid of useless pebbles. The recurrent blue stones were so uncharacteristic of the country that he showed a few of them to his silent partner in the placer, rancher S. S. Hobson of Utica. A schoolteacher friend of Hobson is said to have recognized them as valuable gems of some kind, and a knowledgeable Swiss jeweler at Helena pronounced them sapphires of high quality. Of several colors of sapphire, the most esteemed is a light cornflower blue. This was the color of the Montana stones. An early edition of the Britannica states that some rubies were found among the Montana sapphires, though of a lighter red than the prized Burmese jewels.

The partners now got busy working the gravel. Lean in gold, it was rich in sapphires. Soon a soft, veinlike formation above the creek was investigated and found to contain sapphires in place. At that time, at least, it was the only spot in the world where the gems were found in their native matrix. All other recoveries had been from placer. The mine's output found a ready market in New York and in Europe.

After Hoover had sold his interest and Dr. Bovet had died, the diggings were acquired by a British syndicate, which undertook work on a larger scale. An adjoining deposit discovered by John Ettien had previously been bought and consolidated.

These developments fill up the years until about 1900. Then another occurrence of sapphires was accidentally found higher in the mountains and about three airline miles from the original strike. Operations here were by an American company and they continued, presumably at a profit, well into the 1920s. From then on the sapphire story is obscure. One property became tied up in litigation, another may have been worked out. There is a puzzling lack of information about any further prospecting or about additional discoveries, if any. The U. S. Geological Survey, however, has declared Yogo and its environs to be "the most important gem locality in the country."

The Bitterroot Mountains, separating Montana and Idaho, extend in a twisting, northwesterly course for more than three hundred miles. A number of roads cross them, but there are large segments of the range that always have been difficult of access and remain so today. There are two highways that roughly parallel the mountains for considerable distances. These are Interstate 90, running northwest from Missoula, and U.S. 93, which runs almost due south. On the latter one can visit numerous old mines and camps quite close to the highway. A turn east at Florence leads to the once active Eight Mile Creek district. At Victor a short jog to the west brings one to Curlew and some of the better silver-lead mines, properties that operated for decades, well into the present century. Just before U.S. 93 crosses Lost Trail Pass into Idaho, the traveler is in the Overwich-Hughes Creek district, scene of much optimism and hard work in 1870, but which gave up little gold. The ghost town of Alta will be found at the confluence of Hughes Creek and the Bitterroot River. Some helpful information about this region is available in a report by Uuno M. Sahinen published by the state in 1957, *Mines and Mineral Deposits of Missoula and Ravalli Counties*.

Northwest of Missoula there was more extensive mining. Numerous old camps will be found along Interstate 90, State Route 200, and subsidiary roads. The town of Superior, center of the Iron Mountain region, is the logical base for exploration. Cedar Creek was a stream that lived up to early indications of placer gold and it was being worked well into the 1900s. The luckiest prospectors on Cedar Creek were doubtless the LaCasse brothers. On winding up their operations in 1907, they told the *Wallace Miner* that they had taken more than $500,000 from their claims.

Lode mining developed early in a zone of mineralization that appears to have lain around Superior on all sides. A string of camps flourished along the road that climbed into the mountains to Mayville. Until recently, at least, this road was passable although the mines have long been inactive. Another route that has old workings and ghost camps at frequent intervals is U.S. 10A out of Missoula.

Some of the mines of the general Missoula area were essentially gold mines, but for the most part values were in silver, lead, and copper. Later the importance of zinc in many of the deposits was recognized.

It may be said that on the eastern face of the Bitterroots, in this central part near Missoula, there was a great deal of activity and some profits were made. Since the first virgin placer strikes, however, one looks in vain for any great success stories.

Between the northern Bitterroots and Glacier National Park there occurred the Kootenai boom, along the river of that name. Activity extended north across the Canadian border. The present towns of Kalispell and Whitefish are in this area and are reached by U.S. 93. The most active camp was Libby, on U.S. 2. Originally a gold camp, Libby owes its later prosperity to the mining of vermiculite, a nonmetallic mineral with varied industrial uses.

To examine an area of more exciting mineralization we go to Lewistown in central Montana and the Judith Mountains. The name recalls the Judith River and its sapphire mines, but we are dealing with a separate area. The principal camp of the Judith Mountains boom was Maiden. Joaquin Miller, writing in 1884, said of the district:

"Maiden claims for its mineral kingdom all the peaks and ridges and foothills of the Judith Mountains. Nearly all of them are literally covered with vast quantities of gold float and they must be intersected with numerous veins of gold ore. . . . The Judith Mountains must have been, are now, and must continue to be the paradise of prospectors." The developing poetic license of the "Poet of the Sierras" is apparent.

There were indeed some very rich mines developed to support the poet's vision and the town of Maiden was building up rapidly. Then both town and mines were suddenly involved in a governmental nightmare.

On a map of this heretofore unpopulated country the lines of a military reservation had been laid out, and it was discovered that the new mining district was within its boundaries. In Washington an armchair officer issued a "General Order No. 26," commanding all civilians to remove their buildings, their machinery, and themselves from the reservation within sixty days.

By this time there were transcontinental telegraph lines. The burgeoning mining district and a good part of Montana officialdom went into action. When the effect of the furious protests became known, the local newspaper announced it under the large headline "GLORY TO GOD." In Washington the lines of the military reservation had been hastily redrawn.

Gold or silver ores, sometimes very rich, were the basis of profitable mining at the camps of New Year and Gilt Edge. At the latter place, development was interrupted by fraud and embezzlement, accompanied by a near lynching, in the operation of the Gilt Edge Mine itself. The trouble brought near starvation to the families of unpaid miners during the winter of 1893–94. The unprincipled manager of the mine, a New York lawyer representing absentee owners, appropriated the payroll funds and barely escaped with his life from angry townspeople. Even in this exigency he managed to take the money with him in flight and $25,000 in gold bullion as well. The "colonel," as the manager styled himself, got away safely to the east, but was soon in trouble with New York authorities and served five years in Sing Sing. Under better management the Gilt Edge proved its worth, but by 1909 all the local mines had gotten into ore that could not be mined at a profit and activity came to a halt.

Across a wide depression from Maiden and its neighbors are the North Moccasin Mountains. Remains of the old camp of Kendall tell of successful gold mining, both lode and placer, which continued up to World War I. The strikes were all on the eastern or more accessible side of the range. Here as in other localities we have a question: How well did the old-timers prospect the other side?

In its northwest corner, Montana's mineralization merges into that of British Columbia. The southeast part of the state exhibits something of a gap in mineralization, but in that direction lie the Black Hills of South Dakota. They were the objective of one of America's major gold stampedes.

The Black Hills were the scene of what has been called, on more than dubious grounds, "America's happiest gold rush." The hills rise from the flat South Dakota prairie near the Wyoming state line and extend well into Wyoming. The search for gold has reached as far west as the Powder River, but that stream has been more productive of cowboy fiction than of the precious metal.

The Black Hills may be approached today via Interstate 90. At Rapid City, the metropolis of a wide and sparsely settled region, roads fan out to various points of interest. There are several national parks and monuments in the Black Hills, including that of Mount Rushmore with its colossal sculptures. In the town of Deadwood one may find memorabilia concerning "Wild Bill" Hickok, who was shot dead at a poker table as he held aces and eights, now known over the world as "the dead man's hand." There also are reminders of Calamity Jane, the lady gambler and dead-shot. And near Deadwood is the town of Lead, pronounced Leed, site of the Homestake Mine, whose output is climbing toward the billion dollar mark, the nation's greatest gold mine by a wide margin.

Reports of gold in the Black Hills go back a long way. On rather contradictory evidence, discoveries between 1840 and 1850 are attributed to a Father Pierre Jean DeSmet, a missionary to the Sioux Indians. The yellow metal definitely was found in 1874 by soldiers in the command of Lieutenant Colonel George A. Custer, a force wiped out two years later in the Battle of the Little Big Horn, often referred to as the Custer Massacre.

The usual struggle of Indian versus miner occurred in the Black Hills. Its episodes lie somewhat outside our province, and some aspects of the Indians' troubles are emotionally charged subjects even today. For the mining picture it is sufficient to say that, rightly or wrongly, the Sioux were induced to open up the Black Hills country. A treaty providing a monetary settlement for the tribe was signed in 1876 and the miners streamed in.

This report has cautioned against expecting the old, heavily worked placer streams of the West to yield enough gold today for much more than hobby operations. The caution applies with special emphasis to placer locations in the Black Hills. Not only were the really productive gravels worked to the limit, but large numbers of misinformed men moved tons of sand and delved down to bedrock in places that were utterly barren. Watson Parker, a historian who lived in the region for years, tells in his *Gold in the Black Hills* of some stupendous hoaxes and some deliberate knavery that set off rushes into hopeless territory. The calculated false report was too often the work of local merchants with provisions and equipment to sell.

For example, in 1876, when most of the known productive ground had been staked, such a scheme to set off a new stampede was carefully plotted in Deadwood. News of a rich strike in the "Wolf Mountains" was whispered around. Until that time such a range was unheard of. Insiders let it be known, by the device now known as "leaking" information, that an appendage of the Big Horn Mountains was so named. The distance from Deadwood was considerable, so everyone who started for the new district had to carry ample supplies. As a result, at least $60,000 worth of mining tools, provisions, and other necessities was sold. Some of the smaller merchants who were not aware of the fraud closed out their businesses at a sacrifice and joined the trek to the west. Mountain ranges, several of which might or might not bear the name of "Wolf," were inevitably found. Of gold there was none.

Besides such promotions as this, there were several rushes that originated as hoaxes with no discernible profit motive. The Polo Creek and Rockerville Hill enthusiasms caused no great distress, but the excitement over False Bottom Creek involved men in days or weeks of hard, fruitless labor. Two young men conceived the idea. They managed to convince Deadwood and nearby Central City that the barren, already explored bedrock in a nearby ravine was a "false bottom," and that a true bottom rich in gold lay beneath it. This was proved entirely untrue, but only after much difficult and profitless digging. What retribution if any was visited on these and other hoaxers is not recorded.

It would be inaccurate to say that a plague of grasshoppers inspired the Black Hills gold rush. Equally dubious would be the thesis that the Fenian Movement, an Irish liberation group active in the last century, was behind it. Yet both influences were present, in addition to the lure of gold, when the first major civilian party penetrated deep into the Black Hills. Known as the Gordon expedition, these men, together with one woman and a boy, made a trip that for a long time was beyond the capacity of even the U. S. Army to duplicate.

The grasshopper plague was visited on much of America's farmland for four successive years and its effects culminated in 1873. The loss in crops was well nigh disastrous. A severe business depression resulted, many small merchants were forced to the wall, thousands of men lost their jobs. With scant prospects of finding new ones, it was a time to grasp at any opportunity. The early, inspiring days of the California and Colorado mining had faded. News from the Black Hills fell on receptive ears.

As for the Irish independence agitation, it found both a mouthpiece and an organizer in Charles Collins, editor of the weekly *Times* of Sioux City. The latter settlement in western Iowa was one of the chief organizing and departure points for travel to the Black Hills. It could not have been a very imposing publication, this little frontier paper, but the printed media in those days had an influence disproportionate to their individual circulations. Through "exchanges," as they were labeled on the mailing lists, editors received the papers of many colleagues, and many columns were filled with borrowed material. None was borrowed more freely than news from the gold fields.

Collins, while writing diatribes against the English and pleas for Ireland's independence, conceived the idea of an all-Irish colony in the far Northwest. He hoped that Ireland itself would soon revolt. This northwest base of the so-called Fenian Brotherhood would then be the springboard for an invasion of Canada. Congress authorized the colony in general terms, without learning that the probable locale would be in Indian lands in the Black Hills.

To what extent Collins was personally activated by the gold fever can only be conjectured. It may have been used only as a selling point for his Fenian enterprise. It certainly was the major theme of his recruiting. Two huge tents were raised at Sioux City, and by the late summer of 1874 they were filled with adventurers. The interest in Irish independence was far from unanimous. What was unanimous was an interest in gold.

There were soon a good many defections, as bad news filtered into the big tents. Several lesser prospecting parties had set out and had been promptly turned back by army patrols. The Black Hills were Indian land, ceded by the treaty of Laramie in 1868 and this treaty was still in force. There came a more severe threat. General Phil Sheridan, whose exploits both in the Civil War and in civil disorders were known the country over, sent stern orders to his western forces. Any group heading toward the Indians' Black Hills country was to be intercepted, its wagons burned on the spot, and its leaders, along with any other possible troublemakers, were to be confined in the nearest military prison. When the Collins party left Sioux City it numbered only twenty-four men. One of them, D. G. Tallent, was accompanied by his wife and son.

As a guide they had engaged John Gordon and paid him a fee of $1,000. By subsequent accounts, Gordon appears to have been an unpleasant character who gradually assumed dictatorial authority. Collins fades from notice. Whatever his personality, Gordon got things done. He led his people in evasive action past army patrols, got them into the mountain country and then through unfamiliar, rugged terrain to French Creek. This was where Custer's men had made their discoveries and the fact was generally known.

Aware of the Gordon expedition, the Army doggedly followed Sheridan's orders and set out after the offenders. Three times the soldiers found the rugged trail and supply problem too much for them, and had to turn back. Then, on the fourth attempt, a detachment managed to reach the colony on French Creek and evict the little group. A stockade had been built for defense against Indians. The guide Gordon had gone out to civilization for supplies. Later he was apprehended, put on trial in Omaha, and acquitted. Rank and file were merely escorted out of Indian country and released.

Little gold was brought back by the Collins-Gordon party. Their example, however, proved to many others that gold seekers could penetrate far into the region, fortify themselves against the Sioux, survive winter cold that hovered far below zero, and yet survive and mine. Many miscellaneous bands followed. An undetermined number of such prospectors, possibly quite a large number, lost their lives to Indians or to the harsh winter climate. Others were turned back by army patrols. But the rush could not be stopped. At length Chief Red Cloud negotiated a new treaty for his people. It provided other reservation land, government donations of food, and an immediate cash indemnity. The Black Hills now were open for mining. And as for the determined John Gordon? He evidently suffered the slings and arrows referred to by Shakespeare, for when the boom was at its height, he was encountered in Deadwood, destitute and heavily in debt.

The early placers fall into two geographical divisions. The first mining was in what came to be known as the Central Hills. The Gordon stockade was located here and the nearby camp of Custer became a boom town. Custer still is very much on the map, being served by U.S. 18 and U.S. 385. The Northern Hills were richer, and Deadwood was the center of activity. Other towns in the area, still active as trade centers and all on or near Interstate 90, are Sturgis, Spearfish, and Belle Fourche, plus the Homestake's town of Lead.

The two regions mentioned by no means include all the old placer workings. Between those areas were at least a dozen active camps, scattered along virtually all the flowing streams except Box Elder Creek. As one moves west toward the Wyoming state line it is evident that the gravels thin out in value and gold content eventually disappears altogether.

There were thousands of latecomers to the Black Hills who necessarily sought out new territory. Many fanned out to the west, but the scantiness of mining reports from the region tells its own story. These thrusts led into Wyoming, where a vast country lay ahead. The streams in this northeastern section of the state included the Powder River and its various forks, the Belle Fourche, the many branches of the Cheyenne, and such major creeks as Crazy Woman, Clear, Willow, and Rawhide. Although not promising as sources of gold, they cannot be categorically dismissed, for the record of exploration there is too obscure.

We will look at Wyoming's modest gold and silver resources presently, but will first turn to the later phase of Black Hills operations, hard-rock mining. The old quartz mines appear to offer the best present-day opportunities.

In the severe Dakota winters, placer ground was frozen solid. The Gordon party experienced temperatures of forty degrees below zero. In the Klondike miners learned to thaw the gravel by forcing steam or water into the depths, but this came a quarter of a century later. During their enforced idleness, and when the snow cover permitted, the miners prospected for ore in place. They were not disappointed. Few indeed were the spectacular outcrops of gold-laden picture rock that enliven the mining annals of California and Arizona, but there were many discoveries of ledges that could be mined at a profit.

The general Deadwood area, which had the most productive placers, also had the best lode mines. There was a theory concerning a long, massive north-south gold formation that came to the surface only near Deadwood but must extend far beyond the visible signs. This was true in a way and was exemplified by the massive, continuous Homestake deposit, the Father DeSmet Mine, the Hidden Treasure, and other good producers. Attempts to uncover the supposed formation to the north and south of this belt were unsuccessful.

There was much successful mining unconnected with this theory and unconnected with the Lead-Deadwood district. Few of the mines had ore worth shipping to a distant treatment plant. Typically each mine had its own mill, and when titles were cleared by the Indian treaty ample outside capital appears to have been available.

Silver also was found in the Black Hills, notably on Bear Butte Creek and Rapid Creek. The old camps of Galena, Strawberry Gulch, Bear Butte, Virginia City, and Gibraltar were based on silver discoveries. What remained in the mines when they were abandoned in the 1893 silver collapse is a question that may be worth an answer.

If the end of the Black Hills boom can be associated with a single event, that would be the burning of Deadwood in 1879. Built entirely of wood in a wind-swept location, the seat of so much legendary gaiety and gunplay was quickly consumed, its colorful, dime-novel characters left without a stage. The town was rebuilt and it settled into a quieter existence. Mining was also settling down to a more businesslike operation of the major lode deposits between Lead and Deadwood, now mostly consolidated into the great Homestake Mine. And great it is. During many years the Homestake has made the flat, farming state of South Dakota the nation's leading producer of gold.

14
Wyoming and Utah

Across the Wyoming state line from the Deadwood area there is a wide, untamed expanse of mountain and prairie where several place names will be recognized by followers of western fiction or film. The Devil's Tower, now a national monument, is the principal landmark. The Powder River has been mentioned, and the towns of Sundance, Oshoto, Wildcat, and others have had their brief prominence in tales of cattle wrangling and outlawry. This rather barren region constitutes the northeastern part of Wyoming. The northwestern corner is occupied by the first of our national parks, Yellowstone. South of the park is the Jackson Hole country and the Grand Teton National Park, while a little to the east the town of Cody recalls the career of Buffalo Bill.

The first gold mining in Wyoming is associated with South Pass, near the central part of the state. The point may be reached today by turning off Interstate 80 near Rock Springs and driving north forty miles to the town of Farson and the junction of State Route 28. This paved highway takes one over the Continental Divide via South Pass itself and runs quite close to the old gold camps of Atlantic, Miner's Delight, and South Pass City.

Gold was discovered near the pass in 1842 but mining did not get under way until the 1860s. South Pass has associations of interest aside from mining. It offered a fairly easy way across the Rockies. The altitude is only 7,805 feet and grades on each side are not steep, so it was favored by fur traders from Astoria, Oregon. It was a principal feature of the historic Oregon Trail. Two missionaries to the northwest Indians, Marcus Whitman and H. H. Spalding, crossed the pass in 1836 accompanied by their wives. A stone monument commemorates the ladies' passage at this early date.

The branch of the Rockies that traverses the South Pass area is known as the Wind River range. The Sweetwater River flows from their eastern side, and there was moderately successful placer mining along this stream from 1861 onward. An H. G. Nickerson worked the gravels industriously, but failed to explore the mountain above him. On that slope was a strong fissure vein that was developed into the great Clarissa Mine, which yielded high-grade gold ore for many years.

Once discovered, the Clarissa's riches set off a rush to the district. Today one may have trouble locating the Lone Star, the Oriental, the Carrie Shields, and other old mines, but in their day they were profitable operations. What they have left to offer is a speculative question. A railroad never was built to this mining region, nor did a line ever run close to it. We have seen how this lack of transportation affected certain other camps, causing mines to close down when they got into ore of borderline value. The three old camps are still identifiable, and some accommodations are available because of the hunting and fishing attractions.

For the only other scene of very substantial mining activity in Wyoming, one must go southeast, almost to the Laramie-Cheyenne area. There the North Platte River flows between the Sierra Madre—not to be confused with the much-publicized namesake in Mexico—and the Medicine Bow range.

The Medicine Bows and their subsidiary Snowy range afforded some limited gold mining at Centennial, and also furnished one of the few platinum excitements of American mining. The town of Platinum City was laid out, but nothing commercially productive ever developed from these encouraging lode deposits. C. S. Dietz, deputy state geologist, in a 1929 report, calls the platinum discovery "important" and places it in "the mountain ridge lying due west and south of the old mining village of Centennial." The ore carried more than an ounce of platinum to the ton. The difficulty, Dietz conceded, was that no sizable body of such ore had been found.

As for gold occurrences, they extended west to a town called Gold Hill, which has long since disappeared from maps. The better-known Centennial still exists and may be reached by State Route 130 out of Laramie. Other old camps in this district were Greenville, Gold City, Golden Courier, and Altamont. A few of the outcrops had very rich ore at the surface. Some of the deposits carried additional values in silver, but apparently quite subsidiary to the gold. To quote Dietz again, "No extensive deposit of rich silver ore has been uncovered in Wyoming."

It appears that there was a dearth of working capital at these properties. Hard-pressed miners struggled to finance their operations from current production and in the face of what they felt were exorbitant freight rates on the Saratoga and Encampment Valley railroad, the feeder line to the Union Pacific. Economic troubles were the worst at Gold Hill, which was one of the later developments. All this leads up to a tempting report stated as fact by a careful historian, that large tonnages of ready-mined ore were abandoned at Gold Hill. The statement that this was "high-grade" may be questioned, but at today's gold and silver prices high-grade it may be.

Across the valley of the North Platte were the old camps of Dillon, Rambler, and others, all in the Sierra Madre and on or near the present state routes 70 and 230. They are outside the scope of this report, all the significant production being in copper.

Far to the north, in the Big Horn Mountains, is a lost gold ledge which its founder boasted could "make the whole world rich." His knowledge of the mine was instrumental in getting him confined in what the press of the day called the "state lunatic asylum," but eventually that same knowledge effected his release. The location is in Big Horn County, which adjoins the state of Montana.

The discoverer, a man named Thompson, brought his story to the diggings at Alder Gulch on the Montana side. This happened to be the nearest settlement in 1863.

Thompson was one of four men prospecting together in Sioux Indian country when the strike was made. The quartz outcrop pushed up out of fairly level ground. The man's startling statement about "making the whole world rich" was based on a startling sight, a showing of massy gold occurring liberally through the exposed vein. Its presence showed up so brilliantly and obviously that the partners decided to build a cabin directly over the ledge in order to conceal it. This was not the only instance in which such a stratagem was adopted. There was a widespread belief that the "castle" of Death Valley Scotty stood squarely over his reputed gold mine.

The cabin in the Big Horns was completed and a start made on getting out the ore. Then came the almost inevitable Indian attack. Three of the miners were killed at once in the onslaught. Thompson, who was outside at a little distance, was not hurt, evidently not even observed. He was able "to drop down the slope out of sight," he related, and then to make his way along the base of the mountain while the Sioux took time to reconnoiter the cabin and establish that the intruders in their country were dead. Nowhere in the narratives from different sources is it stated whether the Indians burned the cabin, whether it might still be there as a marker. Thompson, fleeing for his life, could hardly know.

A woman in peril will snatch up her baby, a miner will snatch up his gold. When the terrified Thompson arrived at the booming placer camp of Alder Gulch he therefore had not only his story to tell but the gold to back it up.

He described the outcrop as that of a quartz vein three or more feet wide with streaks or bands of the solid metal running through it. These varied from strands of wire gold, he declared, to a belt that was an amazing seven inches in width. Miners are not easily deceived by ore specimens. Thompson's display of samples was convincing. The pieces plainly had been chiseled or gouged out of a massy occurrence of solid metal. More than a hundred miners were ready to escort him safely back to the site. Their leaders agreed that the original ledge as it existed under the cabin should go to Thompson, plus a reasonable length along it on each side. They tried to convince him that the guaranty would be honored. The object of their solicitations hesitated. He had been through an unnerving experience, and his later, undeniable mental trouble may already have developed. He agreed to the expedition, but insisted that first he wanted to visit San Francisco. A disappointed community of Alder Gulch had to let him go.

It is now eighteen years later. Alder Gulch is all but worked out. One of its miners has drifted to San Francisco. This man has an unexplained interest in abnormal psychology. He is being shown through a mental hospital and the attendant pauses to comment on a curious case.

The poet Christopher Smart, confined to an English madhouse and denied writing materials, is said to have composed his entire *Song to David* by scratching the words on the walls of his cell. This San Francisco inmate has been drawing a map on the floor. When it is occasionally scrubbed out by routine cleaning, he draws it again. Even after the lapse of years and in this strange scene, the visiting miner recognizes the patient. He is the Thompson who once stirred Alder Gulch with his gold discovery.

From hospital records it was now possible to learn something of Thompson's troubles. Soon after arriving in California he had been robbed of his gold. He next fell seriously ill. He recovered physically, but a dementia was developing which at times led to violence. His raving about a fabulous mine did not help him. In 1878 he had been adjudged insane and ordered confined. The staff physician in charge of his case said that between seizures Thompson was rational. He had told the story of his life, and the physician fully accepted its highlight, the account of the rich gold mine.

The visiting Montanan now supplied the background from old Alder Gulch days. As a result, far from both Alder Gulch and the Big Horn Mountains, a search at last was organized for the Lost Cabin Mine.

The first thing was to effect Thompson's release, and this the doctor did by certifying him to be recovered. The physician then arranged a long summer's leave for himself. He became the most sanguine of the gold hunters. The Alder Gulch miner was of course one of the party, and it included the hospital orderly who had unwittingly set events in motion. Possibly this man knew too much to be excluded, possibly he was needed to deal with Thompson should he become violent. The patient already had experienced one such upset during the planning stage and the burly attendant had had to protect the others from attack.

In early June of 1882 the four were in Virginia City, Montana, preparing to head for the Wyoming state line and the Big Horn Mountains. Thompson remained quite rational. He expressed no doubt that he could find his way back to the log cabin, whether burned down or not. The party gave out their destination to be the Yellowstone River country and outfitted and provisioned themselves for a long journey.

After entering the Big Horns they traveled south for three days. When they made camp the night of June 15, Thompson announced that on the following day they would reach the mine. Excitement over the impending discovery was now running high, but unfortunately not high enough to keep the doctor, the miner, and the orderly awake. They slept soundly after their day's exertions. When they woke up, their guide was missing.

It took the anxious men three days to find him. No doubt, in their search, they were as alert for the sight of a cabin as a sight of their companion. The cabin and the bonanza it concealed never were discovered. Thompson was, a body at the base of a high, rocky cliff. There was no sign of foul play. The two men from the hospital agreed that he must have had a sudden relapse and been on a wild, demented rush through the night woods when he suddenly stepped off into space.

The San Franciscans returned home and the Alder Gulch man told his cronies what he remembered of the map that Thompson had drawn and redrawn at the hospital. It depicted mountain country. In the center of the map, a small stream was joining a larger one. A line then extended due south to a mountain peak, and the distance was marked as three miles. From the mountain, the line continued south, but now it was drawn zigzag. Did this mean the route followed a crooked stream, or did it indicate a crooked trail by which one might ascend a mountain? At the end of the zigzag was a cross within a circle, obviously the goal itself.

Only one contribution was remembered from Thompson's guarded conversation long ago at Alder Gulch. He said the original prospecting party was intent on getting higher in the range, and that just before their great discovery they were looking down on a fairly distant grassy meadow. This information may have inspired other hopeful searches but one finds no record of them. The physician, being on leave of absence, presumably returned to his hospital post.

A Wyoming state report on mineral resources names thirteen counties in which gold is known to occur. They are well scattered over the state, being Albany, Big Horn, Carbon, Crook, Fremont, Hot Springs, Johnson, Laramie, Park, Platte, Sublette, Sweetwater, and Teton. "To date only a small per cent of the recoverable gold has been mined in these counties," the report states.

On the south, Wyoming borders on Utah and Colorado, and both states loom large in mining history. Interstate 80, which runs clear across southern Wyoming, crosses into Utah at Evanston, and takes one through the Wasatch Mountains and scenes of long and profitable mining. Across the state to the west, it enters Nevada at the little town of Wendover. It is at this western point that we will start an examination of Utah's gold and silver districts.

The best-known natural feature of Utah is its Great Salt Lake, although from the scenic standpoint the various national parks in the southern part of the state are of greater interest. The lake is the remnant of a vast inland sea that spread into Nevada and Idaho, known as Lake Bonneville. The name honors Captain Benjamin Louis Eulalie de Bonneville, a West Pointer whose ambiguous military-civilian expedition into the West in 1833 was chronicled by Washington Irving. The Great Salt Lake is shallow, having an average depth of only thirteen feet, but it extends seventy-five miles on its north-south axis and is fifty miles in breadth.

Reports of the lake filtered out to the early Spaniards in Mexico. It was known as Lake Timpanogos, and was connected in the lore of the times with the hypothetical Rio Buenaventura, which was reputed to provide a navigable passage to the Pacific from the Rocky Mountains, possibly even from the central plains. Legends that clustered about "the mountain sea," as unearthed by Dale L. Morgan for his *The Great Salt Lake,* told of a white race known as Munchies which inhabited islands of the lake. There also were red men, gigantic in stature, who rode on large beasts improbably assumed to be Asian elephants. Aquatic animals foreshadowing the Loch Ness Monster were seen, and noxious vapors rising from the brine poisoned or suffocated birds in flight. These legends persisted to an extent that had economic effects. When the heavy salt content of the lake was recovered and marketed as common table salt many persons, disturbed by these old tales, would not use it.

The immense, flat expanse of the dried-up section of ancient Lake Bonneville is familiar to all who have traveled by land or air between Salt Lake City and the Pacific Coast. One part has frequently been in the news, the Bonneville Salt Flats, which have been the scene of many automobile speed trials. So flat and featureless is this western Utah desert that between the towns of Wendover and Knoll Interstate 80 runs in an unswerving straight line for forty miles.

If from Wendover you drive not across the flats but south along the western edge of the desert you will come to a trio of mountain peaks about 8,000 feet in elevation. Their names are Dutch, Ochre, and Montezuma. In their midst is the town of Gold Hill, an early mining settlement of rising and falling fortunes. It is in an area of curiously mixed mineralization. You can prospect with some hope of success for gold and silver, but also should be on the alert for lead, copper, bismuth, tungsten, and, if not alienated from it by reading murder mysteries, for the valuable mineral arsenic.

Another old camp, Clifton, at the base of Montezuma Peak, may offer the best promise of finding a workable deposit of the precious metals. It is said that the silver-lead deposits about Clifton never were thoroughly explored. The mines were worked in the early 1870s. At that time the site was so remote that haulage of ore was difficult and expensive and a profit hard to come by. Clifton's more prosperous neighbor, Gold Hill, although now all but abandoned, had a series of modest booms, and a few lingering inhabitants believe that higher metal prices may bring another one.

In the three-peaks region we find the old Alvarado, which was a gold mine with one ore shoot that ran $1,100 to the ton. Another mine of the same period was working a vein running 1,800 ounces of silver per ton, and $400-a-ton copper ore was coming from the Copperopolis. Values declined at depth in these and comparable workings and the mines closed.

Then came World War I. The slopes of Dutch and Ochre mountains were found to contain scheelite, an ore of tungsten. This metal was sorely needed in producing munitions and other wartime manufactures, and there were tungsten booms throughout the West. They quickly collapsed after the armistice of 1918. It took another conflict to bring Gold Hill to life again. With World War II it was not only tungsten but arsenic that was needed, the latter being a component of certain insect sprays employed not only in the South Pacific jungles but also in domestic agriculture. When the government abruptly stopped the purchase of arsenic, Gold Hill was vacated with equal abruptness. Except for a rare watchman's job, an inactive mining camp seldom offers employment.

From Gold Hill it is possible to drive in a wide arc around the southern edge of the Great Salt Lake desert. An inferior road with few water sources must be followed for about ninety miles before reaching pavement at the Air Force's Dugway Proving Grounds. The prospector may wish to pause along here to see what is left of some rich silver deposits. If accounts printed in obscure weekly newspapers of the 1890s are correct, the mines in question were quite rich but were worked under such severe difficulties that owners gave up the struggle, leaving veins of high-grade galena, the silver-lead ore, only partly mined out.

To find these workings, the investigator should establish his whereabouts by driving to the proving grounds, then turn back on his tracks for about twelve miles. This should place him in sight of the Dugway Mountains. Different maps show the Dugways in different locations and some maps do not show them at all. A map dated 1910 labels them the Granite-Dugway Mountains, and there is indeed a Granite Peak in the area which, with an elevation of 9,779 feet, is a desert landmark. If local information can be obtained in this almost uninhabited country it would be highly desirable. We know from the meager records of these mines that the Dugway range extends well into the

Great Salt Lake desert and that operations involved a long haul over the desert flats. The name of the old camp is Bullionville. There is little left to identify it other than the remains of an old smelter, plus such tunnels and dumps as testify to mining activity.

The richest mine at Bullionville was the Silver King. Well up a west-trending canyon is the Black Maria, and just over the top of the ridge the Queen of Sheba. The Sheba's ore was different, with values in gold and copper in addition to silver. The district was opened up in the era of horse-drawn transport, and getting heavy ore wagons through the desert sands was a grueling task. The small smelter was built in order to treat ore on the spot, but because of the scanty water supply it could operate only for short periods.

Bullionville may have collapsed with the collapse of silver prices in 1893, or its owners and miners may have left for other fields during the great Nevada mining revival a few years later. We know that very early in this century, and perhaps even earlier than that, the camp and the mines were deserted. Carloads of good ore are known to have been left on the dump of the Silver King. Has anyone been back and hauled it away?

At the proving grounds one is headed toward Salt Lake City and the focus of Utah's greatest mining activity. The heavily mineralized areas near the Utah capital were recognized and developed largely because of the discovery of the Emma Mine at Alta. This property, often described as "notorious" or as "a swindle," caused serious friction between the United States and Great Britain.

When Brigham Young sighted the Great Salt Lake and announced to his followers, "This is the place," he had an agricultural empire in view. The Latter-day Saints held closely to this objective. Search for the precious metals was discouraged and at times forbidden, but the church sanctioned the mining of such commodities as coal and iron in specific instances. Brigham–Bancroft always refers to the Mormon leader by his first name–told his people, "Coal and iron have made Britain great, gold ruined Spain." Mormon prospectors occasionally ignored church policy and made important discoveries, including, as has been related, the remote and very rich Potosi lode in southern Nevada.

It was non-Mormons, sometimes army men, sometimes roving prospectors, who found and opened up Utah's important mining districts. In 1864 the wife of an army doctor picked up a piece of high-grade silver ore in Cottonwood Canyon, which leads up to what was to become the booming camp of Alta, not far from Salt Lake City. Then a discouraged prospector named J. B. Woodman chanced upon the Emma outcrop, high on the slopes of Mount Baldy. This snowy summit and the resort of Alta are now devoted to skiing, but in 1869 and in ensuing years the people of Alta were unearthing the rich, massive silver-lead deposit that was the Emma lode and finding and working additional veins of similar value. The Emma's first shipment of ore was got off with great difficulty. Woodman and his helpers were operating on a very slender grubstake from skeptical Salt Lake City businessmen. The physical handicaps of operating were extreme. Appeals were made in vain for adequate working funds. When some ore finally was sent out, it had to be got to a seaport and carried around Cape Horn to faraway Wales for treatment. Even after such monumental expense had been incurred, the lot returned a net profit of $189 a ton. This achievement brought adequate new capital flowing into the operation.

In barely three years the Emma's yield was so prodigious that English investors formed a syndicate and bought the mine for $5 million. Half was paid in cash. The new management took over, and as work went deeper and deeper the silver-bearing vein narrowed down between its side walls. Then it pinched out altogether. The mine was worthless.

At first it appeared that deliberate deception had been practiced on the English buyers. High diplomatic officials in Washington and London became involved. An investigative team with both British and American members was formed and their geologic and financial expertise applied to the case. Their finding: At the time of the sale, neither the American owners nor the purchasers' consulting engineers could have known that the ore body was close to exhaustion. It was an unfortunate investment but not a swindle.

High in the sky though Alta was, the crest of the Wasatch Mountains rose higher, forming a dark backdrop behind the headframes and loading bins of the working mines. A man named Herman Budder, who had been at the California mines and learned something from them, roamed about these higher levels a good deal and paid particular attention to the Emma mine and its surroundings. Then he equipped himself for a long outing and climbed over the top of the ridge.

The far side of the Wasatch was at that time unexplored. At a distance of about ten airline miles from Alta, Budder found what he had hoped for, what he more than half expected. Nature had marked a spot where Alta's great zone of mineralization came through to the other side of the range. It was not the individual Emma Lode that extended to this side, but the giant formation of host rock that was threaded with silver-bearing veins. Budder's discovery was the beginning of Park City, which has long outlasted its sister camp and has poured out $250 million, with values not only in silver but in gold, lead, copper, and zinc.

About twenty-five miles south of the Alta-Park City complex are some very old gold or silver workings whose history and values remain unknown. The early Mormon settlers were aware of them, but the church's attitude and the demands of their barely established farms and cattle ranges forbade any exploration. Like so many old diggings they were called "Spanish mines," although if worked by men from south of the border they would be better identified as Mexican. In 1860, thirteen years after the Latter-day Saints had settled beside the Great Salt Lake, a pack train was observed coming from the mines, the mules laden with heavy leather bags. A Mormon farmer wrote in a letter of seeing the party encamped near his place, with armed men standing guard. He thought it best not to inquire into their business. It later was reported that the group had been ambushed and killed by Indians near what is now the town of Nephi. It would be helpful if someone had picked up some of the scattered ore at the scene, but there is no record of this. In later, calmer days the Ute Indians told a little about these expeditions that had come from the south during their fathers' time. It was indicated that the miners had gone up a major creek which flowed into Utah Lake. The latter is a body of water south of the big lake, with Provo, site of Brigham Young University, being the principal population center along its shores. An elderly Mexican in the area confirmed the Indians' directions, and added that the mines were along the first creek flowing into the major stream. Using modern maps, this would point to a route up Spanish Fork Creek–How did it get that name?–then up Diamond Fork Creek and one of Diamond Fork's several branches.

Although not aware of these clues, a hunter named Clark M. Rhoades came upon an old shaft near the confluence of Diamond and Spanish Forks creeks. His son Gale looked into the historical, or legendary, background and did some successful exploring. Writing in *Desert* magazine, he described two inclined shafts just over the ridge from the creek junction and near the top of the ridge. They were about fifteen feet apart and both curved in the same direction at the fifty-foot level. Some scanty remains of rusted spikes and tools gave evidence of some antiquity, as did the heavily overgrown dumps of waste rock.

It is apparent that no effective examination was made as to the nature or value of the vein matter, possibly because of the dangers inherent in such old shafts. Both the younger Rhoades and his father got the impression that there were additional underground workings, possibly large chambers that were concealed by timbers and some rock fill. The size of the overgrown dumps was not stated.

It would appear that these two holes alone, though clearly mine workings and clearly very old, would not account for the successive pack trains that emerged from the canyon of Spanish Fork Creek. Other and larger mines are indicated. The area beyond the Rhoades discoveries invites further prospecting.

Across the wide depression south of Salt Lake City is Bingham Canyon, which quite early attracted placer miners. Bear Gulch was especially rich, yielding some $2 million in gold to those who got there early. Exploration followed its usual course to ore in place. After a good deal of gold and silver had been mined, it was evident that much of the district was a vast, low-grade copper deposit. It was here that open-pit mining on a grand scale was developed, an enormous mechanized version of the old glory hole. The great Bingham Canyon operation was a landmark in mining history. Similar massive excavations, with their ore trains spiraling up to ground level, may now be seen in several western states.

Exploration directly west of Bingham's varied ore deposits contributed little to mining history. The Cedar Mountains are the only range between Bingham and the wide, flat desert we skirted on the way to Gold Hill. Perhaps they deserve another going-over in the light of today's metal prices. Another range, the Lakeside, rises to the north. Interstate 80 runs between these two ranges and the small settlement of Delle, which is about forty miles west of Salt Lake City on Interstate 80, stands equally distant from both mountain chains.

A better takeoff point for the Cedars as a whole is the ghost town of Iosepa, which has a dismal history, not connected with mining. Some organizers, possibly well meaning, managed to establish a colony of native Hawaiians at this place. It was to be sustained by farming and cattle raising. The Hawaiians found the soil and water supply to be far different from those in the islands, as was the climate—especially in winter. Poverty, illness, and premature death thinned the islanders out, and the project, whatever the objective may have been for this curious transplant, was a complete failure.

The camps of Ophir and Mercur, southwest of Salt Lake City, were centers of much activity, inspired by shipments of high-grade ore, both gold and silver, found almost at grass roots. The farthest workings that may be grouped with the Salt Lake City complex were around Eureka, on State Route 36. These constituted the Tintic district, where profitable operations extended to depth and had a long life span.

If few leads for the individual prospector have been presented in the Salt Lake City region it is because mineralization was fairly well concentrated and exploration and development were well financed and thorough. The four Walker brothers, Utah bankers, were early investors. In spite of the Emma disaster, additional capital came from London. Cripple Creek millionaires financed the pioneering Bingham Canyon experiment and, as various mines proved profitable, New York money was available.

No mineral zone as extensive as that of the Salt Lake City area exists anywhere else in Utah, but a number of gold and silver discoveries in scattered places deserve attention. We may exclude at once the immense region of stratified sandstone that spreads over the southeast part of the state. These formations have been weathered to create much remarkable scenery. Large areas are protected in the form of national parks and monuments. Parts of the sandstone country were important during the hunt for uranium, but except for the exceedingly rare placer bar this is not a gold region.

In the extreme southwest corner of Utah is the town of St. George, which for years lived in apprehension of radioactive fallout. When nuclear bombs were being exploded above ground at the Nevada test site, the drift of the upper air was often to the northeast and consequently toward St. George and its environs. Time and again scientists and distinguished visitors to the test site were about to don protective goggles and witness one of the stupendous bomb blasts when some motion in the atmosphere was detected and the detonation postponed. "Save St. George" was the unspoken policy. Now that tests are conducted deep underground, St. George can feel secure and perhaps speculate about the great sandstone reefs that lie just to the north. For while the atomic reactions conform to scientific theory, the sandstone reefs do not. Contrary to both theory and experience elsewhere, they contain silver.

Silver in sandstone may be heresy, yet the great rock formations that can be traced for miles near St. George are undeniably sandstone, and there is a record of $10.5 million in cash payments from smelters that says the metal recovered there is silver.

The reef's silver was discovered in absentia. The sandstone, as sandstone, was quarried in a small way to make building stone and grindstones. One story is that such material was used in a fireplace in a certain Utah home and that when heated little bright globules oozed out of it. A sharp-eyed visitor, William T. Barbee, scraped off some of the exudation and recognized it. He easily traced the quarrymen to their source on government lands, located the first claim of the Silver Reef district, and laid out a townsite.

Another story has to do with a young assayer named Murphy who was doing a brisk business in Pioche, Nevada, which is about forty miles from St. George and the reefs. Business was altogether too brisk. The encouraging values that he reported in almost any ore sample were making the miners suspicious. About this time a grindstone shipped to Pioche had been broken in transit and left in a vacant lot. Someone now had an idea for trapping the assayer. A little group of mining men crushed a piece of the grindstone into a mass of sand so that it would look like a grab sample of "fines" from a mine dump. Then they grimly submitted it to Murphy.

After his test, the young assayer made out a certificate variously reported to have shown two hundred ounces to more than eight hundred ounces of silver to the ton. In any case it was a fantastic value, when by all the rules of mining the sandstone should have proved quite barren. Murphy was yanked out of his laboratory, taken to the edge of town, and warned not to return.

Disgraced but not discouraged, the exile pondered his best course of action. He knew the assay to be correct. With darkness, he crept back into Pioche and through a friend got a clue as to where his nemesis grindstone had come from. It had been delivered by an Isaac Duffin, who was in the business of quarrying and shaping the stones somewhere between Ash Creek and the Santa Clara River. The location was across the Utah border but not a great distance from Pioche. Murphy made the trip, talked to Duffin, and then investigated the puzzling sandstone. The metal was there in nature just as surely as it was in Murphy's assay cupel. The enriching magmas had somehow entered the sandstone, and later there were eyewitnesses to the fact that the metal at times occurred in sheets that could be stripped off the broken sandstone and were almost pure silver. So now we have

Murphy, not Barbee, as discoverer of the anomalous silver formation.

The sandstone reef can be traced for one hundred miles. Bancroft says that the silver-bearing part of the formation was at least fifteen miles long. He describes various profitable operations up into the 1880s, and observes that there remained "hundreds of locations as yet unworked" in the known mineralized section. His report on the Christy Company's mining returns over a four-year period figures out to a silver value of $205 per ton of ore. This is higher than the average along the reef. Much of the old, known profitable area is off limits to the prospector, for it is corporation-owned and much of it is fenced. The principal camp was called Silver Reef. Its site may be reached by turning off Interstate 15 at the hamlet of Leeds.

Young Murphy was vindicated, but we have no information as to whether he profited by his discovery. History is equally silent on what amends were made by the suspicious miners of Pioche.

Besides the reef mines of the St. George region, there were some occurrences of gold and silver in the more familiar form of veins in quartz. Such a deposit was discovered at one of the places—just which one is the question—where the Old Spanish Trail crossed the Santa Clara River. A miner named James Houdon, said to be French or a French Canadian, found the ore so rich that it was worth packing out on burros over a long and dangerous trail. He made two profitable trips. Before his third visit a heavy flood tore up the banks of the river and the mine was lost to Houdon, as well as to several men who were waiting for an opportunity to jump his claim. The general area may be identified today by the bridge that takes State Route 18 across the Santa Clara.

About midway between here and the northern mining districts, on Interstate 15, is the town of Beaver, all of its early houses being picturesque structures of hewn stone. It was once the supply point for quite active mining in the mountains to the northwest, extending as far as Wah Wah Springs. State Route 21 runs through the region. The principal camp was Frisco, now a ghost town.

The Frisco district afforded one of the instantaneous successes that helped sell stock in less fortunate mining ventures. The property was the Horn Silver. The historian Bancroft, writing only a few years after its discovery, states that in its first year of operation, and with workings down only five hundred feet, the mine produced silver and lead to the value of $6 million and that $1.5 million was paid out in dividends.

Much nearer Beaver is an undiscovered gold vein that must be extraordinarily rich. The indicated location is reached only by trail, and is about two hours' ride by horseback into the hills west of town. Float assaying $80,000 a ton was picked up here by a very old miner living in Beaver. In 1930 an Oklahoma oil man and a Denver professor joined forces for a field trip with the patriarch. They found gold disseminated throughout a huge, iron-stained dyke but the value per ton, as shown by a dozen assays, was too low for mining under conditions then prevailing. Nothing corresponding to the handsome piece of float was found. The party admittedly did not make a very sustained search.

In southeastern Utah, but not part of the widespreading sandstones with their arches and other scenic features, are the "unknown" mountains of early-day maps, now bearing the name of the Henry Mountains. Formed by igneous intrusions in the great sedimentary plateau, they reach a height of 11,000 feet. Geologists do not think well of their mineral possibilities, but there have been a few vaguely described discoveries. There was enough gold to cause a small rush in 1890 and the camp of Eagle City flourished for a brief time.

Between World War I and the early 1920s an engineer named F. T. Wolverton tried some serious development. His diary spoke optimistically of a large, medium-grade ore body. He constructed an ore crusher powered by a remarkable water wheel of great size, built by himself with ordinary hand tools. When Joyce Rockwood came upon the place in 1969 during an exploring trip with Lurt and Alice Knee, long-time residents of the canyon country, the wheel was in good operating condition in spite of nearly half a century of neglect. Wolverton suffered injury in a fall from his horse in the early 1920s and complications proved fatal. Whether his confidence in the mine was justified seems never to have been determined.

Some brief but useful information on Utah mines is available in a report published in 1973, *Utah Mineral Industry Operator Directory*. It was compiled by Carlton H. Stowe of the Utah Geological and Mineral Survey. As would be expected under current conditions, most of the listings relate to minerals with industrial uses—oil, gas, coal, sand, and gravel. There are a great many entries, however, showing precious metal and uranium mines, both operating and inactive, with names and addresses of the latest owners of record. Such publication projects in other western states would seem well worthwhile.

15
Colorado: East of the Divide

Colorado is one of the great mining states. Its major gold and silver mines were of astonishing richness and it remains a storehouse of minerals that are less glamorous but none the less essential to any industrialized society. Its reserves of coal are the largest in the nation, underlying at least one quarter of the state's total area. It has more than 2,000 producing oil wells and vast reserves of natural gas, while the petroleum locked in its great shale beds promises a well-nigh illimitable supply when technology and economic considerations make it possible to bring oil shale into production. The radioactive ores used by the Curies in their historic experiments came from Colorado. The little-known but important metal molybdenum adds $80 million or more a year to the state's income, all of it from a mountain near Climax, Colorado, which early gold and silver prospectors passed by.

The searcher for precious metals will immediately divide Colorado into two sections. Somewhat less than half the state, the eastern part, is flat prairie, which merges into the great wheat fields of Kansas. The rest of Colorado, extending west to the Utah state line, is all mountainous. If the Pamirs of Asia are the "roof of the world," Colorado is similarly the roof of the United States.

The capital city of Denver, although it lies on the plains fifteen miles from the mountains, is an even mile above sea level. To the west an immense area of mountain country has a mean elevation of some 10,000 feet, a land of rarefied atmosphere and banks of perpetual snow. Many peaks rise above 14,000 feet, and even the mountain passes, the lowest points at which one can cross the lofty ranges, carry roads high into the sky. The elevation of Mosquito Pass, which you would travel between the gold camp of Alma and the gold-silver-lead metropolis of Leadville, exceeds 13,000 feet and it has close rivals in altitude all along the Continental Divide.

The gold and silver of the Centennial State lie in the mountain country, and often very high in that country. It is curious, however, that the discovery that set off the Colorado mining rush was made on a stream flowing out of the sterile eastern section of the state. The stream is Cherry Creek. It flows into the Platte River in the midst of what is now the business district of Denver, after wending its way through an expanse of low, flat-topped mesas southeast of the capital city. Gold was washed from Cherry Creek sands when this point of confluence was merely a good camping spot. Neither the creek nor the tablelands from which it issues became the scenes of any mining, but before leaving them for more active regions it may be noted that an occasional mining engineer has believed that a great gold-bearing reef of sedimentary origin might be found by exploring the upstream mesas. The gold of Cherry Creek, overshadowed though it was by richer discoveries, had to come from somewhere.

Colorado's first gold discovery was made by a party of six men from Georgia. It was in the early days of the California mining rush. Assisted by some Cherokee Indians, they were driving a herd of horses across the country to the new settlements on the Pacific slope. They planned to make a good profit on the horses, then remain to try their own luck in the California diggings.

When the Georgians, the Indians, and their horses drew near the Rocky Mountains it was late in the season. They had been warned against trying to get through the mountains under winter conditions. The great range ahead was something different from the Appalachians. Great banks of snow already could be seen on the higher peaks, and the party wisely decided to wait out the oncoming winter. The Platte-Cherry Creek site offered good forage for their horses and here they settled down.

There had been limited gold mining in Georgia and some of the six travelers were from Dahlonega, one of the early mining centers. During their long wait, two brothers, William Russell—usually known by his middle name, Green—and Joseph Russell, panned the sands of Cherry Creek. They accumulated a small amount of gold dust. This was stored in quills from the feathers of wild geese and carried on to California when, in the late spring, the party and their stock of horses were able to get through the Rockies.

It is Green Russell who is commonly credited with starting the Colorado gold excitement. The Georgia men had experienced only fair luck in the California placers and had returned to their home state. The memory of Cherry Creek's gold and its implications kept recurring to Green Russell and eventually led him to organize a group consisting of himself and ten fellow Southerners to see what this unexploited land had to offer. Nine years had gone by since the modest placer findings. It was in May 1858 that the old camp was reoccupied and serious prospecting undertaken.

Meanwhile other travelers to and from California or the new Fraser River gold district in Canada had begun to wonder about this intermediate mountain land. It was known that it harbored only a sparse population of Ute and Arapaho Indians, that it was virtually unexplored as to either mineral or agricultural promise, and that most of it was mountainous. Mountains, then and now, at least hint at valuable mineralization.

Officially, Colorado was a vague, distant portion of Kansas Territory.

The interest of the Russells and of several others–soon it would be thousands of others–turned from their river camp to the nearby mountains. Streams flowed out of these lower reaches of the Rockies at intervals of a few miles. What influence caused the first prospectors to choose Clear Creek from among the others is unknown, but Clear Creek was where the gold lay. The creek and its tributaries took them to a gold district that became so rich and self-sufficient that, although officially constituted as Gilpin County, it often was styled "the Kingdom of Gilpin."

Among the Canada-bound travelers who paused in Colorado was John H. Gregory, a competent mining man. Like the early Georgians he came to a halt when he faced the towering Rockies just at the onset of winter. As far as weather permitted, Gregory improved his time by prospecting, and spring found him well in advance of others in one of Clear Creek's little branches. It was here that he came upon the rich placer ground and crumbling quartz outcrops of even greater richness that lay along a rugged ravine known later as Gregory Gulch.

Similar discoveries were made all through the adjacent region. Here, thirty miles from the original Russell camp, soon to build up and bear the name of Denver, the booming towns of Central City and Blackhawk took shape. This was the beginning of Colorado mining, and these towns were also the scene of great advances in technology that benefited metallurgical operations the world over.

Although Gregory's strikes were the most important, they were not the first. The initial discovery in Colorado, aside from the unimportant panning on Cherry Creek, is credited to George A. Jackson, a cousin of the frontier scout Kit Carson. Braving the winter snows, Jackson had entered the lower mountains via Clear Creek Canyon and had thawed out some promising gravel by building a fire over it. Panning showed it to be quite rich. It was a good strike in itself, and it also pointed the way to the more opulent deposits beyond. All this was in 1859, just ten years after the California rush got under way. Jackson successfully developed his claim, with a partner known as Phil the Cannibal, a sobriquet originating with some grim episode in the wilds of Montana.

Discoveries at more distant points followed rapidly. In the wide, lofty basin now known as South Park, extremely rich placers were found at Tarryall. Latecomers who found themselves excluded termed the place "Graball." When they came upon equally good diggings not far away they self-righteously named their new camp Fairplay. Another booming camp of the South Park region was Buckskin Joe, where one Jacob B. Stansell struck it rich. He knew how to spend his money. Minstrel shows were a popular stage attraction at that time, and in his isolated mountain camp Stansell kept an entire troupe of minstrels on the payroll for his personal entertainment.

At the same time that the Central City-Blackhawk discoveries were establishing Colorado as a mining territory, news of similar import was coming from the distant southwestern part of the state. This region was known as the San Juan country, after the principal mountain range.

It was a land of exceptionally sharp, rugged mountains, of long winters, and of difficult access at any season. Even John C. Frémont found the San Juans impassable, and they were the scene of one of the few tragic failures in his career as an explorer. In 1848 "the Pathfinder" was leader of a party charged with laying out a trail through this section of the West. Blizzards can come early and unexpectedly in southwest Colorado. Frémont and his men were trapped by such a storm, and snowfall was so heavy that egress was cut off for many days. Eleven of the band perished. The hardier ones survived, but only by subsisting for days on the flesh of their frozen pack animals.

The earliest explorers of the San Juan were Spaniards, who had been established at Santa Fe, New Mexico, since 1605. It was to be expected that they would scout on northward. One small expedition is recorded as of 1765, and several others followed. From the Spaniards, from French-Canadian trappers, and from certain roving American prospectors came reports of gold and silver lodes, and in the San Juan country Colorado soon had another focus of mining excitement. As to present-day possibilities, this great Alpine fastness presents both promise and puzzle. In the next chapter there will be a more detailed look at it.

Another early, important development was in California Gulch, lying almost at the base of the awe-inspiring bulk of Mount Massive and far to the southwest of the Central City-Blackhawk mines. The gold discoveries here led miners on to the great deposits of the Leadville district, which produced gold, silver, and lead at a rate which for years maintained Leadville as the second largest city in the state.

The great mining operations at Leadville have been largely obscured in the public mind by an avid interest in the ups and downs of one of the town's citizens, Colorado's most publicized mining king, H. A. W. Tabor. Tabor rose to dizzy heights of wealth and prominence. Whether he blundered his way to success in the Leadville mines or was guided by native shrewdness is a question to which there is no satisfactory answer. Lucky he certainly was during much of his career, and it is equally evident that Tabor was bad luck for the women who were closest to him. His first wife, Augusta, who ran their little store at Leadville while her husband ranged through speculative and at first unprofitable mining ventures, was divorced soon after he struck it rich. A daughter, said to have been christened "Silver Dollar Tabor" but usually referred to as Rose, lived out a short, sordid, and unhappy life in Chicago. Baby Doe, Tabor's second wife, belle of a lavish wedding reception attended by President Chester A. Arthur, was left penniless at her husband's death. His fortune had been dissipated in rash investments and he left only a mine of uncertain value.

Having been adjured by Tabor to "hold on to the Matchless," for the unproved mine was so named, Baby Doe did just that. Moving to a small cabin at the deserted mine shaft near Leadville, she guarded the supposed treasure of its depths for many years, existing in abject poverty but occasionally being interviewed for recurring feature stories on the Tabor saga. At last, in one of the bitter winters that Leadville experiences, she was found dead. The charmer and coquette of younger days, for whom Denver social circles showed little regard, had by this time become a heroine of sorts.

On the favorable side, Tabor's luck included these episodes:

Out of Augusta's grocery stock, he grubstaked two prospectors, with specific instructions as to where he wished them to explore. The provisions included a quart of whiskey. When the men had gone only a short way, they sat down, drank themselves into confusion, and, disinclined to further climbing, dug a prospect hole right where they were. This became the Pittsburgh Mine, which poured millions into Tabor's purse.

During the first years of mining success, the Tabors continued to operate and expand their store. Several substantial businessmen in Denver, connected with wholesale grocery firms, authorized Tabor to buy a mine on their behalf. He made such a purchase, putting up $40,000 of his own funds, pending reimbursement. The mine proved to have been salted with high-grade silver ore stolen from a better property. The deceit was soon apparent. The Denver syndicate, understandably (but as it proved unwisely), refused to honor its commitment. Tabor appeared to have lost $40,000. Never averse to throwing good money after bad, he went ahead and had the mine developed at depth. The silver was there. One report, possibly exaggerated, states that this $40,000 fiasco eventually gave Tabor a return of $17 million.

Tabor's fortune was lost in investments chiefly unrelated to mining. A great promotion involving mahogany forests in Honduras was among his speculations. The bankrupt miner, however, did not die in poverty as sometimes reported. Influential men in Denver felt that something should be done for him. After all, the fine Tabor Grand Theatre had been built in his days of affluence and Denver was still enjoying its use. He finally was settled into a comfortable income as Denver's postmaster.

Mining engineers who have expressed an opinion about the Matchless Mine do not think at all highly of it. Certainly Baby Doe was unable to interest any capital during her long, hopeless guardianship. One factor perhaps was the continued depressed state of the silver market. Silver prices are much higher now, and on the record most of Tabor's hunches on strictly mining matters were good ones. Some entrepreneur of romantic bent and with money to spare may yet wish to find out what lies deeper down in the Matchless.

The scenes of the Tabor legend lie on the western side of the Continental Divide. This geographical feature, represented by a sinuous line on the map, runs north and south along the crests of the highest mountain ranges. Prospectors, always able to surmount any natural barrier, quickly crossed this one. The difficulties posed by the Divide, however, and by equally forbidding highlands beyond, figure importantly in mining operations and their cessation. Access to some abandoned or neglected areas is not easy even today. Nevertheless, modern highways bring one much closer to them, and on a selective basis there are individual deposits and even whole mining districts that can be recommended for fresh examination.

All the mining camps that lay well back from the front range of the Rockies had to be reached by mountain passes. The passes figure prominently in Colorado history. They were high, rough, and windswept and for much of each year were quite unapproachable. At times contemporary writers almost personified the passes as actors—in the villain's role—in dramas of hardship and heroism among pioneer miners. Among the worst was high and dangerous Boreas, which one had to cross to get into the lofty Golconda known as the Breckenridge district.

The town of Breckenridge has survived. It is the county seat of little Summit County, whose boundaries zigzag along the surrounding peaks and which is the most irregularly shaped county in the United States. Around Breckenridge, principally to the east, an even forty old mining camps have been identified by the Colorado historian Perry Eberhart. Each of these settlements represented discoveries of gold or silver. Some were immediate bonanzas. In French Gulch, for example, a lucky glance by roving Harry Farncomb fell upon an outcrop of quartz heavily laced with wire gold. He became rich, as did others who located claims on Farncomb Hill. In less spectacular cases, strong veins of medium-grade ore were discovered and were developed in a businesslike way. There was much sustained large-scale mining and substantial towns grew up. The mines at Park City for a time supported a population of 10,000, and a private mint turned out gold coins.

The fact that forty camps came into existence around Breckenridge pretty well proves the intensity of early-day exploration. Nevertheless there are several rich lodes in the area that, once discovered, were promptly lost. The one which has attracted most searchers became known as the Tenderfoot's Lost Mine. As readily surmised from the name, it was found by a newcomer to the gold country, one who had not yet been impressed with the importance of marking his location and carefully observing landmarks along his trail.

At some lofty point from which he could see the Warrior's Mark Mine buildings, the amateur prospector knocked loose a piece from a protruding ledge of quartz. The broken piece failed to fall to the ground. The man looked, and found that the fragment was held by several strands of wire gold. This meant immediate returns. The ore he was able to dislodge with his small pick and carry away brought him about $400. As in many other situations, the winter now intervened. Spring came, and a confident return trip was undertaken with two companions. It ended in confusion and failure. Nothing answering the description of his outcrop was found in all the Breckenridge district's later history. Best information points to its being on Bare Mountain, not to be confused with the similarly named Mount Baldy, a 13,964-foot peak which is nearby.

Baldy itself is the supposed location of a lost ledge of high-grade that is mentioned by the historian Bancroft. It was mined in the summer of 1860 and 1861 by two taciturn men from Missouri. They regularly left the camp of Lincoln at night and returned to sleep during the day. In spite of their attempt to keep their business to themselves, it became known that they were bringing in and shipping small quantities of very rich gold ore. When the Civil War broke out the Missourians left the district, saying they were returning to their families in their home state. There were immediate searches for what was assumed to be a nearby bonanza. These hunts proving fruitless, a belief grew up that the nocturnal miners had no mine of their own, that actually they were stealing from one of the Farncomb properties after the day shift had left.

Although Breckenridge had profitable gold mines, much of the production was of silver and the district was hard hit by the collapse of 1893. It is some of these inactive silver properties that might become paying operations at today's quotations. There are several areas, however, where exploration for virgin strikes or abandoned, slightly worked prospect holes also can be suggested. Collier Mountain has been shown to be highly mineralized. It rises above the old camp of Montezuma. South of Breckenridge is the site of Conger. Mineralization here was widespread, but early reports indicate that Conger, like some of the Nevada camps noted, had too many prospectors waiting for investors and not enough men willing to get out ore. Some grass-roots assays were very high, both in gold and silver. Conger appears to have flourished only during the summers of 1879 and 1880. How much of value was left behind remains to be seen. It is along a small stream called Swan River that one finds the largest number of abandoned workings.

The possibilities of the Breckenridge district today are doubtless confined to lode mining. The placers were rich, but ordinary placer mining was followed by hydraulicking, then by extensive dredging, leaving little or nothing in the way of placer for the individual miner. Today one does not face the chill winds of the old pass to get to Breckenridge. Interstate 70 runs north of the town, and it may be reached by only a short drive down connecting State Route 9. There is one little ghost town along here that should not be overlooked by anyone with some metallurgical skills. It was named Quandary, for the reason that miners never were able to understand the nature of the heavy, obviously mineralized rock which outcropped all around them. It did not yield gold by standard recovery methods. An analysis of Quandary's deposits by modern spectographic procedure would prove interesting and possibly of profit.

Another mining region in which camps proliferated in amazing style may be roughly described as the mountain country back of Boulder. This town, at the base of huge, picturesque rock formations that face out to the plains, is the seat of the University of Colorado. Boulder Creek issues from the mountains at this point, and prospectors had not gone far up the stream until they found both gold and silver. The fact that the region also contained some of the most concentrated occurrences of tungsten in the world was unknown to them and would have been of little appeal in the explorations of 1859 and onward. When World War I caused a heavy demand for this alloy and a sharp rise in prices, there was a tungsten rush that, in the number of men involved, rivaled the earlier boom in precious metals.

From the first gold camps, Red Rock and Boulder Diggings, miners pushed back into the range and up to higher and higher elevations. There seemed no end to the mineralization. The entire area up to the crest of the Continental Divide and towering Arapaho Peaks was dotted with settlements. There probably is not a snow-fed stream between these summits and the town of Boulder that does not exhibit old mine dumps and other relics of early-day activity. The principal towns were Ward, Jamestown, and Gold Hill, but there were many others, Wall Street, Caribou, and Magnolia among them, whose annals were brief but spectacular and which live on in Colorado history.

When the inevitable decline set in, many small miners were kept going through an unusual enterprise conceived and financed by the

Boulder business community and supported also by some of the large mining interests. This was the Boulder Sampling Works. To this plant the miner could bring his ore, however small the quantity, have it assayed, and be paid at once for its metal content. A percentage was deducted for operating costs and freight and smelter charges. The Sampling Works was equipped to sort and store the ore and ship it in carload lots for reduction.

The system's benefits for the small miner are obvious. The sampler operated well into the twentieth century, eventually handling tungsten as well as gold and silver. Such a facility in other districts could have kept many mines going.

The important mining center of Leadville may be taken as a point of reference as we trace a zone of mineralization southward. The Arkansas River has its source near Leadville. Dozens of mining camps lay on or near this stream along its southerly course. Today the country can be traversed by U.S. 24, but most of the ghost camps and mines will be reached only by unimproved roads, many of them unmapped and possibly more difficult than they were when ore and supply wagons creaked along them in the 1880s and 1890s.

The substantial little city of Salida is the largest place along this route. Here the Arkansas turns east and the highway that parallels it is U.S. 50. Founded as a railroad town, Salida became the supply center for mines over a wide area. The name Calumet and Hecla, once familiar on the New York Stock Exchange, refers to mines in the Salida region that tapped vast deposits of magnetite.

For gold and silver possibilities the most promising spot of record is old St. Elmo at the base of Mount Yale in the 14,000-foot Collegiate Range, the other peaks being named Harvard and Princeton. The most productive mine in this so-called Alpine district was the Mary Murphy, named after a Denver nurse who had cared for one of the owners. Hundreds of men were employed and the great ore body held out for decades. The mine closed down in 1926 and its own town of Romley along with it.

A number of settlements in this region grew up around mineral springs, which had a more popular appeal about the turn of the century than they have today. In looking up ghost or near ghost towns in the Arkansas River country, the prospector should be sure he selects mining camps instead of old health resorts. Among the most active of the former were Harvard City, Hortense, Alpine, Whitehorn, Turret, Crazy Camp, Garfield, Babcock, and Monarch. The long ravine of

Chalk Creek has well established mineralization clear up to its source. Cottonwood Creek, flowing along the base of Mount Yale, is of similar but somewhat lesser promise.

To the southwest of Salida, across some formidable mountains, is a string of houses and stores in a deep, narrow canyon that make up a town widely known from the refrain in some popular verses:

> It's day all day in the daytime,
> And there is no night in Creede.

The lines were published in Creede's heyday by Cy Warman, a Colorado writer who later became an editor in the Munsey magazine empire. They hint at the twenty-four-hour activity he witnessed in one of the West's most populous and wildest mining towns. Creede was an early stamping ground of Skagway's notorious Soapy Smith and the scene of the final and fatal swaggering of Bob Ford, the killer of Jesse James. Creede's violent history has been chronicled elsewhere and we will look only at its mines.

Early prospectors somehow failed to recognize the riches at this northernmost bend of the Rio Grande. The ore deposits lay in two narrow canyons between the river and the towering bulk of San Luis Peak. It was a discouraged, middle-aged prospector named N. C. Creede who, in 1890, found a rich surface outcrop on East Willow Creek. His exclamation "Holy Moses" became the name of the mine. As in many other mining districts, the initial discovery proved to be one of the best, but the wealth from the Holy Moses failed to bring happiness to its finder. For unknown reasons Creede committed suicide in 1897.

Nearly a dozen smaller camps surrounded Creede. Most of the profitable workings were to the north, along the small creeks descending from San Luis Peak. If one can locate the sites of Bachelor, Weaver, or Sunnyside, he will be in heavily mineralized country. Besides gold and silver, some deposits carried important values in lead and zinc.

Such a large number of prospectors roamed the Creede district that the present-day possibility of finding a bonanza overlooked in the 1890s would appear slim. One also is skeptical of exploration upstream from Creede on the Rio Grande. Stage routes ran that way and several small settlements developed, but there is little record of mining. Apparently nothing of value could be found. Any current opportunities around Creede may be summed up in the word "silver." The oft-repeated story of mine shutdowns in 1893 applies to Creede also, and the substantial ore bodies that were abandoned in some workings may now invite inspection and reappraisal.

In following the gold and silver trail down to Creede we have come close to the San Juan country, a silver empire with a fabulous past and with riches that are by no means exhausted. Before exploring the San Juan, however, note must be taken of an amazing concentration of gold confined to a small volcanic crater near Pike's Peak, namely, the Cripple Creek district.

In its most active period, about the turn of the century, Cripple Creek was the greatest gold-mining center in the world. No picture of Colorado mining could omit Cripple Creek, but it is not submitted as a possible objective for the prospector or small mining enterprise. Transactions and yields were in terms of millions of dollars. The area of mineralization is quite circumscribed and has been thoroughly explored, so thoroughly, in fact, that continued, hopeful probings deep into the earth have resulted in huge losses. At Cripple Creek the only possible direction of search is downward, and in these deeps a combination of lower-grade ores and subterranean water flow has long since closed the great mines. The changing position of gold in the world's economy may lead to a revival of Cripple Creek, but it would be achieved only by massive outlays of capital.

Poor "Crazy Bob" Womack, the cowboy-prospector who discovered Cripple Creek's gold in an innocent-appearing cattle range, immediately went on a drunken spree and sold half his claim for $500. Nor did he benefit substantially later on, although the Gold Coin, as the property was called, yielded $7 million under new owners. Winfield Scott Stratton, a carpenter who, like Womack, had patiently prospected the unpromising surface of the cattle country, fared better. Beaten to the original discovery, Stratton acted quickly in following it up. On July 4, 1891, he made a location under the name of the Independence. After taking out $4 million in rich ore he rather unwillingly sold the mine for $7.5 million. By this time he had other properties operating and went on to become possibly the richest man in the state. He also was one of the most interesting characters. His heavy drinking, his open-handed largesse to anyone in need, especially miners, and his quixotic financial deals made him good copy for the Colorado press as long as he lived. Much of his estate went to founding a home for elderly mining men down on their luck.

After Womack's discovery there was an immediate stampede to the new field. Cripple Creek was barely seventy-five miles from Denver and only a few miles from the city of Colorado Springs, although of much higher elevation. The scenic journey down to Colorado Springs by an early mining railroad was described by Theodore Roosevelt as "a trip that bankrupts the English language." The old right of way is now an automobile road. As miners swarmed in from the older camps, a town quickly grew up and by 1900 Cripple Creek had a population of 35,000. Near Stratton's Independence the sizable town of Victor took shape. In a well-defined oval that took in these two population centers and a few smaller camps lay all the ore bodies. Outside of that, nothing of consequence.

At places the Cripple Creek ores were fantastically rich. Allusion was made earlier to the "bridal chamber" in the Cresson Mine. This little underground hollow, technically a "vug" in the formation, was lined with golden incrustations that totaled $2 million in value. Highest production was achieved by the Portland, Colorado's richest mine and one of the great gold mines of the world. This giant operation was started on a very small vacant area located by a tenderfoot. He is said to have found the unclaimed little parcel between mines by measuring distances with a clothesline.

The story of Cripple Creek's successes and of a few cases of abysmal failures would fill a book but would not contribute materially to a report on present-day possibilities. An interesting minor success concerns the Jackpot Mine, a rural sheriff, and possibly a jug of hard cider being passed around in the distant town of Creston, Iowa. Learning that the Jackpot property could be bought for a small down payment, a group of temporarily enthused farmers put up the money and sent the local sheriff to run the mine. Wages at Cripple Creek were $3.00 a day. It did not take much money to keep a small crew at work, but even so the lawman-miner had to keep pleading with the dubious farmers to meet each successive payroll. At depth, the Jackpot proved to have a rich, if limited, ore body. At the final accounting the Iowans had invested $13,000 and had received dividends exceeding $400,000.

Now, in a long leap to the Southwest, we come to the so-called Empire of Silver, "the San Juan country."

16
Colorado: San Juan Country

The southwest part of Colorado is probably best known to the public for the Mesa Verde cliff dwellings, now set aside as a national park. They are in the far corner of the state, where the country is merging into the barren sandstone formations of northern Arizona and southern Utah. The region of mineral interest lies northeast of here. The San Juan country of early-day excitement embraces six or seven present-day counties. The rivers that were followed into the highly mineralized mountains are the San Juan, the Piedra, the Pine, the Animas, and La Plata.

Even the great Rio Grande rises in these mountains, having its source in tiny Hinsdale County, population about two hundred. And Hinsdale County recalls a tragic, winter prospecting party and the curious case of Alfred Packer, "the Man Eater."

Statistics on the cannibal practices in Borneo and New Guinea are sketchy, but in the United States Packer's consumption of five adult human beings is a record that has not been challenged. It is backed up by the legal archives of Lake City, Hinsdale's tiny county seat. The victims, or successive entrées on Packer's menu, were Israel Swan, George Noon, Frank Miller, James Humphreys, and Shannon Bell. These five were the more adventurous members of a party of some twenty mining men who, in the winter of 1873–74, were intent on getting a share of the San Juan's rapidly unfolding wealth in silver and gold. They started out as a well-organized exploring party. The Indian chief Ouray had warned them not to venture into the higher mountains in winter weather, and when storms gathered the majority of the men heeded his advice. There were five who decided to push ahead.

With the party from the start was Alfred Packer, a local man who knew the country and had been hired as a guide. When the party split up, Packer elected to remain with the five who were continuing, and consequently to remain on the payroll. The guide was described as a man "of very peculiar appearance," but we find little amplification of the phrase. He had a large head and the forehead sloped back abnormally from his eyebrows. He was smooth shaven except for a small, trimmed beard.

Beautiful Lake San Cristobal, formed when a natural landslide created a dam across the Gunnison River, is a central feature of Hinsdale County. Testimony at Packer's trial showed that the winter travelers had camped at this body of water and continued on. Ahead was an ascent now known as Slumgullion Pass. They had started toward this pass when they were completely immobilized by a heavy snowstorm.

The fact that the little band had remained in the higher mountains was of course known in the settlements. Those of the expedition who had returned safely waited with anxiety for their companions to appear. As weeks went by and snow piled deeper in the canyons, it seemed beyond question that the sojourners had perished. Five of them had.

After the thaws of the new season, the guide Packer appeared—alone—at Los Pinos Indian Agency. He then went on to the town of Saguache. There he spent much of his time in saloons, paying for his drinks from a large roll of currency. During his drinking he told different stories of his party's wanderings and of how he became separated from his five companions. There was nothing to prove or disprove his accounts. Then chance sent an eastern artist to Lake San Cristobal and the highlands beyond. The man was on assignment from *Harper's Weekly*. Near Slumgullion Pass he came upon an empty campsite and scattered about it the remains of five mutilated bodies.

Packer was placed under arrest. He readily admitted knowing of the carnage but asserted that it was the work of Shannon Bell, one, it will be recalled, of the "final five" of the expedition. The man had gone insane and slaughtered his companions, Packer said, and it was only he himself, armed with a hatchet, who had got the best of Bell. Authorities were convinced that Packer had killed the men, robbed their lifeless bodies, and subsisted on their flesh through the winter. An Indian later brought in some strips of human skin left from the grisly proceedings.

The now notorious Man Eater managed to escape from the Lake

City jail. He was at liberty for nine years. Then, recognized in Wyoming—by his unusual, high-pitched voice, it is said—he was brought back to Lake City, tried for murder, and convicted. He escaped the death penalty but served a long term in prison.

An inseparable part of the Packer story, though possibly fictitious, concerns the serious view of the man's behavior voiced by the trial judge.

"Dammitall, Packer," his honor complained, "we only had twelve registered Democrats in Hinsdale County, and now you've eaten five of them."

In examining mining and mineral exploration in the San Juan country, the point should be made that we are dealing with the highest region in the United States and with its most precipitous mountains—with parts of Alaska possibly excepted. The Colorado historian Robert L. Brown notes that this great triangular uplift contains hundreds of peaks of more than 13,000 feet elevation. Of the mountains of all North America rising above 14,000 feet, the San Juan country contains one fourth. This land in the sky extends over an area equal to Massachusetts, Connecticut, and Rhode Island combined.

As to minerals, there have been gold placers, now largely worked out, and some lode mines whose values were almost exclusively in gold. The great wealth of the San Juan, however, has been in silver. Lead, a frequent concomitant of silver ore, also has been of some importance.

Considering the ruggedness of the country and the short working season, it could be assumed that only the richer, "shipping-grade" deposits would be worked. This had been the case in many similar situations. In the San Juan, however, there was more energy and resolution, and doubtless more capital available in this period of general mining enthusiasm. Somehow the heavy components that make up a stamp mill or a small smelter were hauled up to the distant mines and assembled. Ore could be treated where it came out of the ground. Veins of medium-grade could show a profit. If a rich pocket was encountered it provided a welcome entry on the right side of the balance sheet, but what was most wanted was a continuing supply of ore of dependable average value. The pioneers in the erection of these treatment plants must have set a successful example, for a great many were in operation from the mid-1870s on to the 1893 debacle already noted, the demonetization of silver.

Along with this businesslike, fairly stable mining there also occurred rich surface strikes, and, in the rough country of the San Juan, such a discovery too often turned into a lost mine. The clearing house for both mine and lost-mine information is Durango, the principal supply and population center of Colorado's far southwest. Durango may be reached from Denver or any eastern point by U.S. 60 or from the south or southwest by U.S. 550. It was a railroad point, founded later than the initial miners' influx, but soon became the "big town" of the region.

In Durango you may pick up clues to the Mine of the Three Skeletons, where some bags of extremely rich gold ore await you in addition to the treasure of the ledge itself. As put together by Perry Eberhart for his highly informative *Colorado Ghost Towns and Mining Camps,* the facts are as follows:

And old prospector comes into Durango in 1905 with a bag of ore in which the gold is almost equal in amount to its quartz matrix. He tells of a discovery somewhere around Bear Creek. It is a very old tunnel, with three skeletons just inside the portal. There also are sacks of "picture rock," and he has carried as much of this into town as he can. He cashes it in and purchases supplies. Having so equipped himself, he proceeds into the mountains and out of the story.

There are several Bear Creeks in Colorado. The one closest to Durango is a tributary of the Rio Grande River and the quite productive camp of Beartown flourished in its upper reaches in the 1890s. A landmark on the trail is the modern Rio Grande Reservoir in Hinsdale County. Just upstream from the reservoir the ruins of old Junction City mark the turnoff to Bear Creek and presumably to the lair of the three skeletons.

Then, as the story continues, the same mine is found again in 1918 by a Pedro Martinez, who also brings in as much of the gold-spangled ore as he can. The first great influenza epidemic is sweeping the nation at this time and Pedro Martinez dies of the disease.

Now, we are on the eve of World War II and the third discovery is made. An elderly sheepherder appears in Durango with the same story and with the same rich ore. This time there are alert people ready to go into partnership with the sheepherder, doubtless with the skeletons themselves if it will lead to the now celebrated bonanza. A search is organized, and a flattered and befuddled old man does his best for his newfound friends. His best is not enough. The mine remains lost.

There were other convincing cases of rich lodes found and lost in the San Juan, but after the passage of nearly a century the names of discoverers and searchers and, more important, any intelligible directions have largely faded away. An instance in the 1930s may be related for whatever moral it carries.

A young man who was out of work during the great depression appeared at Newton's Cafe in Durango. He felt he could as well be hunting gold as hunting nonexistent jobs, and he hoped for a grubstake arrangement. Impressed by the youth's sincerity, Newton outfitted and provisioned him. The luck of the amateur appeared to have descended on the young prospector. In two weeks he was back in the restaurant, pouring onto the counter a bagful of yellowish, obviously metallic rocks. Where these came from there were many more, he assured his backer, describing what appears to have been the broken, weathered surface of a strong vein.

Whatever optimism the two felt was quickly extinguished. An early cafe patron was an old miner of recognized experience. Shown the ore samples, he smiled grimly and pronounced them pyrites, the iron ore, often bright yellow, which is laughed off as fool's gold. A disconsolate young man left the San Juan, presumably returning to the fields of trade and industry. A deflated Newton heaped the rocks on a shelf back of his lunch counter.

In due course another cafe customer was sitting at the counter and his eye rested on the display. He asked for a closer look, remarking that he was a professional mining engineer. He absently confirmed the fact of the clearly visible pyrite, but continued his examination. "You might have something in addition here–something pretty good," he concluded. Newton now sent a representative sample of the ore to be assayed. It was worth nearly $5,000 a ton.

From Durango one may drive north through striking scenery to Silverton and Ouray. Both are early mining centers and now the county seats of San Juan and Ouray counties respectively. Along the route, U.S. 550, are the old camps of Hermosa, Burro Bridge, a group of camps in the Red Mountain area, then Yankee Girl and Guston. At Silverton lesser roads fork off to the north and reach the once important mines around Gladstone, Howardsville, Eureka, and Animas Forks.

The advice has been heard time and again in the San Juan country to keep your eyes on the sides of the cliffs, the precipitous canyon walls, in search of mineralization. Many important discoveries were made in this way and such locations made the mining easier. A horizontal adit on the vein itself is probably the ideal operation, and it appears that a great many of the San Juan ore bodies permitted development along this line.

The very stress laid on such lower locations by mining reports raises an interesting question: Have the summits and ridges of the San Juans, the San Miguels, and La Platas ever been adequately explored? Prospecting at these heights involves some hard climbing, but chances of a virgin discovery appear greater than at lower levels. At the same time a caution should be offered against a widely held belief that rich lodes are most likely to be found at high elevations. This is not supported by mining history.

Ouray is the turnoff point for Mount Sneffels, the site of extremely rich deposits of both gold and silver. The initial discovery at the Sneffels mine ran $40,000 a ton, and up to 1919 this lode and nearby properties had produced $27 million. Rich Mount Sneffels has been overshadowed, however, by the even richer and much more highly publicized Camp Bird Mine.

The Camp Bird has been kept in the public eye largely through news of the developer's daughter, Washington hostess Evalyn Walsh McLean. Mrs. McLean has been written up as possessor of the fateful Hope diamond, and she was very much in the news at one time when she financed a private search for the kidnapped Lindbergh baby. It was a well-intentioned and very expensive enterprise which proved to be based on false information. Mrs. McLean herself took pen in hand for a time and told the family story in a book, *Father Struck It Rich.*

Thomas Walsh came from Ireland at the age of nineteen. He worked in various western mining camps and by middle age was owner of a small but prosperous smelter at Silverton. Some siliceous ore being required for flux, Walsh or possibly an employee went to examine the dump of an inactive silver mine not far from the booming Sneffels mines. Walsh saw something in the discarded material which the original miners had not detected. The values were not in silver, which had been sought for, but in gold. The casual samples taken from the dump assayed $3,000 a ton in the yellow metal.

Walsh spent $20,000 acquiring the old mine and surrounding claims and started a businesslike development. His confidence was well repaid. In Colorado the Camp Bird production is second only to that of the great Portland Mine at Cripple Creek. Walsh was soon a millionaire and spent some of his gains providing his four hundred miners with living quarters which, by the standards of the day, could be called luxurious. In 1902 he sold the Camp Bird to an English syndicate for $5.2 million, receiving most of the sum in cash. Some tentative interest in mines of the Belgian Congo occupied him in later life, but he lived to enjoy his success for only seven years.

The resourcefulness of the San Juan miners in making the most of their properties has been noted, but like any such statement it cannot be taken as all-inclusive. There is more than suspicion that many promising mining claims did not share in the vigorous development. F. V. Hayden, who headed a government survey in 1874, stated that few of the miners were actually mining. If Hayden's observations were correct, the familiar pattern of locating and opening up a vein, then waiting for a buyer or backer, prevailed even in the active San Juan. Under this procedure many claims would inevitably be abandoned, to await development in later years, if ever. There is some hint of this in propositions made to various Denver businessmen over the years. In the 1930s in particular a number of mining men from the San Juan were looking for capital for mines which, on the basis of verifiable engineering reports and assays, appeared to be of real value. What success they had, either in financing under depression conditions or in mining itself, is difficult to determine.

The story of the San Juan may be concluded by reviewing the first and possibly the richest gold discovery of them all. The account emerges in fragments from the days when Colorado was entirely Indian country, and the sparse white settlements to the south in what is now New Mexico were entirely Spanish. The time was the late 1700s, when the American colonies on the Atlantic seaboard had just gained their independence.

The first general knowledge of the early-day events came from a double-page historical feature in the Denver *Post* prior to World War I. Old miners were getting a friendly reception in Denver newspaper offices in those days, and were often rewarded with a tongue-in-cheek item about their mines or prospects. From one of these visitors an editor heard the improbable tale of a large, official French reconnaissance party that had penetrated the San Juan country in the long ago. The trip had turned into more than a reconnaissance. The French had remained for three years. What had detained them was a remarkable ledge of gold ore, so rich that they were crushing the vein matter by hand, washing out the metal, then melting the gold and casting it into bars. It was said that the accumulated bullion had a value of at least

$5 million. Predictably, later accounts kept raising this estimate and the sum was once printed as $33 million.

The old man's story could not be ignored. A writer was sent by the newspaper to the little southern Colorado town of Alamosa. He found that the tale had been known and accepted locally for a long time. People could even name men who had searched for traces of the French expedition. Asa Poor was one of the first, and two of the partners he took were A. T. Stolsteimer and Leon Montroy, both of them citizens of some standing. Alamosa residents could point to an imposing peak named Treasure Mountain. The reason for this name? The French had been unable to carry away their gold. The millions still lay buried somewhere on Treasure Mountain and not far away, doubtless skillfully concealed, would be the mine itself.

We are dealing with a period before the Louisiana Purchase. The vast territory west of the Mississippi was a French possession, and some authority in the mother country had decided that it had something to offer in the way of mineral resources. The enrichment of Spain by its New World conquests must have been very much in mind. The published date of this enterprise, 1790, probably is incorrect, for this would place it in the first turbulent years of the French Revolution.

Whenever it was organized, the expedition was said in some accounts to be an ambitious one. It consisted of three hundred men, and numbered geologists, surveyors, and practical miners among its personnel. The point of departure was a French outpost in what is now eastern Kansas. The crossing of the prairies, then overrun with buffalo and with many flowing streams, would present few difficulties. What is mystifying is that such a ponderous train of men and animals should cross range after range of the lofty Rockies to get to the distant San Juan country. Did they have a definite objective there from the start?

So much for the more pretentious story.

A more plausible version of the French enterprise leaves Paris officialdom out of it. The newspaper writer who pieced together old accounts from residents of Alamosa has the mining party made up chiefly of French-Canadian trappers, a breed that we have met before in various parts of the early West.

The misty traditions of the San Juan give us little information about the actual mining operations. We know that they were conducted only during the summers and that when snow closed in the Frenchmen went south to the milder climate of the Spanish settlements. We hear of them in Taos, then an important trading place, and in Santa Fe.

They were well treated by the Spaniards. Presumably well financed, they were able to outfit properly each spring and return to the mine.

The Ute Indians of the gold country either found the French party too strong to attack or they were temporarily tolerant of the visitors. But as the third summer drew near an end the Indians attacked in force. Figures from an untraceable source say that seventeen French fled from the workings but that twelve of them were overtaken and slain by pursuing Utes.

The original published account deals at some length with the trials of the five survivors, but here we get into some suspiciously colorful writing. It seems that the men constructed a raft on the Purgatoire River, and from that stream drifted and poled along the Arkansas toward civilization. Whether some starved, died of wounds or disease, or fell victim to later Indian attacks may never be known. Of the whole original party only one man, named Le Breu, came out alive.

The slaughter of these invading whites has been handed down in Indian tradition. In describing the valor of their forefathers, the Utes have accented the fighting and have been singularly indifferent to the cause of the combat, that is, to the topic of gold. What is known about the mine and the treasure therefore comes from the survivor Le Breu, or more precisely from relatives and friends to whom Le Breu talked.

The Denver press and writers in general were, as related, quite tardy in discovering and telling the story, but it was well known in the San Juan country. Many searches were made after Americans settled the region. There was one attempt organized in France. It may have been ordered by Napoleon or by one of the monarchies that succeeded him, it may have been a private enterprise of the Le Breu family. About thirty men participated and their purpose was known to Spanish authorities in Santa Fe. The hunt was dogged but without success. A Bernardo Sanchez of Taos, who for a time acted as their guide, related a little about their efforts but was disappointingly vague as to where the search was made and about what map or other directions the French party possessed.

Although real evidence is lacking, there has been a conviction among mining men of the San Juan that the French mine was near Summitville. This is an old mining camp perched high above the Alamosa River, and almost at the crest of the Continental Divide. As for the great treasure in gold bars, the local tradition continues strong that it is buried on Treasure Mountain. Old graves have been found there, and Indians have picked up metal parts of harness, buckles, and rusted tools. A favored spot, which has seen some scattered modern excavation, is a small natural basin near the summit, what in the Colorado mountains is called a "park."

The ingots may be in several lots and buried separately. Each summer's yield is said to have been put underground before the winter migration to the south. The final concealment was made in anticipation of the Indian attack and, if the oft-repeated narrative is true, the location was carefully mapped by the leader of the group. This individual perished in the ensuing fight.

A map often mentioned but never seen by narrators may well be something that Le Breu drew from memory. It is believed to have been the inspiration of the second French party. The final Gallic note in the story carries the date 1910. In that year a Parisian with little English at his command appeared in Durango. He stayed, hired horses, tramped the hills, and ineffectually asked directions through a long and arduous summer. Then he sadly departed, to be seen no more.

If we stay in the southern part of Colorado but move some distance to the east, we come to where the Rockies face out on the great central plains. The principal ranges are the Sangre de Cristo and the Culebra. We are still in high country. An irregular rectangle lying immediately north of the New Mexico state line contains some of the loftiest peaks in the state. Here there is a great L-shaped area of at least eight hundred square miles without a single hamlet large enough to be entered on a state road map, if indeed anything more populous than a summer sheep camp exists. It is discouraging country to an explorer. A towering peak will be followed by a deep, almost inaccessible canyon, then another rocky summit, another sheer cleft and so on until the country begins to flatten out along the lonely, well-named Purgatoire River.

Just across the New Mexico line are the old gold workings about Taos, including the Mystic Mine, a proved gold district. But we do not have to cross out of Colorado to have proof of gold, for out of the little known "L" came a young Alex Cobsky with a batch of glittering ore that assayed more than $40,000 a ton. The burros he led up to the smelter in Pueblo carried about six hundred pounds of it and Cobsky was paid $12,500. That same year he loaded a larger pack train and received about $30,000.

The unusual youth had no wish to enjoy a splurge in Denver, New York, or London along the lines of other lucky Colorado miners.

Though quiet in his conduct, he could not remain unknown. The time was 1901, when the second great Nevada mining boom was under way and Colorado itself was still a little dizzy from its own great bonanzas. Cobsky's name was in newspapers throughout the country, but the fact left him unimpressed. He took up a pattern of life that apparently had the sole objective of guarding the secret of his amazing mine. This course turned him more and more into a mountain hermit, and gradually he took on the trappings of the hermit stereotype, an unkempt beard, trousers of deerskin, and ragged flannel shirts. Yet the one photograph ever taken of him shows sensitive, almost handsome features and unusually large eyes, staring frankly and seemingly with a little amusement into the camera. The rest of the snapshot is filled with whiskers, rough clothing, the figures of two black burros, and, significantly, a long-barreled rifle. Cobsky was known as a dead shot. The rare hunter or fisherman who chanced near the miner's hideout obeyed his warning to keep away. As it developed later, there were a number of men in the nearest town, La Veta, who knew exactly where Cobsky lived and where he supposedly mined his rich ore.

In this latter assumption, regarding the mine, they were mistaken.

It was necessary for Cobsky to come out of the mountains for supplies from time to time. On such an errand in 1937 he was struck by an automobile. He died in a hospital at Pueblo, the scene of his startling ore delivery many years before. He appeared to be about eighty years of age.

As the fatally injured man lay in the hospital he said nothing about his mine and uttered little else that was intelligible. After his death people at La Veta began to piece together what little they knew of his bonanza.

It developed that the recluse once had the company of a teen-aged boy forced upon him for a short time. The youth had become separated from his father's hunting party. Cobsky had fed him and in due course had guided him to a trail that led homeward. Inside the cabin, which was built against a hillside, the boy had seen a stout, padlocked door set in what was clearly the mouth of a tunnel. Before sending him on his way, Cobsky had given him some grim information to pass along. No one but the hermit could safely enter the tunnel. It was, in today's terminology, booby-trapped. At one spot a shotgun would fire. At another a bundle of dynamite would explode. A section of the tunnel floor would collapse, plunging any prowler into a water-filled shaft. Various additional hazards were implied. To the boy it was all quite convincing, and after Cobsky's death it remained convincing to many a prospective treasure hunter who valued his own life.

At length Undersheriff Carl Swift of Huerfano County and a fellow officer risked an investigation. They found a box of dynamite in the cabin, something to be expected at any miner's place and not in itself a danger. With all possible safety precautions, the tunnel door was forced open. Nothing exploded. Flashlight beams revealed no wires, no shotgun trap, nothing but the walls of native rock. It still was touchy business, but the two men explored the opening and emerged safely from what had proved to be a very short tunnel. Thankfully there were no infernal devices, but to the dismay of the officers' fellow townsmen neither was there any indication whatever of gold ore. Or ore of any kind. The worthless character of the excavation was confirmed by a mining engineer sent to the scene by some Cobsky family connections living in Denver. Where was the mine itself? The answer was no nearer.

Although the country still is difficult, a search for the hermit's gold would be made easier with modern transport. State Route 111 penetrates the mountains north and south of La Veta and there are local roads into most of the country to the east. Westward from La Veta the best that can be hoped for is a few dirt trails of limited extent. Penetration to the higher ground would have to be by foot or horseback, probably the former, for one is in a major mountain uplift, culminating in peaks rising well above 13,000 feet elevation. To decide whether such a search is reasonable, one must choose his own explanation of the Cobsky saga.

One theory would be that the man struck what has been referred to earlier as a "blowout," an isolated body of rich ore that is quickly mined out. There is no record of whether the mining engineer believed that this is what had occurred in the now barren little tunnel. His assignment was only to determine whether there was any basis for further mining, not to reconstruct Cobsky's past operations.

The alternate explanation is that the tunnel from the cabin's interior was a blind, that the hermit's excessive protectiveness used this means to divert attention from a really great gold deposit at another location. He could well have found this short tunnel left by some earlier, disappointed prospector. Such an adjunct to a cabin affords ready-made storage space, and miners at many places in the West have chosen to build their abodes against an old tunnel. In such a case, Cobsky might well have built up a picture of the cabin and its tunnel as the actual source of his gold, a deliberate misdirection. Where then, and how distant, is the real mine likely to be? Geology and old trails supply no answer. Cobsky's secret perished along with his rugged old frame.

Searchers in this region, however, should carry in mind another lost mine. If an elusive little stream can be positively identified as the "North Veta Creek" of Civil War days, you would do well to go to its headwaters, establish camp, and do some serious prospecting. Not only is a certain quartz outcrop rich in free gold, but most of the metal recovered by the man who found the lode is said to be secreted nearby.

In the town of La Veta the story was widely known in post-Civil War days. The lucky miner was understood to have fled from an eastern city to escape conscription into the Union forces. The fact that he was a wanted man but also the sole owner of a very rich mine put him in a position that could at least be described as frustrating. He seems to have placed his personal safety and freedom above the temptation of luxurious living, for he remained on remote "North La Veta Creek." His story, but not his name or home city, came to light after his death. Relatives appeared in La Veta, looked at maps, asked guarded questions, and made several rather bewildered excursions into the hills.

At length these family connections had to take local people into their confidence. The aging man, on a trip to Denver for supplies, had willed the mine to his sister and written down directions to it. The waybill, meaningful to the miner himself, was all too vague to any second party. Which creek was which, what mountain was meant? Townspeople were willing to help, and after the kinsmen had left in disappointment were even more ready to search on their own behalf, for the claim, necessarily left unpatented by the fugitive, was now available for location by anyone. As of this writing it is still available.

17
A Look at Alaska

In turning to mineral prospects in Alaska we are dealing with new dimensions. The airline distance from frozen Point Barrow on the Arctic Ocean to Ketchikan on the southern tip of the Alaska panhandle is roughly 1,400 miles. Alaskans remind us that the Aleutian Island chain is part of their state, so Alaska's east-west dimension, from the Yukon border out to lonely Attu Island, is well over 2,000 miles, not including the southeasterly extension of the panhandle. This is two thirds the distance from New York to San Francisco.

The great mileage involved in exploration must be coupled with the fact that most of Alaska is wilderness, with very limited means of access. Fairbanks, the only interior place of any size, can be reached by both railroad and highway. These transportation links were costly and difficult projects, and have not been duplicated to reach other sections of the interior. The only other railroad in the state, the White Pass and Yukon, is of use only for reaching Canadian territory. In summer there is navigation on the great arc of the Yukon River and its tributaries, and there are of course roads pushing relatively short distances inland from the coastal population centers. How useful the roads and other facilities connected with the great oil pipe line will be is something that remains to be seen. To reach any chosen prospecting territory, one will likely rely on Alaska's justly famous "bush pilots," who bring their little planes down on almost any open space of land, on lakes and rivers and even on the icy surface of glaciers. Making an informed and definite choice of one's intended field of exploration is desirable in any of the western states. In Alaska a precise objective would seem imperative.

Men have penetrated much of Alaska's forbidding terrain and they have found gold, sometimes gold in great quantity. Aggregate production figures have no great meaning for the individual miner, but a few may be cited to show the principal fields of activity. They are from a report by Gordon Herreid, a state geologist, published in 1961.

Down the southerly coastal strip, gold production totaled $148 million. Entering the vast Alaska peninsula proper, the mountains north of Prince William Sound afford reports from various districts that all run into the millions. The gold yield, very largely placer, in the Fairbanks area aggregated $197. 9 million. The Nome placers, which will be described in some detail, produced $104.9 million.

It may be noted that the most celebrated rush to the north country, the Klondike stampede, was not to Alaska. Hopeful thousands were *in* Alaska briefly as they struggled over Chilkoot Pass or White Pass from the port of Skagway, but the rich gold fields to which they were headed were in Yukon Territory, a part of Canada.

The year 1741 is the date accepted for the discovery of Alaska, in the sense of its being visited by a European who left an official record of the event. Fur trappers, American, French-Canadian, and British, must already have been there or at least very near. The discoverer of record was Vitus Bering, a Danish navigator in the service of the Russian czar. He had sailed east from Kamchatka to scout a still further extension of Peter the Great's conquests and colonization in Siberia. On a second expedition, weakened by scurvy, cold, storms, and Indian attack, Bering and many of his crew perished on the island that now bears his name.

After this, it was more than fifty years before Russia developed a stable colony in Alaska. The incentive was the immensely profitable fur trade. Success was achieved under the hard-drinking but capable Alexander Baranov, the first effective governor, whose fame is preserved in many Alaskan place names. In another half century, with many of the fur-bearing animals exterminated, the Russian crown had decided that its distant colony was profitless and unmanageable. In 1867, as most schoolboys know, Alaska was sold to the United States for $7.2 million.

Furs were the basis of Alaska's economy under the Russians, but for most of the years of American rule the paramount industry has been commercial fishing. The harvest of the sea far exceeds in value the output of the mines. It now appears that oil will be a major resource. The fact that the great Arctic petroleum pools went undiscovered, perhaps unsuspected, until about the time of World War II may be significant in relation to gold and silver. Are there also unsuspected districts where the precious metals abound?

The Russians had found gold as early as 1832. The discovery was on the Kuskokwim River, which even today is a lonely, largely untraveled stream. The few Russian colonists failed to get excited and it was years before meager information on the find reached the outside world.

By the time of the next authenticated discovery, Alaska was American territory. Quartz veins were found near the old Russian capital of Sitka, on Baranov Island. This lies in the relatively mild coastal climate, not a great distance from Juneau, Alaska's present capital. The year was 1877 and two claims were located under the federal mining act of 1872. With a stamp mill shipped north from San Francisco, the owner, George Pilz, conducted an unexceptional little operation which doubtless should be recognized as the first mining in Alaska.

The Pilz mine was soon dwarfed by a single giant enterprise on Douglas Island just to the north. Douglas is situated across Gastineau Channel from Juneau and today is the scene of extensive suburban development. The property took its name from John Treadwell, a carpenter sent north on a speculative trip by a group of San Francisco capitalists. The company that was promptly formed developed the great Alaska-Treadwell Mine, which operated for more than three decades and produced $21 million. The ore was low-grade, but mining and milling on a massive scale under efficient management produced a profit. The mining turned out to be altogether too massive. The continuing excavations at last extended too close to the shoreline, the sea broke through, and the mine was hopelessly flooded. A similar low-grade operation, backed by the financier and presidential adviser Bernard Baruch, was the Alaska-Juneau. For years it ran successfully on gold ore even leaner than that of the Treadwell. The small miner is inclined to regard such enterprises as big industrial operations rather than mines as he conceives them.

The first strike of consequence in the interior of Alaska was not made until 1893. It was far up the Yukon River, only a little more than a hundred miles from the boundary of Yukon Territory and Canadian soil. Where Birch Creek flowed into the river, rich placer ground was discovered and the town of Circle City grew up. By 1897 it had a population of 3,500 and was known as the largest city in the world made up exclusively of log cabins.

In the short season when the Yukon was free of ice, miners could reach Circle City in comparative comfort on shallow draft steamers, embarking at St. Michael on Norton Sound. It was a winding, time-consuming trip of more than a thousand miles. Many men preferred to reach the interior by way of Chilkoot Pass, later famous as a gateway to the Klondike.

While the hardships of interior Alaska have been graphically described in terms of freezing climate, isolation, and exorbitant prices, the principal complaint of the Circle City miners is rarely mentioned, namely mosquitoes. The place swarmed with them during the summer months. The thawing muskeg over an area of thousands of square miles supplied breeding pools for billions of the insects. They would settle on the exposed forearm of a miner in such numbers that the arm appeared black. Veils of mosquito netting over the face were not uncommon. Some men wore gloves throughout the summer. A smaller pest than the mosquito, called the "no see 'um," inflicted an equally irritating bite. Accounts of returned miners dwelt feelingly on the misery they endured, and numbers of men were rendered seriously ill by the cumulative amount of venom absorbed. The winter, for all its severity and interruption of work, was welcomed by many because it terminated the insect plague.

The placers around Circle City were lucrative enough to keep men working in this inhospitable place, but few of the miners became rich. The same may be said of the gravels at Rampart, several hundred miles farther down the Yukon. The real wealth was far away on the Seward Peninsula facing Bering Sea, on whose southern side occurred perhaps the wildest, most disorganized gold rush in the annals of American mining. This was at Nome.

As this settlement attracted attention, a map of the district was undertaken in faraway Washington. There was no designation for the new town, so one cartographer penciled the query "Name?" on a rough draft. Another government man read this as "Nome." Nome it became on this and all succeeding maps, and Nome itself was soon a ragtag, uncomfortable boom town of 20,000 souls. Among its earliest and more prosperous citizens was Erik Lindblom.

Lindblom was a Swede, a tailor, who in 1897 was working at his craft in San Francisco. Although forty-four years of age, fires of adventure were smouldering behind his uneventful daily routine. To make his way to the rumored wealth of Alaska he joined the crew of a whaling vessel. Once arrived in the far north, he jumped ship and soon was associated with two much younger Scandinavians, John Brynteson and Jafet Lindeberg.

September of 1898 found the three men at a Swedish missionary station at Golovin Bay, not far from the beach where in a short time Nome would come into being. There had been the merest hint of gold in this region, occasional colors having been panned by half a dozen prospectors. The trio set out in the waning summer weather and immediately struck pay dirt. Lindblom in particular became a symbol of prospector's luck, an impoverished, deserting sailor in 1898, a millionaire in 1899.

The Snake River enters the sea near Nome and both the river and the creeks draining into it were rich in placer gold. Many other early arrivals made fortunes. Largely through the level-headed management of Jafet Lindeberg, a Norwegian and at twenty-one the youngest of the three partners, the discoverers consolidated the best claims, defended them against litigation and a corrupt federal court, and took out an average of $15,000 a day over a long period of time.

New and surprising opportunities developed for the thousands of people, men and women, who were well-nigh monopolizing West Coast shipping in the rush to Nome.

A soldier, Jack Clunin, tried the unheard-of experiment of panning the sand of the ocean beach. He recovered more than $1,000 in his first day's work. The shoreline soon became a mob scene. The gold values extended for a great distance along the beach, and contemporary accounts say that at least 7,000 persons worked the tiny parcels to which, by decision of a miners' council, their claims were restricted.

Some $5 million was taken out of the ocean sands before they were worked out. The belief that the tide was continually bringing in gold proved an illusion. The metal had been washed down from the interior.

In 1902 a new type of deposit was uncovered. Back from the ocean and at a higher elevation, men dug down to a layer of gravel resting on a hard clay formation. Gold dust and nuggets occurred there in quantity. This higher deposit was called "the Second Beach." A Third Beach was found in 1904, and a year later came the final discovery of the Nome district, the Intermediate Beach.

There was ample capital for any promising operation at Nome. When the deposits played out for the hand laborer and were abandoned, mechanized equipment was moved in and larger scale recovery was attempted. Results were not spectacular but a steady profit was returned, and some dredging activity continued for many years.

The only substantial population center in the interior of Alaska, Fairbanks, owes its beginnings to placer gold discoveries on the Tanana River in 1903. The early arrivals inexplicably failed to find workable ground and hundreds of men left in disgust. Later it was shown that the gold-bearing gravels were not only extensive but in places were very rich. Dome Creek in particular paid off handsomely. How the first wave of prospectors failed to open up the opulent Fairbanks field is one of the unexplained mysteries of mining. This was one, very nearly the only one, of the Alaska placer districts where a successful search was made for ore in place. The resultant lode mining gave Fairbanks some lasting stability, and this remote spot was chosen as the site of the University of Alaska. Fairbanks is in the news each spring with its far-north version of the Irish Sweepstakes, a huge lottery based on the hour, minute, and place of the clock's second hand when ice breaks up in the Nenana River.

In 1906 there was a large rush to Iditerod, on a tributary of the lower Yukon. Quite a little city grew up but by 1910 the workable gravel was all but exhausted.

A number of such strictly placer camps flourished and faded. It may be conjectured that the strikes were in locations where it was not possible, as in California for example, to discover where the gold had come from and to start mining the gold-bearing veins themselves. Mining after the turn of the century was largely by experienced men. Nome was the last place to see a rush of *cheechakos*—newcomers, tenderfeet. Later there were hundreds, probably thousands, of good miners to whom Nome, the Klondike, or any of the western states had failed to bring any success. Having accustomed themselves to the rigors of the country, they spread out into the great unknown and brief, spotty accounts of placer diggings and a few remote lode mines over a wide area tell their story.

The preliminaries of an Alaskan prospecting trip are not difficult. The state wants tourists, and information will be readily, even enthusiastically, furnished by the official Alaska Travel Division at Juneau, or by airlines, shipping lines, and travel agencies. You can go in comfort to any of Alaska's principal cities and towns and even be routed on limited exploration trips through country that is rugged, scenic, and possibly mineralized.

Getting into parts of the interior that have been scantily prospected, or conceivably not prospected at all, is another matter. Alaska has twice the area of Texas but has very few roads. The unique service of the bush pilots between the populated sea coast and points in the hinterland has therefore grown up, not specifically in connection with mining or any other industry but to meet transportation needs in general. The bush pilot can be a valuable ally for the prospector.

18
Good-bye and Good Luck

People who have taken to the outdoors as contemporary Americans have need no instructions on the hiking and camping involved in mineral exploration. The hunter, the fisherman, the rock hound, and the sportsman or hobbiest in various other fields are well prepared to look after themselves. If the experiences of mine hunters as related in this book have anything further to contribute, it might be an accent on just two things: Carry ample water in the desert, and have an ample respect for early snowstorms in the mountains.

One aspect of planning a trip that is more important to the prospector than to his fellow outdoorsmen is the importance of preliminary information. Once an area of investigation has been decided upon, he should read all he can about its known mineralization and its mining history. As Dr. Samuel Johnson elaborately puts it, "He who would bring home the wealth of the Indies must carry the wealth of the Indies with him," and for the benefit of simpler souls he goes on to explain that this means equipping one's self with advance knowledge.

By far the most valuable and readily accessible sources of such knowledge are official mining reports. On the federal level, the U. S. Geological Survey and the U. S. Bureau of Mines have produced separate series of reports extending back many years and covering virtually every mining locality in the United States. Whatever anyone may think of the perennial avalanche of publications carrying the government imprint, it must be conceded that the federal mining and geological reports are very much worthwhile. They contain a wealth of information about individual mines and even about many prospects that so far have been unproductive. The individual write-ups, in nearly all cases, are based on a personal inspection of the property by a staff mining engineer. They are commonly quite detailed.

The standard reports by the states follow the model of the federal reports, in most cases even in typography. There is duplication of subject matter, but considerable amplification may often be found in the state version.

In consulting these reports, it takes time and patience to extract adequate information. They seem to be thoroughly indexed—the sheer bulk of the index volumes is impressive—but the same subject, say a certain mine, may have to be followed through the reports for a number of years. This is because there have been successive entries at various dates.

In the West, at least, the larger public libraries have complete sets of the federal reports and those of their own states in bound volumes. Besides the annual (or monthly or quarterly) publications, there are many special ones issued, some of which can be of great value. *Geology and Ore Deposits of the Blank Quadrangle* would be a typical title of such a special study. If you encounter a special report covering your intended field of prospecting, you will have a real asset.

The full-fledged report on a mine by a professional engineer in private practice is a different matter. Such a report would ordinarily be commissioned as a guide to further development, to assist in some financing, for the information and encouragement of the stockholders, and so on. It is not often printed or available for public scrutiny. If a present-day gold hunter becomes seriously interested in an old mine he should learn whether such professional inspection and findings were ever made and endeavor to trace the resulting document down.

The sources noted so far, although easy enough to read and understand, would be described as technical. Books on mining directed to a more general readership can be helpful for orientation in the subject itself and for background. These for the most part deal with mining history. Some contain quite detailed information on certain districts and can be of direct assistance. Of particular interest are the smaller volumes in which some miners, looking back on their careers, have set down in their own way their recollections of a particular gold rush, gold district, or individual gold mine. Usually privately printed or brought out by small specialty publishers, these are hard to come by. If you can find one relating to your own intended field of exploration you will be at least interested, perhaps helped.

In the course of this report a good many informative books have been mentioned by author and title and they need not be listed here. The references to *Desert* magazine, however, may be amplified a bit. The current issues of this monthly and the back files if available will be of genuine help to anyone prospecting in the Southwest. Featured throughout the years of its publication have been detailed maps of certain localities, sometimes printed in connection with a lost-mine narrative, sometimes as a guide to rock-hunting parties.

In the matter of maps, almost any experienced miner or prospector will recommend carrying maps of the U. S. Geological Survey. Get one or more as required to cover your area of exploration. These are sometimes referred to as contour maps, because their lines, commonly representing intervals of forty feet of elevation, show the contours of the land. Another term is quadrangle maps, since each one depicts a quadrangular area to which the Survey has applied a name, for example, "Ivanpah Quadrangle." These are the most detailed maps available. One caution should be given against accepting them on a literal, current basis: Look at the date. The number of quadrangles mapped is so great that it is financially impossible to keep updating the maps from year to year. The old Reno area map, for instance, shows much open land close in to the city. The physiographic features shown have not changed from what the map shows, but large sections of the quadrangle are now covered by residential developments.

Data from geologic maps has not figured in our account, chiefly because of the specialized terminology involved, partly because the maps deal mostly with large areas rather than specific mineral locations. Much excellent work has been done in this field, but not every region of interest has as yet been mapped as to its geology. Where such a map is available, it should by all means be obtained, studied, and carried.

This report has limited itself to assistance in *finding* a gold or silver deposit. To go beyond discovery and on to the development of a mine would be to enter an additional field of considerable complexity. It is one that has its own literature, necessarily technical, somewhat scanty as regards lode mining but with quite adequate publications for the placer miner.

We may encroach on the post-discovery situation just far enough to say a word about property rights, for this is something to be kept in mind from thc beginning. To reduce the matter to its simplest form, you would not wish to spend time, money, and energy searching for a gold-bearing vein and then, if successful, learn that your bonanza was on a privately owned ranch (though you might, of course, be able to negotiate a satisfactory working arrangement). There is lots of government land where your mine would be your own. Unless a particular tract has been withdrawn or reserved for presumably good reason, the national forests, under the Department of Agriculture, and the nonforest public domain in general, under the Department of the Interior, are open to the filing of a mining claim and development of a mine.

The recommendations in this report concerning known mines, now inactive, which may be of value under present conditions, involve another kind of determination. This would take you to the county seat. If the mining claim has been patented, it is legally like any other piece of real estate and you must deal with the owner. If it is not patented, the question is whether the claimant has kept up his assessment work, the $100 worth of labor each year that is required by law. If not—and the county records will give the answer—you are at liberty to "jump" the claim. This is a reprehensible action in western fiction and sometimes in fact, but if the claim simply has been abandoned your takeover is quite legitimate.

A privately owned, patented mine that has been idle for years is often found to be tax delinquent. If so, you have a good chance of getting it put up for sale at competitive bidding. Policies in this regard vary from state to state and county to county, so no over-all directions can be given here. It may resolve itself into how well you get along with the county tax collector.

This report opened with a note on luck and it will close with a similar one. There will be additional gold and silver strikes made in our Western states, but not every searcher is going to make one. A sign in a Reno gambling house warns patrons that EVERYBODY CANNOT WIN. This applies equally to the prospector roving over nearby Sun Mountain or through the desolate Monte Cristo range that lies just beyond. But if a reward in terms of legal tender eludes him, it is to be hoped that the intangible reward will not, a memory of days spent in a portion of the earth that remains primitive and unspoiled, and in a quest that throughout human history has had an aura of mystery and romance.

Index